CONTAINER MOLECULES AND THEIR GUESTS

Monographs in Supramolecular Chemistry

Series Editor: J. Fraser Stoddart, *University of Birmingham, UK*

This series has been designed to reveal the challenges, rewards, fascination, and excitement in this new branch of molecular science to a wide audience and to popularize it among the scientific community at large.

No. 1 Calixarenes
By C. David Gutsche, Washington University, St. Louis, USA

No. 2 Cyclophanes
By François Diederich, University of California at Los Angeles, USA

No. 3 Crown Ethers and Cryptands
By George W. Gokel, University of Miami, USA

No. 4 Container Molecules and Their Guests
By Donald J. Cram and Jane M. Cram, University of California at Los Angeles, USA

Forthcoming Title
Molecular Assemblies and Membranes
By J.-H. Fuhrhop and J. Köning, Freie Universität Berlin, Germany

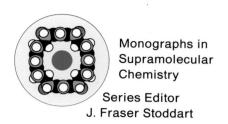

Monographs in
Supramolecular
Chemistry
Series Editor
J. Fraser Stoddart

Container Molecules and Their Guests

Donald J. Cram and Jane M. Cram
University of California
Los Angeles, USA

ROYAL
SOCIETY OF
CHEMISTRY

ISBN 0-85186-972-6

A catalogue record of this book is available from the British Library

© The Royal Society of Chemistry 1994

All Rights Reserved
No part of this book may be reproduced or transmitted in any form or by any means – graphic, electronic, including photocopying, recording, taping or information storage and retrieval systems – without written permission from The Royal Society of Chemistry

Published by The Royal Society of Chemistry,
Thomas Graham House, Science Park, Cambridge CB4 4WF

Typeset by Computape (Pickering) Ltd., Pickering, North Yorkshire
Printed and bound by Black Bear Press Ltd., Cambridge, England

Preface

Stimulated by the progress made in reducing biochemistry and molecular biology to organic chemistry in the late 1950s, I was inspired by the extraordinary ability of working biological entities to bind specifically certain organic compounds, and to regulate their syntheses, transport, and disposal in living organisms. Since the physical laws governing biological and non-biological chemistry are fundamentally the same, it seemed possible to design, prepare, and study synthetic systems that might mimic aspects of the evolutionary biological systems.

By the late 1960s, crystal structure determinations along with mass and nuclear magnetic resonance spectra were refined enough to provide the tools necessary to the success of a systematic study of synthetic complexing systems. The presence of 100 crystal structures in this volume attests to their importance to our progress. Likewise, Corey–Pauling–Koltun molecular models based on crystal structures of biologically important compounds became available for use in designing spacial relationships between complexing partners. When in 1967, C.J. Pedersen published his crown ether complexation of alkali metal ions, I recognized that this work provided an entrée to the field of complexation phenomena. My research group started to make chiral crown ethers in 1970, and by 1974, my wife (Dr. Jane M. Cram) and I published our first general paper summarizing our first thoughts, methods, and results.

This book traces the evolution of host–guest complexation chemistry in the minds and hands of my co-workers at UCLA over the 25-year period from 1970–1994. To accommodate space and time limitations and provide continuity, we review here other groups' research results only to the extent that our work was inspired or helped by them. Because some of our early work was reviewed in preceding volumes in this series, we will emphasize results obtained since 1987. We hope others will build on the results summarized here.

This book is essentially a chronicle of the more successful parts of the journeys my research group took into the wilderness of complexation phenomena. Described are the results of an estimated 375 person-research years. Key people who contributed centrally to this research on an ongoing basis were Drs. Roger C. Helgeson, Carolyn B. Knobler, Emily F. Maverick, and Professor Kenneth N. Trueblood. Ms. June O. Hendrix turned our many hundreds of crudely sketched structures into computer-drawn art pieces. The NIH, NSF, and DOE provided most of the direct support for the work, and UCLA provided the laboratories and infrastructure so important to this endeavor.

Drs. Robert Thomas and Christophe von dem Busche generated the colored art work, and we warmly thank them. I express my gratitude to the Nobel Prize Committee whose 1987 award extended my career, and provided great motivation for me to prove that this prize can act as a stimulant rather than a sedative to research careers. My most heartfelt thanks go to the profession of teaching research in organic chemistry in a stimulating environment. No other profession is endowed with such a rich landscape, draws inspiration from so many fields of science, exercises the hands and mind in so many different ways, offers such opportunities to employ creative instincts, and mixes ideas, theory, and experiment on a daily basis. Hurrah for the science of organic chemistry, and for the joy it brings those who play the research game.

Donald J. Cram
Los Angeles, California
October, 1993

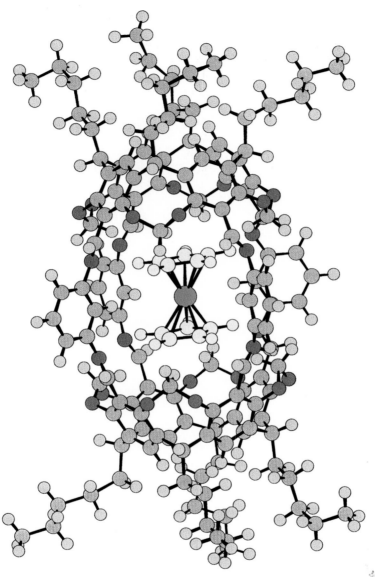

*This drawing is of ruthenocene guest incarcerated in octaimine host **5** of Section 9.2, and is based on the crystal structure of **5**·ruthenocene.*
Courtesy of E.F. Maverick, unpublished.

Contents

Chapter 1   Contexts, Conceptions, Corands, and Coraplexes   1

     1.1   Origins of Host–Guest Chemistry   1
     1.2   Early Types of Hosts, Guests, and Complexes   2
     1.3   Molecular Modules   7
     1.4   Structural Recognition in Complexation   10
     1.5   The Crystal Structures and Molecular Modeling Connection   12

Chapter 2   Spherands, Spheraplexes, and Their Relatives   20

     2.1   Spherands and Spheraplexes   20
     2.2   Syntheses of Spherands   22
     2.3   Crystal Structures for Spherands and Spheraplexes   25
         2.3.1   Non-bridged Hosts   25
         2.3.2   Bridged Hosts   26
     2.4   Hemispherand and Hemispherand Crystal Structures   28
         2.4.1   Three Preorganized Ligands   28
         2.4.2   Four Preorganized Ligands   30
         2.4.3   Cryptahemispherands   31
     2.5   Correlation of Structure and Binding   33
         2.5.1   Determination of Binding Power   33
         2.5.2   Corand and Anisyl Hemispherand Binding Comparison   34
         2.5.3   Cryptahemispherand Binding and Specificity   37
         2.5.4   Spherand Binding and Specificity   37
     2.6   Principles of Complementarity and Preorganization   39
     2.7   Illustration of the Effects of Preorganization on Binding   42
     2.8   Rates of Complexation and Decomplexation of Spherands and Hemispherands   44

	2.9	Cyanospherands	44
	2.10	Chromogenic Ionophores	46

Chapter 3 Chiral Recognition in Complexation 49

3.1	Hosts Containing One Chiral Element	49
3.2	A Chiral Breeding Cycle	52
3.3	Hosts Containing Two Chiral Elements	53
3.4	An Amino Ester Resolving Machine	55
3.5	Chromatographic Resolution of Racemic Amine Salts	56
3.6	Failure of a Magnificent Idea Guides Research	57
3.7	Chiral Catalysis in Michael Addition Reactions	59
3.8	Chiral Catalysis in Methacrylate Ester Polymerization	61
3.9	Chiral Catalysis of Additions of Alkyllithiums to Aldehydes	63

Chapter 4 Partial Enzyme Mimics 65

4.1	A Partial Transacylase Mimic Based on a Corand	66
	4.1.1 Kinetic Acceleration	66
	4.1.2 Competitive Inhibition	67
	4.1.3 Chiral Recognition	68
	4.1.4 Abiotic and Biotic Comparisons	69
4.2	Hosts Containing Cyclic Urea Units	70
	4.2.1 Preorganization of Cyclic Urea Units by Anisyl Unit Attachment	71
	4.2.2 Binding Properties of Hosts Containing Multiple Cyclic Urea Units	71
	4.2.3 Crystal Structures	72
	4.2.4 Binding Power Dependence on structure	73
	4.2.5 Highly Preorganized Hosts	75
	4.2.6 An Unusual Color Indicating System	76
4.3	An Incremental Approach to Serine Protease Mimics	77
	4.3.1 Kinetics of Transacylations	78
	4.3.2 Rate Enhancements Due to Complexation	79
4.4	Introduction of an Imidazole into a Transacylase Mimic	80

Contents   xi

	4.4.1	Synthesis	80
	4.4.2	Transacylation Kinetics	81
	4.4.3	Dependence of Rate Enhancement Factors on Structure	82

Chapter 5 Cavitands   85

- 5.1 Origins of the Cavitand Concept   85
- 5.2 Systems Based on Cyclotriveratrylene Ring Assemblies   87
- 5.3 Octols From Resorcinols   87
- 5.4 Bowls from Octols   91
  - 5.4.1 Syntheses   91
  - 5.4.2 Crystal Structures   93
- 5.5 Cavitands with Cylindrical Cavities   97
- 5.6 Cavitands Containing Two Binding Cavities   99
  - 5.6.1 Syntheses   99
  - 5.6.2 Crystal Structures   101
  - 5.6.3 Complexations   101
- 5.7 Hosts Based on Fused Dibenzofuran Units   102
  - 5.7.1 Cavitands Containing Two Clefts   102
  - 5.7.2 Cavitands Containing A Single Cleft   103

Chapter 6 Vases, Kites, Velcrands, and Velcraplexes   107

- 6.1 Vases, Kites, and Temperature   107
  - 6.1.1 Synthesis   107
  - 6.1.2 Conformational Analysis   108
  - 6.1.3 Rationalization of Temperature Effects on Conformation   110
  - 6.1.4 Crystal Structure   111
- 6.2 Design and Synthesis of Kite-shaped Systems   112
  - 6.2.1 Structural Basis for Dimer Formation   112
  - 6.2.2 Synthesis   114
- 6.3 Evidence That Velcrands Form Velcraplexes   116
  - 6.3.1 Crystal Structures   116
  - 6.3.2 Solution Dimerization   120
- 6.4 Free Energies of Formation of Velcraplex Dimers in Solution   121
  - 6.4.1 Determination   121
  - 6.4.2 Correlation of Binding with Structure   122
  - 6.4.3 Importance of Preorganization to Binding   123
  - 6.4.4 Binding in Homo- *vs.* Hetero-dimers   123
  - 6.4.5 Effect of Solvent Changes on Binding   125
- 6.5 Enthalpies and Entropies of Complexation of Velcrands   127

	6.5.1	Partition of Driving Forces Between Two Parameters	127
	6.5.2	Solvolytic Effects on Dimerization	127
	6.5.3	Balancing of Entropic and Enthalpic Dimerization Effects	128
6.6		Thermodynamic Activation Parameters for Association of Velcrands and Dissociation of Velcraplexes	129

Chapter 7 Carcerands and Carceplexes — 131

7.1	Conception		131
7.2	The First Closed Molecular Container Compound		132
	7.2.1	Synthesis	132
	7.2.2	Characterization	133
7.3	Soluble Carceplexes with CH_2SCH_2 Connecting Groups		135
	7.3.1	Syntheses	136
	7.3.2	Characterization	138
	7.3.3	Guest Rotations in Carceplexes	139
7.4	Carceplexes with OCH_2O Connecting Groups		140
	7.4.1	Syntheses	140
	7.4.2	Carceplex Crystal Structure	142
	7.4.3	Guest Movements in Carceplex	144
7.5	Inner Phase Effects on Physical Properties of Guests		146
	7.5.1	Molecular Communication Through the Shell	146
	7.5.2	Possible New Type of Diastereoisomerism	147
7.6	Comparisons of Carceplexes, Spheraplexes, Cryptaplexes, Caviplexes, Zeolites, and Clathrates		147

Chapter 8 Hemicarcerands and Constrictive Binding — 149

8.1	Hemicarcerand Containing a Single Portal		151
	8.1.1	Synthesis	151
	8.1.2	Crystal Structure	152
	8.1.3	Characterization	153
	8.1.4	Decomplexation at High Temperatures and Complexation at Ambient Temperatures	154
	8.1.5	Complexation at Elevated Temperatures	156

8.2	Hemicarcerand Containing Four Potential Portals		158
	8.2.1	Synthesis	158
	8.2.2	Crystal Structure	159
	8.2.3	Guest Variation in Hemicarceplexes	160
	8.2.4	Rotations of Guests Relative to Host	161
	8.2.5	Proton Magnetic Resonance Spectra of Guests in Hemicarceplex	162
	8.2.6	Mechanism of Guest Substitution of Hemicarceplexes	162
	8.2.7	Dependence of Decomplexation Rates on Guest Structures	163
	8.2.8	Constrictive and Intrinsic Binding	164
	8.2.9	Driving Forces for Intrinsic and Constrictive Binding	166
8.3	Chiral Recognition by Hemicarcerands		167

Chapter 9 Varieties of Hemicarcerands 170

9.1	An Octalactone as a Hemicarcerand		170
9.2	An Octaimine as a Hemicarcerand		172
	9.2.1	Synthesis and Characterization	172
	9.2.2	Complexation	173
	9.2.3	Decomplexation	174
	9.2.4	Crystal Structure	175
9.3	An Octaamide as a Hemicarcerand		176
	9.3.1	Syntheses and Characterization	176
	9.3.2	Crystal Structure	177
	9.3.3	Complexation	180
9.4	A Near Hemicarcerand Based on [1.1.1]Orthocyclophane Units		180
	9.4.1	Syntheses	181
	9.4.2	Crystal Structures	182
	9.4.3	Characterization	183
	9.4.4	Complexation	184
9.5	Rigidly Hollow Hosts That Encapsulate Small Guests		185
	9.5.1	Synthesis	185
	9.5.2	Crystal Structures	186
	9.5.3	Complexation of Tritosylamide Host	187
	9.5.4	Complexation of Triamine Host	188
9.6	A Host with a Large Cavity		189
	9.6.1	Synthesis	190
	9.6.2	Complexation	190
	9.6.3	Correlation of Decomplexation Rates with Guest Structures	192

	9.7	A Highly Adaptive and Strongly Binding Hemispherand	193
		9.7.1 Synthesis	194
		9.7.2 Complexation	194
		9.7.3 Structural Recognition in Complexation	196
		9.7.4 Crystal Structures	196
		9.7.5 The Unusual Guest Structure in $52 \cdot 6H_2O$	199
Chapter 10		**Reactions of Complexed Hosts, of Incarcerated Guests, and Hosts Protection of Guests from Self Destruction**	**202**
	10.1	Energy Barriers to Amide Rotations in the Inner Phase of a Carcerand	203
	10.2	Acidity of Amine Salts in the Inner Phase of a Hemicarcerand	204
	10.3	Hemicarcerand as a Protecting Container	206
	10.4	A Thermal–Photochemical Reaction Cycle Conducted in the Inner Phase of a Hemicarcerand	206
	10.5	Cyclobutadiene Stabilized by Incarceration	208
	10.6	Oxidations and Reductions of Incarcerated Guests	211
	10.7	Reductions of the Host of Hemicarceplexes	212
		10.7.1 Octaimine Reductions	212
		10.7.2 Reduction of Hemicarceplexes with Four Acetylenic Bridges	213

Subject Index 217

CHAPTER 1

Contexts, Conceptions, Corands, and Coraplexes

'An investigator starts research in a new field with faith, a foggy idea, and a few wild experiments. Eventually the interplay of negative and positive results guides the work. By the time the research is completed, he or she knows how it should have been started and conducted.'

1.1 Origins of Host–Guest Chemistry

An examination of the receptor sites of evolutionary biological molecules reveals them to have concave surfaces to which substrates with convex surfaces bind. This generalization was the starting point in our thinking about complexation between synthetic entities, and it inspired the coining of the terms *host* and *guest* and the expression *host–guest complexation*.[1–3] Hosts are defined as organic molecules containing convergent binding sites, and guests as molecules or ions containing divergent binding sites.[2,3] Complexes are defined as hosts and guests held together in solution in describable structural relationships by electrostatic forces other than those of full covalent bonds or of ionic crystals.[2,3] These forces are of a pole–pole, pole–dipole, and/or dipole–dipole variety that attracts host to guest. In solution these enthalpic attractions are supplemented by the desolvation of host and guest binding surfaces to turn somewhat organized solvent into randomly oriented 'drowned' solvent. This decollection, disorientation, and randomization of solvent provides entropic driving forces for complexation which compensate somewhat for the entropically expensive process of collecting, orienting, and freezing out degrees of freedom of host and guest molecules as they come together to share a common surface. This so-called *solvophobic effect* is always a factor when complexes form in solution, but is frequently outweighed, particularly for complexes with small shared surfaces dissolved in non-protic media.

Of the millions of synthetic organic compounds known in 1970, very, very few molecules possessed enforced concave surfaces. In contrast, concave

---

[1] D.J. Cram and J.M. Cram, *Science*, 1974, **183**, 803.
[2] D.J. Cram, R.C. Helgeson, L.R. Sousa, J.M. Timko, M. Newcomb, P. Moreau, F. de Jong, G.W. Gokel, D.H. Hoffman, L.A. Domeier, S.C. Peacock, K. Madan, and L. Kaplan, *Pure Appl. Chem.*, 1975, **43**, 327.
[3] E.P. Kyba, R.C. Helgeson, K. Madan, G.W. Gokel, T.L. Tarnowski, S.S. Moore, and D. J. Cram, *J. Am. Chem. Soc.*, 1977, **99**, 2564.

surfaces were abundant in naturally occurring biological compounds. The cyclodextrins were natural products put to extensive use in pioneering complexation studies by Cramer,[4] whose work was far enough developed by 1958 to have inspired faith in us that the field of solution complexation chemistry was viable. Cholic acid is another example of a simple molecule with a shallow, enforced concave surface whose complexing abilities were widely recognized. The naturally-occurring inorganic polymers known as zeolites have large concave surfaces in the shape of canals useful for absorbing in the solid state a variety of small organic compounds. However, the richest sources of concave surfaces are found in the binding and active sites of enzymes and nucleic acids, which offer a wealth of binding site types for appropriate substrates.

In considering the possibility of studying complexation with designed compounds in the 1950s and 1960s, we found the absence of enforced concave surfaces in synthetic organic molecules remarkable, and somewhat daunting. Although some of our larger paracyclophanes possess concave surfaces of molecular dimensions in certain conformations (*e.g.* **1**,[5] **2**,[6] **3**[7]), we never observed solid complexes containing solvent. In their complexes with tetracyanoethylene, this strong π-acid was never located between the aryls, but appeared to complex the outside faces of the aryls.[6,7]

1.2 Early Types of Hosts, Guests, and Complexes

In 1967, C.J. Pedersen published his seminal work on crown ethers,[8] which demonstrated that a rational approach to host–guest complexation was possible. He synthesized a series of macrocyclic polyethers based mainly on CH_2CH_2O repeating units, of which 18-crown-6 is the prototype. This molecule strongly binds K^+ because of the stereoelectronic complementarity between host and guest. The K^+ guest nests in the center of a wreath-like complex ligated by six oxygens, as is suggested by structure **4**, and by **5**, the drawing of a Corey–Pauling–Koltun (CPK) molecular model[9] of **4**. Structure **6** illustrates the ability of 18-crown-6 to bind NH_4^+ and RNH_3^+ salts. In CPK models, $CH_3NH_3^+X^-$ beautifully hydrogen bonds to alternate oxygens of 18-crown-6 to

---

[4] F. Cramer, *Rev. Pure Appl. Chem.*, 1955, **5**, 143.
[5] H. Steinberg and D.J. Cram, *J. Am. Chem. Soc.*, 1952, **74**, 5388.
[6] D.J. Cram and R.H. Bauer, *J. Am. Chem. Soc.*, 1959, **81**, 5971.
[7] L.A. Singer and D.J. Cram, *J. Am. Chem. Soc.*, 1963, **85**, 1080.
[8] C.J. Pedersen, *J. Am. Chem. Soc.*, 1967, **89**, 7017.
[9] W.L. Koltun, *Biopolymers*, 1965, **3**, 665.

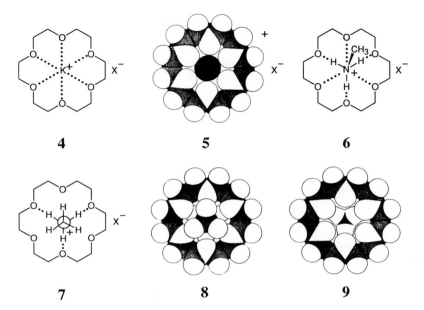

form a complex represented by **6**, by Newman projection **7**, or by drawings **8** and **9**, which show front and rear faces of CPK models of **7**. We refer to **6–9** as containing a tripodal, perching-type structure.

This early work of Pedersen provided us an entrée into the field of host–guest complexation chemistry. Between 1970 and 1974 we had designed and prepared systems which incorporated one or more 1,1'-binaphthyl units into crown ethers (**10** and **12**) for complexing chiral amine salts, such as 1-phenylethyl-ammonium salts (**11**), enantioselectively. Through examination of CPK molecular models of the complexes (*S,S*)-**12**·(*R*)-**11** and (*S,S*)-**12**·(*S*)-**11** (structures **13a** and **13b**), the former tripodal, diastereomeric complex was selected prior to experiment as the less sterically encumbered and therefore the more stable diastereomer.[10] The gratifying feature of the results of complexation of (*S,S*)-**12** with racemic (**11**) in CDCl$_3$ was that they upheld the previous choice of stable diastereomer based on the models.

The binaphthyl unit was also useful for attaching the first side chains ('arms') to crown ethers, which might aid in complexing various ions by intramolecular ion-pairing. Examples of such complexes of binaphthyl hosts with metal ions as guests are **14–18**, in which the charges on host and guest could be matched, as in **14** and **16**. Both complexes exhibit parent ions in their mass spectra.[11] Complex **14** formed serendipitously when the host scavenged Sr^{2+} from bulk Ba(OH)$_2$, thus exemplifying a high selectivity. Two plus charges of guest Ba^{2+} were matched by the simultaneous complexing of two host molecules each having one arm with a terminal carboxylate group. The two binaphthyl host molecules

---

[10] E.B. Kyba, K. Koga, L.R. Sousa, M.G. Siegel, and D.J. Cram, *J. Am. Chem. Soc.*, 1973, **95**, 2693.

[11] R.C. Helgeson, J.M. Timko, and D.J. Cram, *J. Am. Chem. Soc.*, 1973, **95**, 3023.

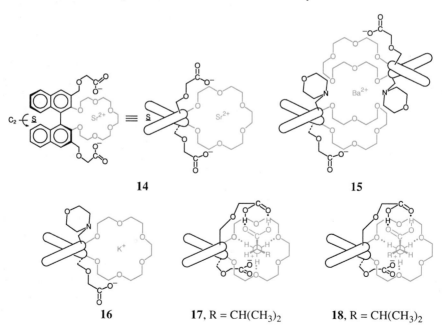

provided a lipophilic skin for the Ba^{2+} ion, rendering the complex soluble in organic solvents, stable to chromatography and to sulfuric acid dissolved in methanol–water.[10]

The binaphthyl host of complexes **17** and **18** with two terminal carboxylates is able to bind both the carboxylate and amino groups of the guest valine. The racemic host was completely resolved by a continuous liquid–liquid extraction involving (S)-valine dissolved in AcOH–H$_2$O adsorbed on silica gel as the stationary phase, with racemic host in the C$_6$H$_6$–AcOH mobile phase. In a one-plate experiment involving CHCl$_3$–H$_2$O–AcOH as solvent, (S)-(CH$_3$)$_2$CHCH(NH$_2$)CO$_2$H and the two enantiomers as a racemate of this host were distributed between the aqueous and organic layers. As predicted based on a comparison of the steric strain associated with CPK models of complexes **17** and **18**, (S)-valine complexed the (S)-binaphthyl host more strongly that the (R)-host in the aqueous layer by a factor of 1.7. In models, the R group (CH$_3$)$_2$CH– of valine of **17** in the (S)·(S)-diastereomer has more room than in **18**, the (S)·(R)-diastereomer.[12]

In a further survey in our early work on complexes with different kinds of guests, the guanidinium tetraphenylborate complex **19** of benzo-27-crown-9 was designed with the help of CPK models and prepared.[1] This was the first characterized complex of guanidinium ion. Later its crystal structure was determined and found to have its predicted nesting structure (**20**).[13] In the synthesis of the host, the guanidinium ion was used to template the ring closure of the host as indicated in **21**.[3] Finally, examination of CPK models suggested that binaphthyl-crown **10** and 18-crown-6 should complex and solubilize ArN≡N$^+$BF$_4^-$ salts in non-polar media. This hope was realized by the formation of complexes **22** and **23** in organic media. Later, others[14] obtained a crystal structure (**24**) of **23** which turned out to have the structure we originally assigned.[15]

The evidence we developed that the N$_2^+$ group was inserted into the cavity of host **10** to make complex **22** was as follows. Complex **22** was prepared by extraction of an aqueous solution of C$_6$H$_5$N$_2^+$BF$_4^-$ with a CHCl$_3$ solution of **10**. When a methyl was substituted in a position *ortho* to the N$_2$ group of the guest, the guest failed to be extracted. Substitution of two methyl groups in the 3,4-positions of the guest did not inhibit extraction, but produced colors suggesting π–π charge-transfer binding between host and guest that supplemented the pole–dipole attractions.[15]

The beauties of the 1,1'-binaphthyl unit for the introduction of chiral elements or of functional groups into hosts are manifold. (1) The diameter of the hole of such hosts is slightly adjustable, because of variability of the aryl–aryl dihedral bond angle, which acts as a hinge. Moreover, this unit nicely replaces an ethylene unit without much change in the spacing of oxygens attached to the 2,2'-positions of the 1,1'-binaphthyl. (2) Although chiral, the unit contains a C$_2$ axis, which renders the two faces of a macrocyclic host identical. Consequently, in the formation of 'perching' complexes such as that of (S,S)-**12** with amine salt (R)-**11**, (structure **13a**), the same complex results by

---

[12] R.C. Helgeson, J. Koga, J. M. Timko, and D. J. Cram, *J. Am. Chem. Soc.*, 1973, **95**, 3021.
[13] J.A.A. de Boer, J.W.H.M. Uiterwÿk, J. Greevers, S. Harkema, and D.N. Reinhoudt, *J. Org. Chem.*, 1983, **48**, 4821.
[14] B.L. Huffman and J.C. Huffman, 4th Symposium of Macrocyclic Compounds, Provo, Utah, August, 1980.
[15] G.W. Gokel and D.J. Cram, *J. Chem. Soc., Chem. Commun.*, 1973, 481.

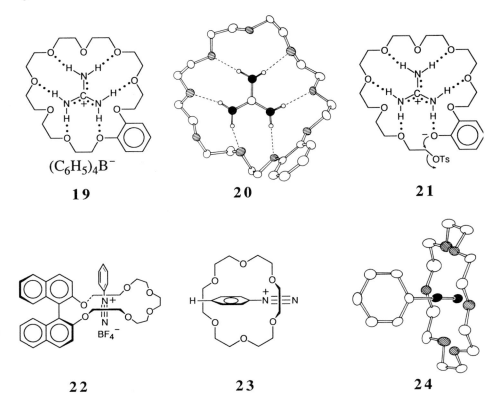

attachment of the guest to either face of the host. (3) Chains attached to the 3,3'-positions extend the chiral barrier, and converge on guests such as amino acids, one over each of the two faces as in **17** and **18**. (4) Chains attached to the 5,5'-positions diverge from the binding site and can be used for controlling solubility of hosts as in **25** and **26**,[16] or for attaching them to solid supports for chromatographic separations as in (R,R)-**27**.[17] (5) Catalytic hydrogenation of 2,2'-dihydroxy-1,1'-binaphthol gives the corresponding tetralin system whose outer aryl groups are reduced. These bitetralyl units are similar to the binaphthyl with respect to geometry and enantiomer stability. Hosts containing the bitetralyl unit (see [R]-**28**) tend to be more soluble than their binaphthyl counterparts. Notice that like the binaphthyl, the chiral barrier of the bitetralyl unit can be extended by A groups substituted in its 3,3'-positions. Also like binaphthyl host (S,S)-**22**, chiral bitetralyl host (R,R)-**29** possesses three C_2 axes to provide overall D_2 symmetry with its attendant advantages.[18]

The hosts of this chapter contain many different kinds of ether and pyridine binding sites built into a wide variety of sterically confining groups which

---

[16] L.J. Kaplan, PhD Thesis, 1978, UCLA.
[17] L.R. Sousa, G.D.Y. Sogah, D.H. Hoffman, and D.J. Cram, *J. Am. Chem. Soc.*, 1978, **100**, 4569.
[18] D.J. Cram, R.C. Helgeson, S.C. Peacock, L.J. Kaplan, L.A. Domeier, P. Moreau, K. Koga, J.M. Mayer, Y. Chao, M. G. Siegel, D. H. Hoffman, and G.D.Y. Sogah, *J. Org. Chem.*, 1978, **43**, 1930.

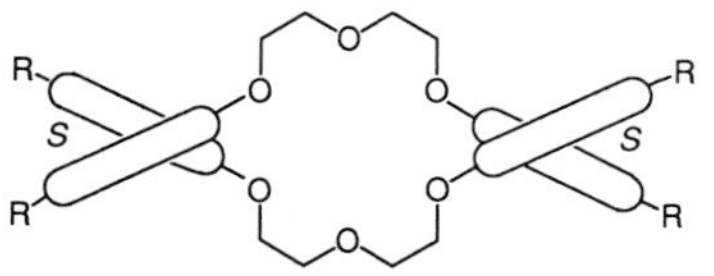

Contexts, Conceptions, Corands, and Coraplexes

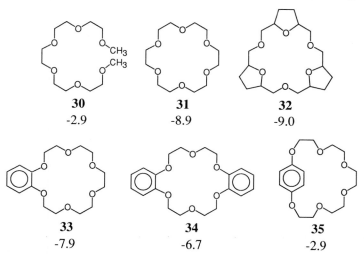

25, R = SO₃H
26, R = C(CH₃)₂(CH₂)₁₀CH₃

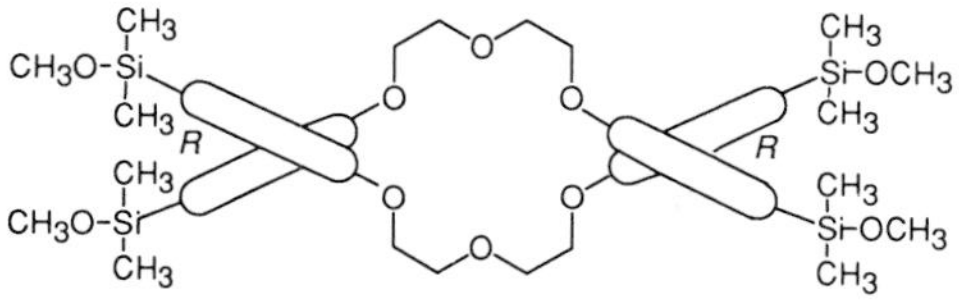

(R,R)-27

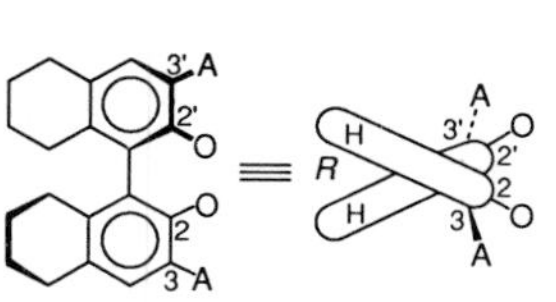

(R)-28

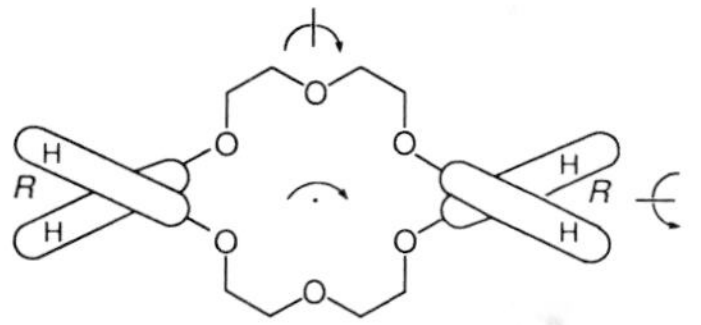

(R,R)-29, D₂ symmetry

impart many different kinds of symmetry to the hosts. Most of them do not have crown shapes. Accordingly, we suggest the family name *corands*, for these hosts, and the family name *coraplexes* for their complexes.

1.3 Molecular Modules

Just as Pedersen[8] found that substitution of a OCH₂CH₂O by a 1,2-C₆H₄O₂ unit in a crown system preserved the general binding and selective properties, we explored the effect of substituting a variety of other modules on the binding power of the cyclic polyethers. Examples of hosts that were prepared are shown in **30–44**. Beneath the compound numbers are given the approximate free energies (kcal mol⁻¹) of binding of the compounds toward $(CH_3)_3CNH_3^+CN^-$ in CDCl₃ saturated with D₂O at 24 °C.[19] The binding scales involved distributing $t\text{-BuNH}_3^+X^-$ salts between D₂O and CDCl₃ and measuring spectroscopically, in the presence and absence of host, the amount of guest in the

30   31   32
-2.9 -8.9 -9.0

33   34   35
-7.9 -6.7 -2.9

[19] D. J. Cram and J. M. Cram, *Acc. Chem. Res.*, 1978, **11**, 8.

CDCl$_3$ layer. From the results, association constants (K_a) between host and guest in CDCl$_3$ were calculated (see Equation 1.1), and from the K_a values were calculated the ΔG^0 values (see Equation 1.2).[19]

$$\text{host} + t\text{-BuNH}_3{}^+ \text{X}^- \underset{\text{CDCl}_3}{\overset{K_a}{\rightleftarrows}} t\text{-BuNH}_3{}^+\cdot\text{host}\cdot\text{X}^- \qquad (1.1)$$

$$\Delta G^0 = -RT \ln K_a \qquad (1.2)$$

Host 18-crown-6 binds t-BuNH$_3{}^+$CN$^-$ better than its open-chain counterpart **30** by 6 kcal mol^{-1} due to the better organization of binding sites in the host *before* complexation. In the perhydrotrisfuranocycle **32** (mixture of the two all-*cis* isomers) the six oxygens lie in an enforced, inward-turned conformation, an arrangement that provides only a slight enhancement of the binding free energy (~0.1 kcal mol^{-1}) over that of the simple crown **31**. Interestingly, **32** was prepared from sucrose as the only organic starting material. Substitution of one 1,2-C$_6$H$_4$O$_2$ for an OCH$_2$CH$_2$O unit, as in **33** compared to **31**, decreased the binding free energy by 1 kcal mol^{-1}, while two 1,2-C$_6$H$_4$O$_2$ for two OCH$_2$CH$_2$O units, as in **34**, decreased the energy by 2.2 kcal mol^{-1}. The attached aryl groups decrease the basicity of the oxygens by delocalization of the electron pairs in the hosts. In **35**, a 1,4-C$_6$H$_4$O$_2$ unit is substituted for a 1,2-C$_6$H$_4$O$_2$ unit of **33** for a reduction in binding free energy of 5 kcal mol^{-1}. Clearly in **31–34** a tripodal arrangement of hydrogen bonds is possible in the complexes, whereas **35** possesses an enforced oxygen arrangement which allows only two hydrogen bonds to form simultaneously in the complex.[19]

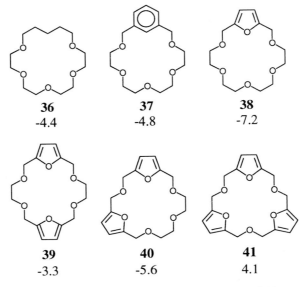

36	37	38
-4.4	-4.8	-7.2
39	40	41
-3.3	-5.6	4.1

In 18-crown-5, **36**, substitution of one CH$_2$ for an oxygen of 18-crown-6, **31**, decreased the binding free energy by 4.1 kcal mol^{-1}. Although a CPK molecular model of the tripodal complex of **36** appears to be sterically feasible, it involves unfavorable *gauche* $(CH_2)_5$ conformations. Furthermore, such a complex contains only two $(CH_2)_2O$----N^+ attractions, as compared to three in the tripodal complex of **31**. In **37**, a substitution of a 1,3-$C_6H_4(CH_2O)_2$ unit for $O(CH_2)_2O(CH_2)_2O$ unit of **31** decreases the binding free energy by 4.5 kcal mol^{-1}. Examination of the CPK model of the perching coraplex of **37** shows host and guest to be sterically compatible, but again in this complex one $(CH_2)_2O$----N^+ attraction present in that of **31** is missing. An *ab initio* calculation of the energies of binding of NH_4^+ and $HOCH_2CH_2OH$ in various conformations suggested that an NH^+----O attraction is about three times that of an N^+----O.[19] Although furano-18-crown-6 hosts, **38–41**, are more preorganized for tripodal binding than is **31**, the aromaticity of the oxygens of the furan units renders them much less basic than those of the $(CH_2)_2O$ units of **31**. This electronic effect appears to greatly outweigh the preorganizational effect, since substitution of each $(CH_2)_2O$ unit of **31** by a furanyl oxygen as in **38–41** results in a decrease of 1.6–2.8 kcal mol^{-1}, depending on whether alternate or non-alternate oxygens are replaced (*e.g.* compare the binding by **40** *vs.* that of **39**).[19]

Structures **42–44** are of hosts in which 2,6-$C_5H_3N(CH_2O)_2$ units replace $O(CH_2)_2O(CH_2)_2O$ units of the parent host **31**. Only trivial differences in binding free energies result. Interestingly, decorative host **43** can give rise to two possible isomeric complexes **45** and **46**, of which we selected **45** as being the more probable.[19] Years later, others demonstrated by ^{15}N NMR that **45** was indeed the correct structure.[20]

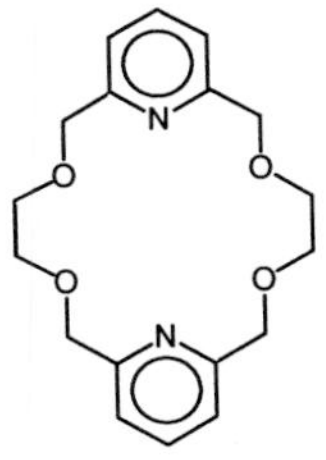

42
-8.4

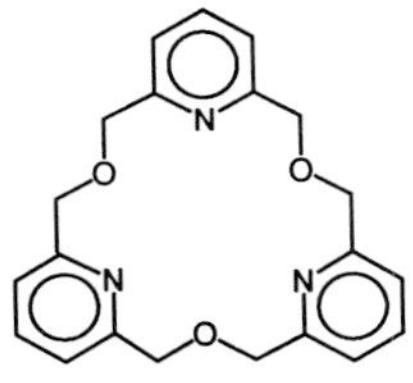

43
-8.8

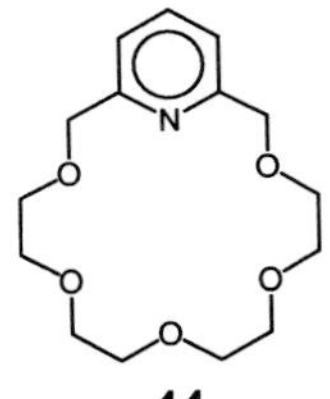

44
-8.9

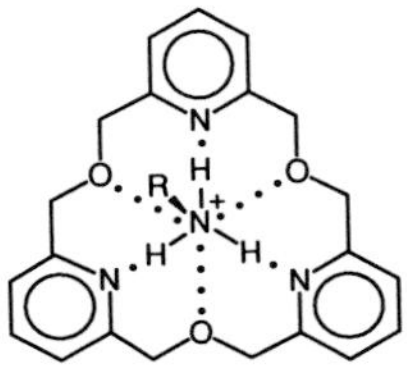

45

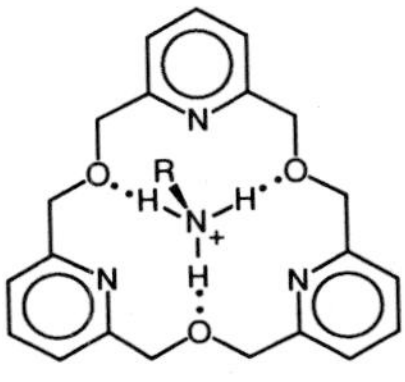

46

Another interesting comparison is that of a binaphthyl macrocycle **48** and its podand **47** binding *t*-$BuNH_3^+SCN^-$. The difference in binding favors the more preorganized cycle **48** by 1.6 kcal mol^{-1}, much less than the difference of 6 kcal mol^{-1} in binding observed for 18-crown-6, **31**, *vs.* its open-chain counterpart,

---

[20] H.G. Förster and J.D. Roberts, *J. Am. Chem. Soc.*, 1980, **102**, 6984.

30. We attribute the smaller difference between the binding of **47** and **48** as compared to **30** and **31** to the fact that the rigid 3,3'-dimethyl-1,1'-binaphthyl unit imposes conformational constraints on its attached polyethylenoxy 'arms' in **47** that are not present in **30**. Thus podand **47** is more preorganized for binding than is podand **30**.[21]

47
4.8

48
6.4

1.4 Structural Recognition in Complexation

Several highly structured three-dimensional modules were incorporated into corands to give hosts that were then tested for their abilities to exhibit structural recognition in complexation. The differences in binding free energies of each host with NH_4^+ picrate and $t\text{-BuNH}_3^+$ picrate were employed as a measure of the steric inhibition of tripodal binding of the more space-demanding $t\text{-BuNH}_3^+$ guest. The extraction method used involved $CDCl_3$ and D_2O as media at 24 °C. [See Equations 1.1 and 1.2 of Section 1.3.] Host **49**, 2,3-naphthaleno-18-crown-6, was selected as a standard because its naphthalene module made it distribute only into the $CDCl_3$ layer, and the crystal structure of its $t\text{-BuNH}_3^+$ complex demonstrated it entered into tripodal binding (Section 1.5).

Several rigid three-dimensional modules proved to be particularly interesting in host design. The layered structure of [2.2]paracyclophane **50** has sixteen substitutable sites, eight of which orient attached groups in a semiconvergent direction and the other eight in the opposite direction. The three ditopic hosts of complexes **51–53** (incorporating **50**), have splendid symmetry properties. The host of **51** has D_2 symmetry, is chiral, and is non-sided. Those of **52** and **53** have C_{2h} symmetry, are achiral, and are sided. Examination of CPK modules of these complexes indicates that steric effects severely limit the CH_2 conformations of the $ArOCH_2$ moieties to those distant from both the transannular ring and the CH_2CH_2 groups that bridge the aryls. This steric effect forces the binding, unshared-electron pairs of the $ArOCH_2$ oxygens to orient toward one another enough to inhibit somewhat the tripodal binding of RNH_3^+ groups in **51** and **53**, and accounts for the reduced binding in **51** and **53** of NH_4^+ as compared to the naphthaleno host **49** by 1.8 and 2.5 kcal mol^{-1}, respectively. The same steric effect operates in **52** to orient these oxygen electron pairs somewhat in the direction of the transannular aryl ring which is more compatible with tripodal structures. An interesting consequence of these enforced

---

[21] R.C. Helgeson, G.R. Weisman, J.L. Toner, T.L. Tarnowski, Y. Chao, J.M. Mayer, and D.J. Cram, *J. Am. Chem. Soc.*, 1979, **101**, 4928.

49
NH$_4^+$, -9.64
t-BuNH$_3^+$, -7.66

50

51
NH$_4^+$, -7.8
t-BuNH$_3^+$, 3.01

52
NH$_4^+$, -9.04
t-BuNH$_3^+$, 3.80

53
NH$_4^+$, -7.1
t-BuNH$_3^+$, -3.1

orientations is to limit tripodal binding to those structures drawn in **51–53**, in effect making the hosts non-sided. These allowed tripodal structures force the R attached to the NH$_3^+$ group to occupy sterically encumbered positions. This arrangement accounts for the ≥ 4 kcal mol^{-1} stronger binding of NH$_4^+$ than of *t*-BuNH$_3^+$ in **51–53**, which compares with the difference of only 2 kcal mol^{-1} in the two non-compressed complexes of model host **49**. That complexes **51–53** are formable was shown by the fact that each host in CDCl$_3$ solubilized just two moles of *t*-BuNH$_3$B(C$_6$H$_5$)$_4$. Many of the above conclusions were supported by ^{1}H NMR spectral comparisons.[22]

The high structural recognition of 5.2 kcal mol^{-1} observed for the difference in binding of NH$_4^+$ and *t*-BuNH$_3^+$ in complexes such as **52** suggests that a host of structure **54** should exhibit high chiral recognition toward the enantiomers of racemates possessing the general structure LMSCNH$_3$ClO$_4$, where L, M, and S are large, medium, and small groups. Models of **54** suggest that diastereomer **55** should be several kcal more stable than the other diastereomer. Attempts to prepare compounds such as **54** were unsuccessful.[23]

Although **51–53** are formally ditopic (contain two separate arrays of binding sites), all attempts to prepare one-to-one complexes failed with diamine guests of the type $^+$H$_3$N(CH$_2$)$_n$NH$_3^+$, where *n* was appropriate in CPK models to span the distance between the two NH$_3^+$ groups to complex the two binding sites intramolecularly. Only polymeric complexes were obtained.[22] The first ditopic host to be prepared in which the two sets of binding sites behaved cooperatively was **56**, which contains a C_2 axis and is chiral. In CPK models, the two 'jaws' were complementary to polyfunctional salts **57–60**. Experimentally, a solution

---

[22] R.C. Helgeson, T.L. Tarnowski, J.M. Timko, and D.J. Cram, *J. Am. Chem. Soc.*, 1977, **99**, 6411.
[23] H. Nakamura and D. J. Cram, unpublished results.

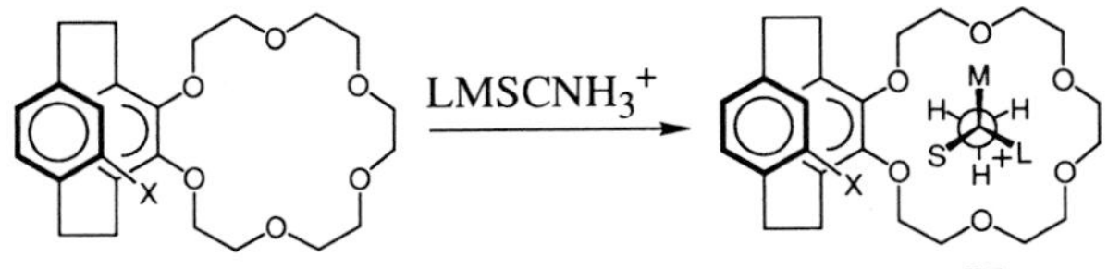

of **56** in CDCl$_3$ extracted about one mole of each of these guests dissolved in D$_2$O to provide complexes in which two remote functional groups of the guest were each bound to one of the jaws of the hosts. A crystal structure of complex **56·57** was obtained and is discussed in the next section.[24]

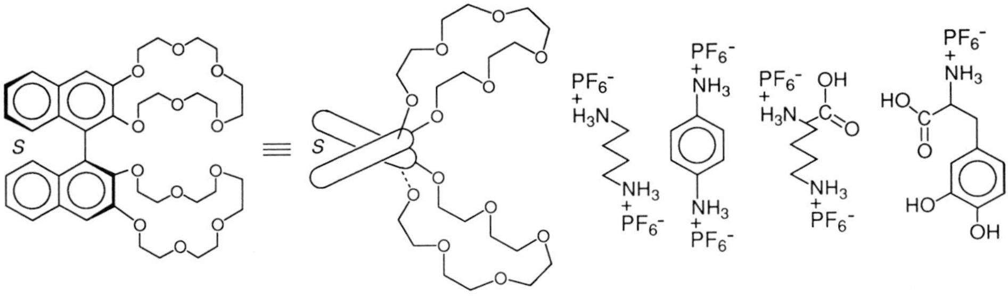

1.5 The Crystal Structures and Molecular Modeling Connection

Although much can be inferred about the structures of complexes from NMR spectra, elemental analyses, solubilization, extractability, structural recognition experiments, and CPK molecular modeling, crystal structure determinations of complexes provide the most convincing and substantial evidence for how hosts and guests are bound to one another. When crystal structures of complexes and of hosts alone are compared, the degree of conformational reorganization of the host upon complexation can be assessed. The crystal structures of over 150 different hosts and complexes have been determined from 1975 to 1992 at UCLA by Drs. K.N. Trueblood, C.B. Knobler, E.F. Maverick, and I. Goldberg. This accomplishment has been of inestimable importance in providing a substantial structural foundation for host–guest complexation chemistry, and in building confidence in the CPK molecular modeling approach to complex design. Examples of crystal structures of some of the complexes prepared early in our research in this field are presented here.[25] Those prepared later are integrated into the text in appropriate places.

The crystal structures of coraplexes **61** and **63** exemplify the manner of binding of t-BuNH$_3^+$ to corands containing six binding sites distributed symmetrically in 18-membered rings. Both hosts produce tripodal perching

---
[24] T.L. Tarnowski and D.J. Cram, *J. Chem. Soc., Chem. Commun.*, 1976, 661; R.C. Helgeson, T.L. Tarnowski, and D.J. Cram, *J. Org. Chem.* 1979, **44**, 2538.
[25] D.J. Cram and K.N. Trueblood, *Top. Curr. Chem.*, 1981, **98**, 43.

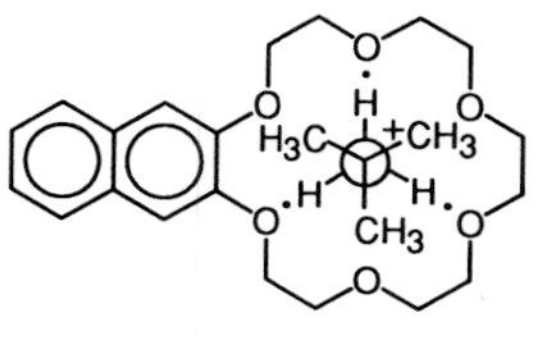

61

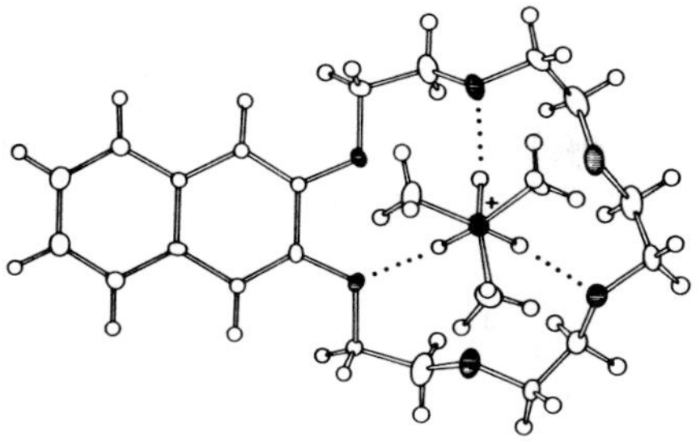

62

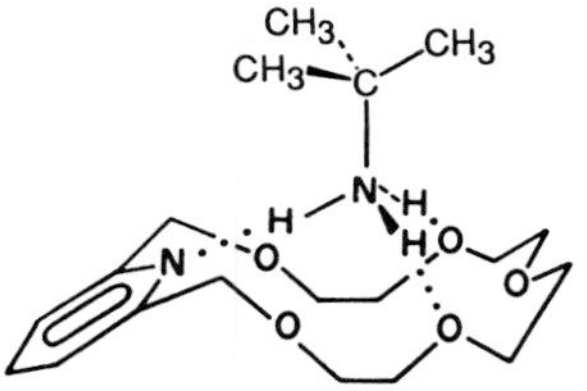

63

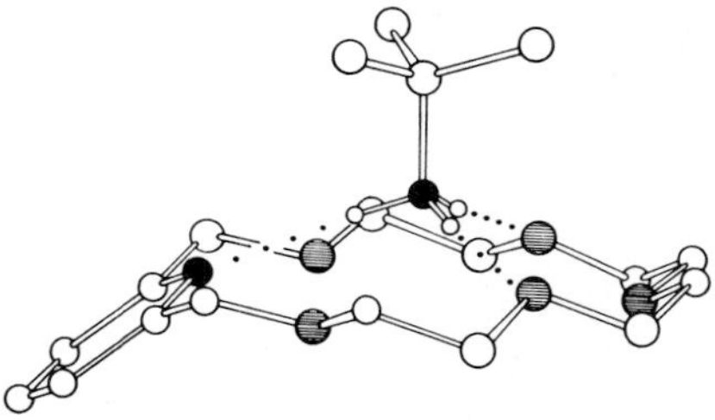

64

complexes in which three hydrogen bonds (O----HN$^+$ or N----HN$^+$) are formed between host and guest with three close O----N$^+$ contacts supplementing the binding. In these and many other complexes involving host–ammonium type binding, the C—N of the guest is nearly normal to the plane of the three hydrogen-bonded heteroatoms of the host. This is exactly what is predicted from examination of CPK models of **61** and **63**.[25]

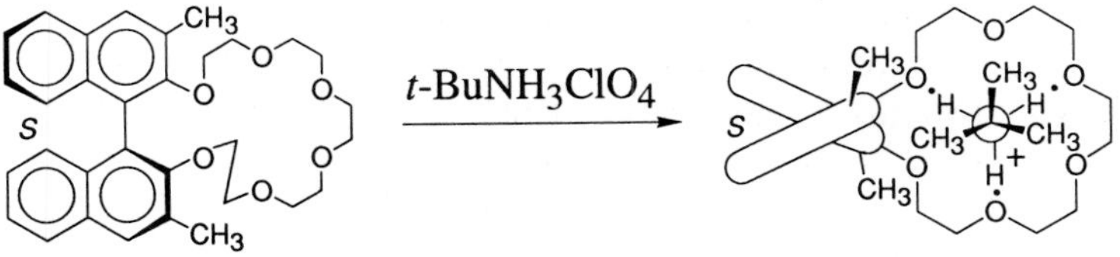
t-BuNH$_3$ClO$_4$

65 **66**

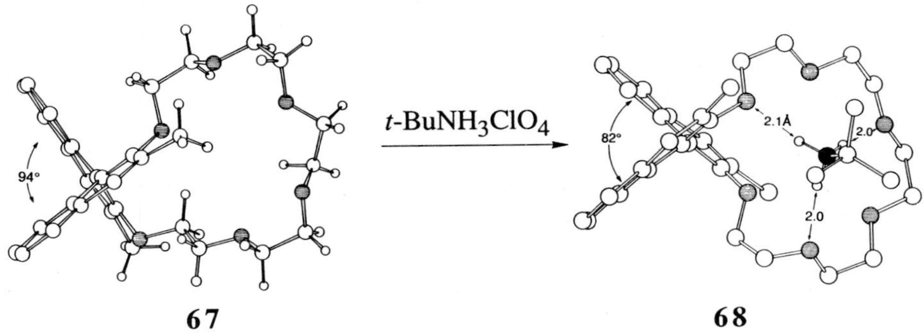
94°
t-BuNH$_3$ClO$_4$
82°
2.1Å
2.0
2.0

67 **68**

In CPK models of binaphthyl host **65**, one methylene can turn inward to fill the corand hole, as is drawn. Models of coraplex **66** show that this methylene must relocate to provide the tripodal binding of **66**. Crystal structures **67** of the corand and **68** of the coraplex indicate the guest conformationally reorganizes the host as suggested by the models. Noteworthy in **68** are the following facts: (1) the methyl group attached to the 3-position of the binaphthyl unit occupies a position between the two methyls of the $(CH_3)_3CNH_3^+$ guest; (2) the dihedral H—N—C—CH$_3$ angles are about 60°; (3) the C—N bond is normal to the plane of the three hydrogen-bonded oxygens. The arrangement described in (2) and (3) was observed repeatedly in coraplexes of t-BuNH$_3^+$.[25]

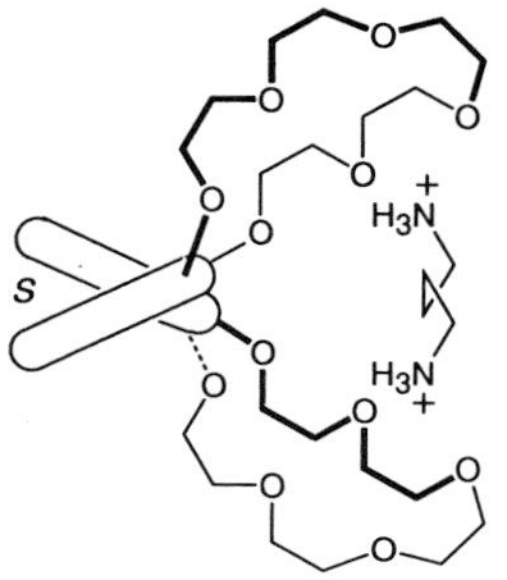

69 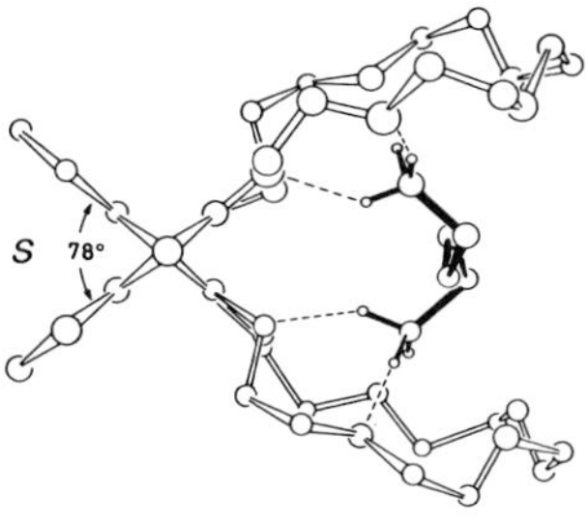 **70** 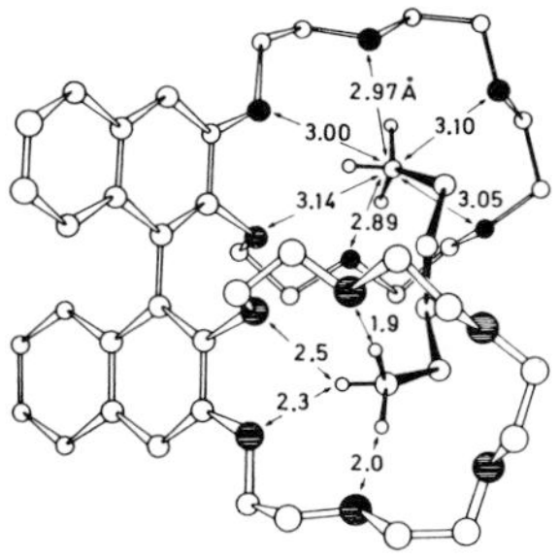 **71**

A molecular model of the ditopic host **56** complexing $^+H_3N(CH_2)_4NH_3^+$ resembles **69**. The two views **70** and **71** of the crystal structure show two of the six hydrogen bonds in **69** are bifurcated, a phenomenon sometimes encountered in coraplexes involving $R_3CNH_3^+$ guests.[24]

In CPK models of host **72**, the carboxyl beautifully hydrogen bonds a transannular oxygen atom, and the C=O carbon is coordinated with two oxygens, one on each CO_2 face. Crystal structure **74** confirms such a structure. When treated with t-BuNH$_2$ host **72** forms a salt, which in models possesses the perching structure **73**, and whose crystal structure turned out to be as predicted

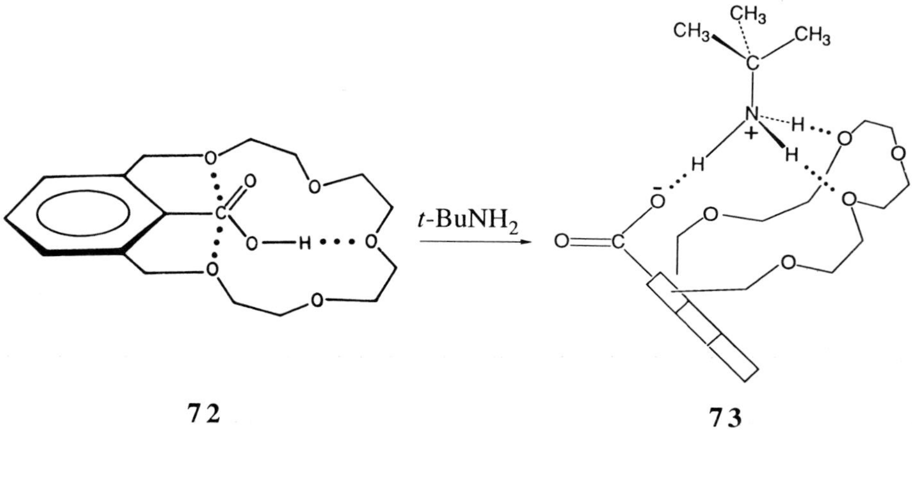

72 **73**

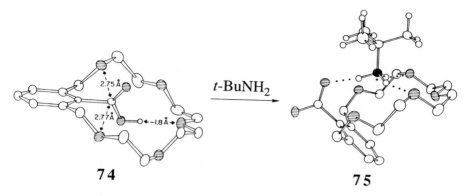

74 **75**

(see **75**). Additional examples of intramolecular complexes involving carboxyl groups hydrogen bonding either a pyridine nitrogen (**76**) or one another (**78**) are illustrated in crystal structures **77** and **79**, respectively. Again, the correspondence between CPK models and the crystal structures is excellent.[25]

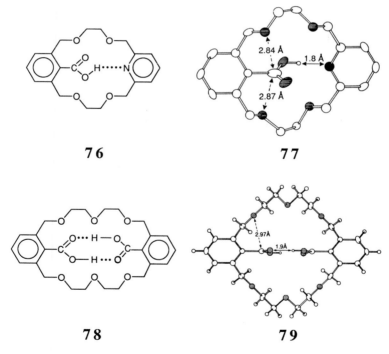

76 **77**

78 **79**

A beautiful example of a complex which exhibits reciprocal hydrogen bonding is found in the unanticipated formation of a corand complex of water **80**. The crystal structure **81** exhibited four close contacts, one between the hydrogen of an inward-turned methylene and a phenolic oxygen, the other three involving OH----O hydrogen bonds. A general characteristic encountered in both hosts and their complexes is that, whenever possible, structures assume

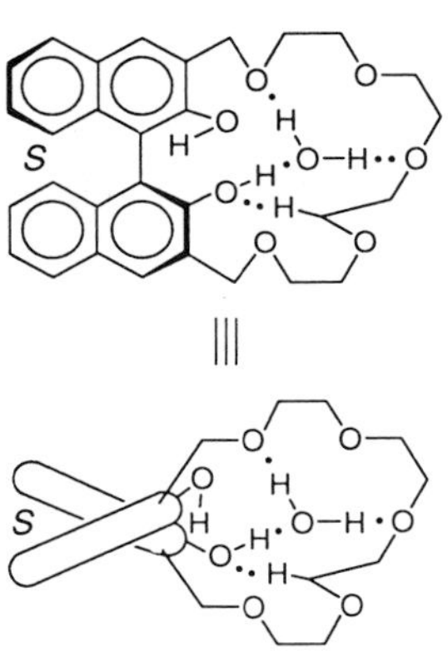

80 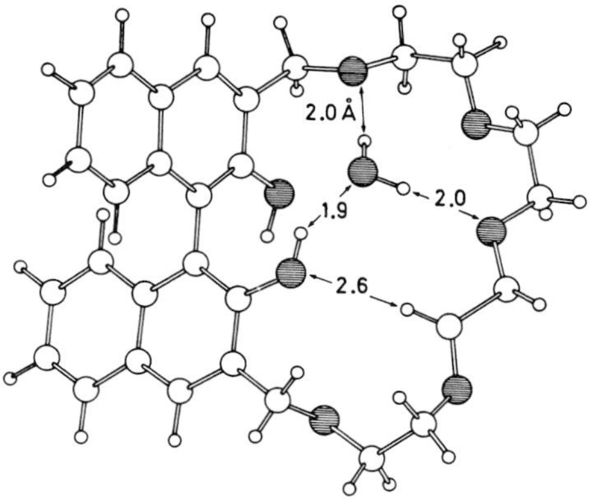   **81**

conformations which avoid empty space. Coraplex **80** illustrates this generalization.[25]

Dunitz's crystal structure of Pedersen's 18-crown-6 **82** and Weiss's crystal structure of Lehn's [2.2.2]cryptand **84** provide important examples of how these free hosts fill their own cavities with inward-turned methylene groups. When treated with K^+, both hosts conformationally reorganize to form coraplexes with K^+ whose crystal structures are pictured in **83** and **85**.[25]

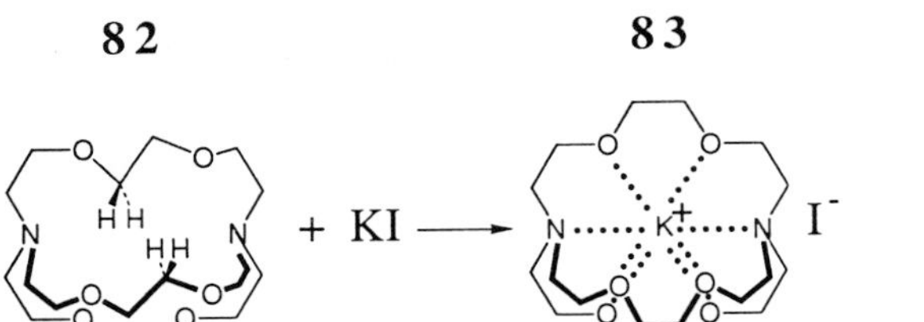

82   **83**

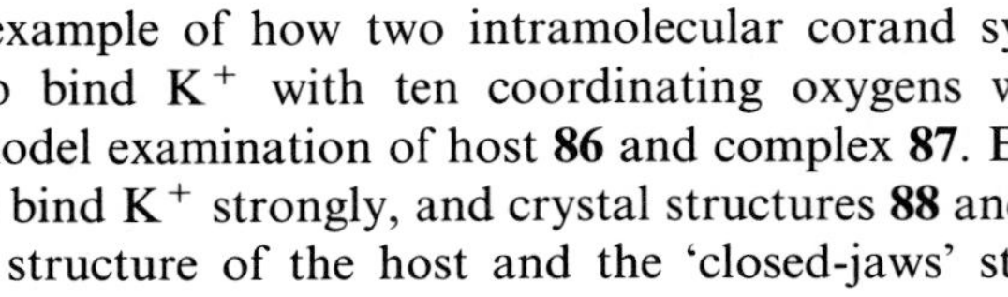

84   **85**

An unusual example of how two intramolecular corand systems operate cooperatively to bind K^+ with ten coordinating oxygens was envisioned through CPK model examination of host **86** and complex **87**. Experimentally, **86** was found to bind K^+ strongly, and crystal structures **88** and **89** confirmed the 'open-jaws' structure of the host and the 'closed-jaws' structure of the coraplex.[25]

The use of two $CH_2PO(OEt)_2$ arms attached in the 3,3'-positions of a

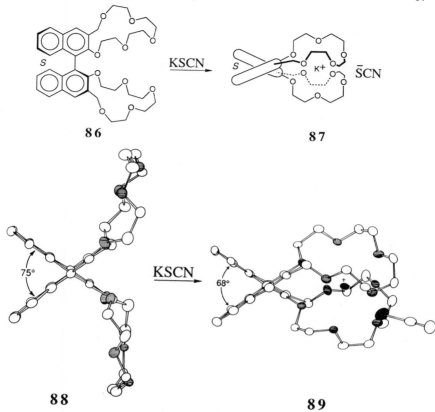

binaphthyl corand to supplement the binding of K$^+$ by ether oxygens is found in **90**, a system envisioned *via* CPK models prior to synthesis. The crystal structure **91** shows how beautifully the eight oxygens contact the K$^+$, the two P→O oxygens being located on opposite faces of the macroring to provide an almost linear P→O----K$^+$----O←P arrangement. As expected, the host of **90**, turned out to be a strong binder of K$^+$.[25]

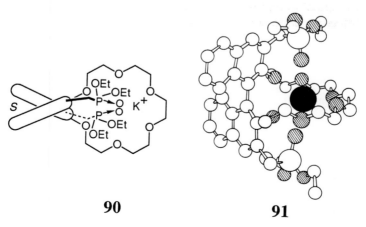

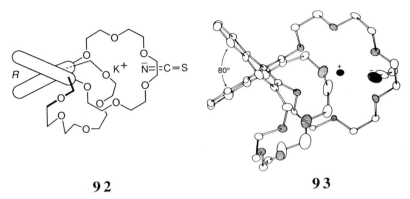

92 **93**

In the host of coraplex **92** a O(CH$_2$CH$_2$O)$_5$ bridge between the 2,2'-positions of the 1,1'-binaphthyl module is combined with a CH$_2$O(CH$_2$CH$_2$O)$_4$CH$_2$ bridge spanning the 3,3'-positions. In CPK models, the second bridge locates one oxygen in a position to coordinate a K$^+$ guest, which supplements the binding provided by the six oxygens of the first bridge. The crystal structure **93** of coraplex **92** gave the expected structure.[25]

The above examples illustrate the *importance of CPK molecular modeling to the design of host–guest complexes*, and of crystal structure determinations of hosts and complexes to the test of predictions of structure. Chart 1.1 summarizes the relationship between the biotic and synthetic systems linked together through the use of CPK models. These mechanical models were developed under the National Institutes of Health, Education, and Welfare, the National Science Foundation, and the American Society of Biological Chemists 'for constructing macromolecules of biological interest'. They are based on the crystal structures of biologically important compounds.[9] The atoms of the models summarize a large amount of data, which are additive when the model atoms are assembled to make molecules. Hence models of complexes of molecular weights of several hundred or a few thousand daltons contain an enormous amount of mechanically stored structural information applicable to either biotic or abiotic systems. When a new complex is modeled, then prepared, and its crystal structure determined and found to be that predicted by the CPK model, a cycle of operations is completed which acts as a test of the

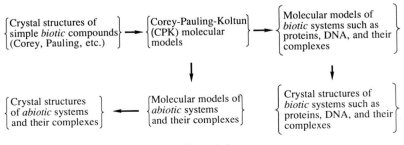

Chart 1.1

validity and usefulness of the procedure. Most of the hosts and complexes of this book could never have been envisioned without CPK models. The fact that over 150 cycles of this kind have been successfully completed has led us to have tremendous confidence in this type of leverage of the imagination.

In 1974, in reviewing our initial results, we wrote,

'Molecular evolution provides chlorophyll, hemoglobin, and vitamin B_{12} as instructive examples of complexes composed of metal ions and an array of ligands and counterions structured into cages by covalent bonds. Although the chemist lacks the time span of nature to produce equally interesting substances, he [or she] does have certain advantages. He has nature's example. He is not limited to functional groups that are stable to water. He can perform experiments at a variety of temperatures. He might know in advance what specific tasks his compounds are designed to perform'.[1]

CHAPTER 2

Spherands, Spheraplexes, and Their Relatives

> 'The aesthetic appeal of the structures of certain organic compounds not only embellishes the satisfaction of working with them, but frequently contributes to the motivation for their design and synthesis.'

An examination of the structures of naturally occurring alkali metal ion binders, ionophores, indicates them to be rich in heteroatoms embedded in five- or six-membered ring systems. Often these systems are incorporated into macrorings which provide for an orderly spacing of oxygen or nitrogen ligating sites whose electron pairs can face inward in low energy conformations. The numbers of their conformations are severely limited by the ring systems and by an artful use of steric effects. Chemical evolution of these ionophores has provided a series of ion-specific compounds preorganized to possess 6–10 ligating sites which sequentially replace water of solvation by conformational reorganization to give nesting or capsular complexes.

This chapter describes the results of investigations designed to create synthetic systems in which preorganization for binding is carried to an extreme. The object of this work is to determine how binding free energies, ion specificities, and rates of complexation and decomplexation correlate with the degree of preorganization of hosts, the nature of the ligating sites, and the character of the guest ion.

2.1 Spherands and Spheraplexes

Crystal structures in Section 1.5 illustrated how 18-crown-6 and [2.2.2]cryptand must undergo conformational reorganization during complexation with K^+. The free hosts do not contain cavities, but their methylene groups are turned inward filling the space ultimately occupied by K^+ in the corresponding coraplex and cryptaplex. In solution, small stereoelectronically complementary guests, such as H_2O, might occupy the cavity and would have to be replaced in the cavity by K^+. The free energy costs of such molecular reorganizations must come out of the free energy of complexation with wanted guests, such as K^+. Much higher reorganizational costs must be paid during the complexation of the K^+ by podand $CH_3O(CH_2CH_2O)_5CH_3$ than by 18-crown-6 (Section 1.3).

A group of systems of ligands was designed to be preorganized prior to complexation so that orbitals of unshared electron pairs of the ligating sites line a roughly spherical cavity enforced by a support structure of covalent bonds. These preorganized systems were named *spherands*. A spherand must be organized for binding during its synthesis rather than during its complexation. Spherands must be rigid enough so that their parts cannot rotate to fill their own cavities, which should be empty and little solvated. The crystal structures of a free spherand and of the host part of a spheraplex should be very similar. These conditions were fulfilled with spherands **1–5**,[1] **6** being a podand which is the open-chain counterpart of **1** in which two hydrogen atoms take the place of an Ar—Ar bond in **1**.[1]

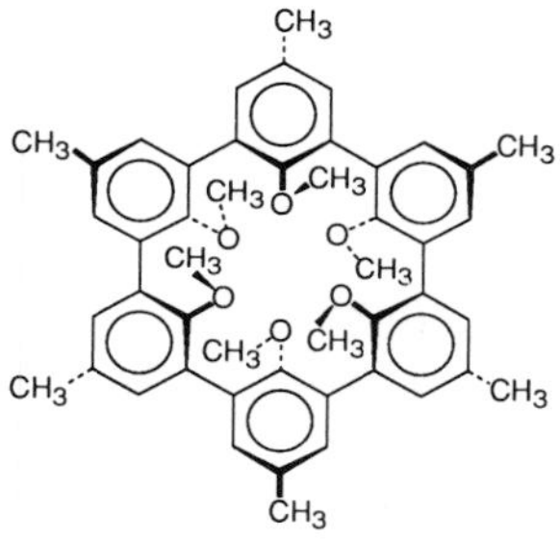

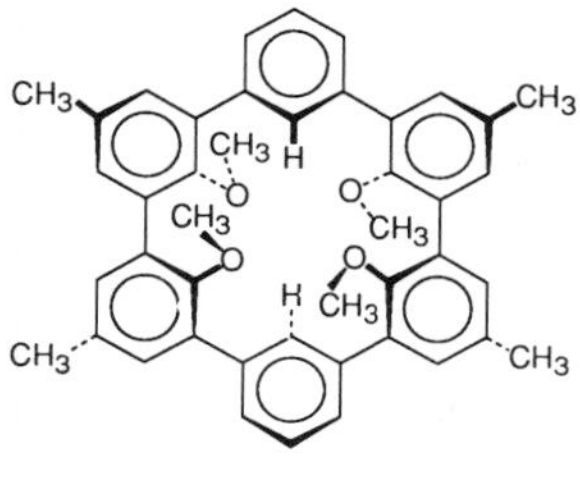

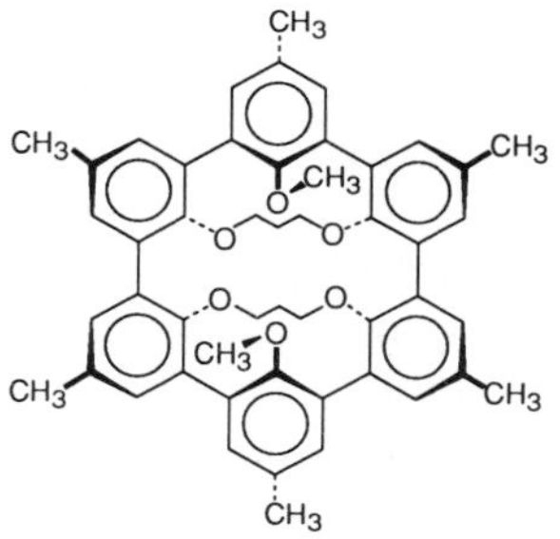

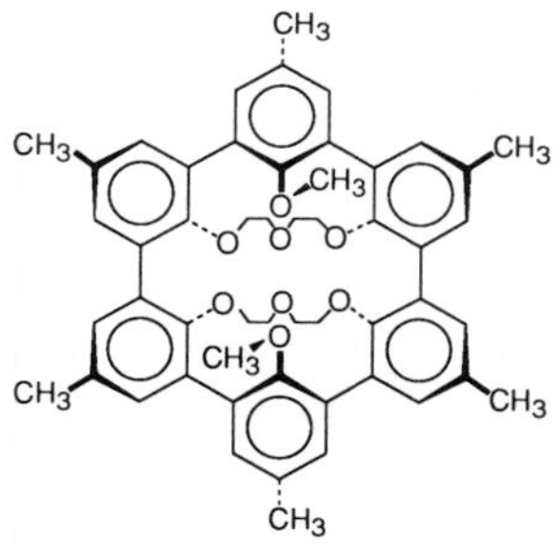

[1] D.J. Cram, T. Kaneda, R.C. Helgeson, S.B. Brown, C.B. Knobler, E.F. Maverick, and K.N. Trueblood, *J. Am. Chem. Soc.*, 1985, **107**, 3645; K.N. Trueblood, E.F. Maverick, and C.B. Knobler, *Acta Crystallogr.*, 1991, **B47**, 389.

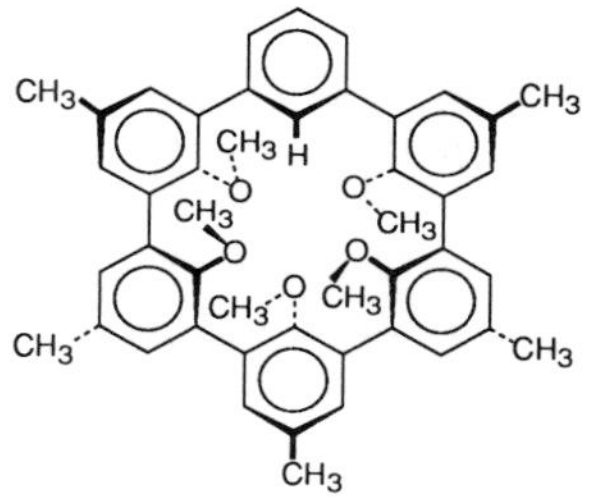

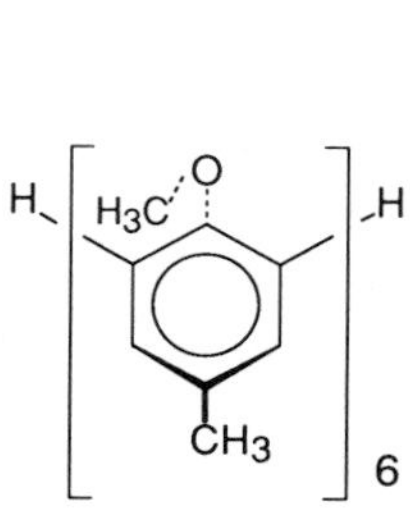

The spherands are at the end of a progression of ligand structures which vary in the extent to which guest cations organize their coordination shells. Thus spherands > cryptands > corands > podands > solvents in respect to the degree of preorganization of the host. Guest cations must exert maximum organization when solvent or solute molecules have to be collected and oriented. In podands, such as **6**, the molecular parts are already collected, but to obtain the roughly octahedral organization of oxygens found in $(OMe)_6$-spherand (**1**), about 500 conformations must be frozen out. Moreover, the oxygens of podand **6** must be desolvated during complexation, while those of **1** are shielded from solvation by the six aryls and six methyl groups.[1,2]

2.2 Syntheses of Spherands

The critical ring-closures in the syntheses of **1**·LiCl, **13**·LiCl, and **15**·LiCl required invention of a new reaction for converting aryl bromides to biaryls. The aryl bromide is lithiated, and the resulting aryllithium is oxidized with $Fe(AcAc)_3$ to produce aryl radicals which couple. When this procedure was applied to **7**, **9**, and **8**, the yields of spheraplex **1**·LiCl were 28, 7.5, and 3%, respectively. As expected, the yields decreased as the number of bonds being formed increased. Decomplexation of **1**·LiCl was driven by phase transfer by heating the spheraplex in 4:1 methanol–water at 125 °C. The free spherand crystallized from the medium, in which it is very insoluble. Similarly $(OMe)_5$-spherand **2** and $(OMe)_4$-spherand **3** were prepared in 6% and 12% yields respectively by cross-coupling one mole of triaryl **7** and 4 moles of triaryl **10** as starting materials. Bridged spheraplexes **13**·LiCl with two $O(CH_2)_3O$ bridges and **15**·LiCl with two $OCH_2CH_2OCH_2CH_2O$ bridges were prepared from triaryls **11** and **12** in 13% and 6% respective yields. They decomplexed to give **13** and **15** making use of phase transfer to drive the reactions, similarly to the decomplexation of **1**·LiCl.

When heated to 200 °C in pyridine, **1**·LiCl lost one methyl group to give **17**·Li, a salt of the spherand anion and lithium, which when heated in acid gave monophenol **18** (70% overall). This demethylation is undoubtedly an S_N2 reaction in which Li^+ acts as an electrophile and Cl^- or pyridine as a nucleophile. When monophenol **18** was treated with $NaOH-(CH_3)_2SO_4$, **1**·$NaCH_3SO_4$ was produced as an intermediate which was then methylated to produce **1**·NaCl.

Attempts to encapsulate K^+, Mg^{2+}, and Ca^{2+} resulted only in **1**·NaCl, the Na^+ being scavenged from bulk sources of the other bases. Thus $(OMe)_6$-spherand **1** appears to be unique in its ability to complex only Li^+ and Na^+.[1]

When the monophenol **18** was treated with CH_3MgI, methane was generated along with the mixed salt **21**, spherand·MgI, which was unstable. Attempts to recrystallize **21** gave spherand^{2-}·Mg^{2+}, **22**, characterized only by its ^1H NMR spectrum.[3] Apparently Mg^{2+} is a strong enough Lewis acid in its carceplex

---

[2] D.J. Cram and K.N. Trueblood, *Top. Curr. Chem.*, 1981, **98**, 43.
[3] C.B. Knobler, E.F. Maverick, and K.N. Trueblood, *J. Inclusion Phenom.*, 1992, **12**, 341.

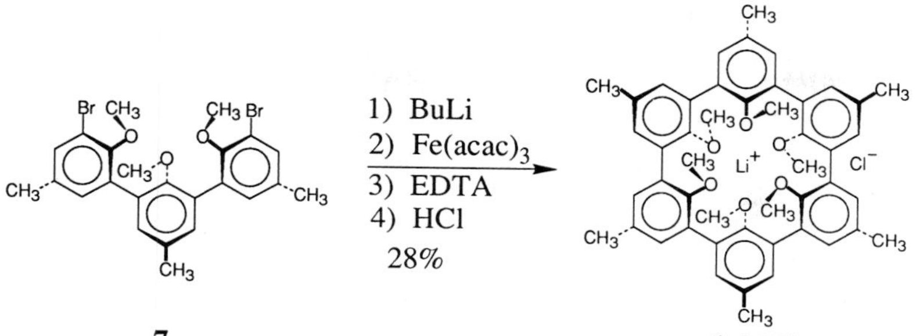

Spherands, Spheraplexes, and Their Relatives

7 → 1) BuLi 2) Fe(acac)$_3$ 3) EDTA 4) HCl 28% → **1**·LiCl

7

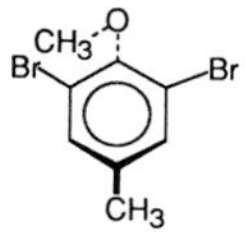

8 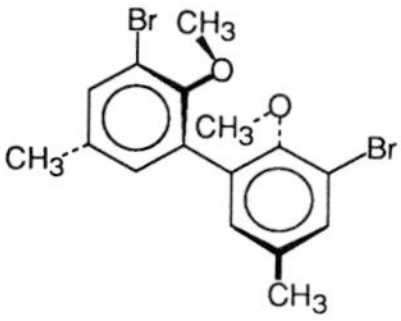 **9**

1·LiCl $\xrightarrow{\text{H}_2\text{O, CH}_3\text{OH}}_{125 \,°\text{C, 89\%}}$ **1**

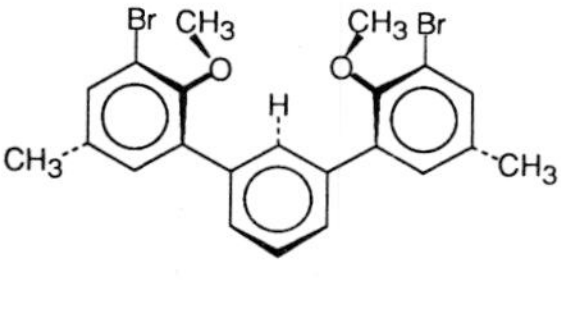

10 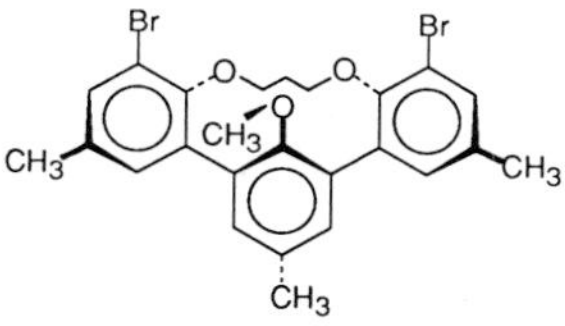 **11** 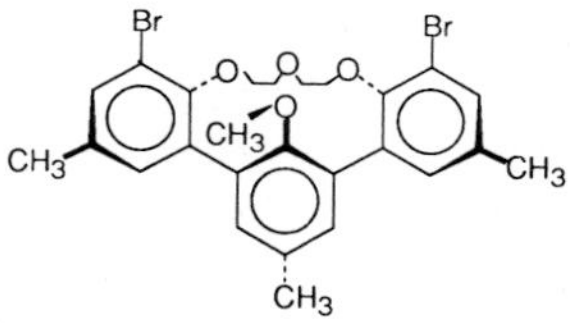 **12**

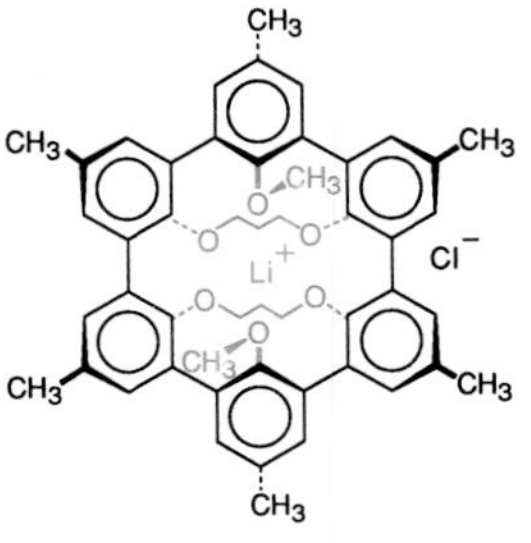

13·LiCl 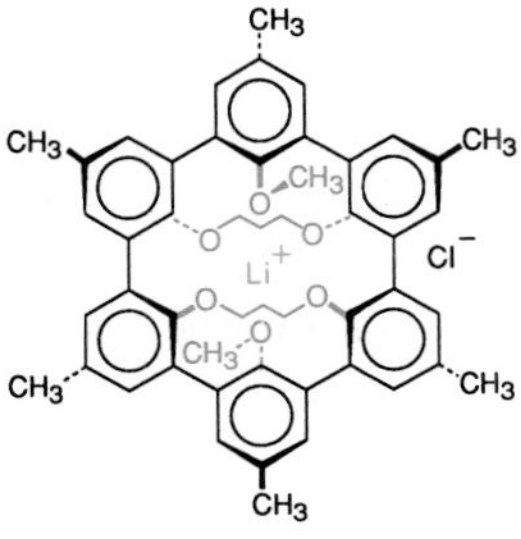 **14**·LiCl 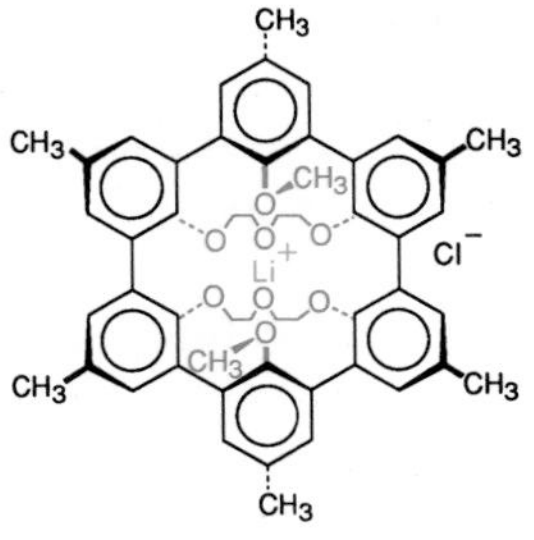 **15**·LiCl

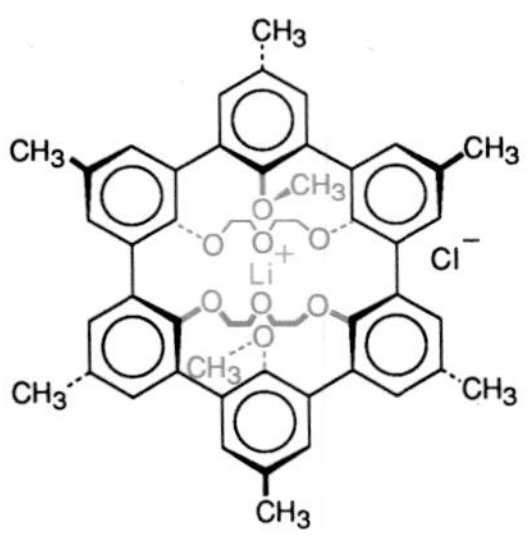

16·LiCl 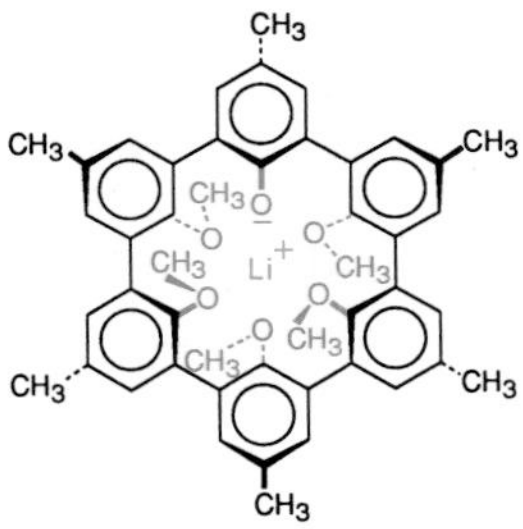 **17**·Li 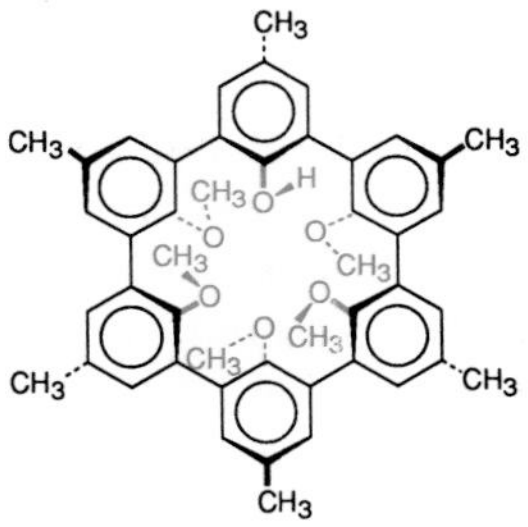 **18**

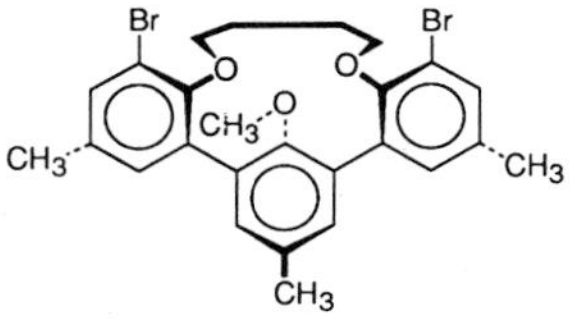

19 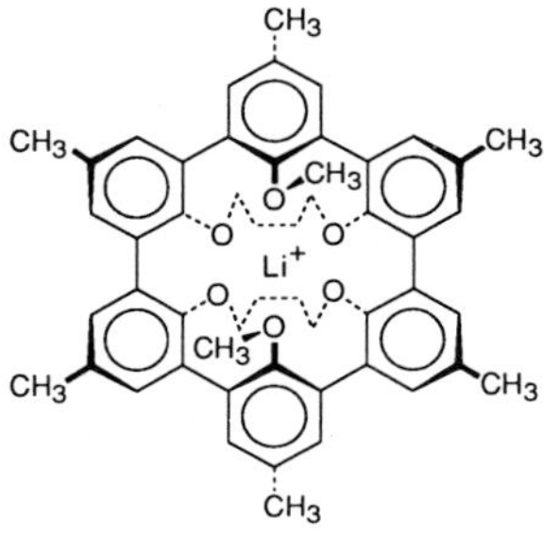 **20**

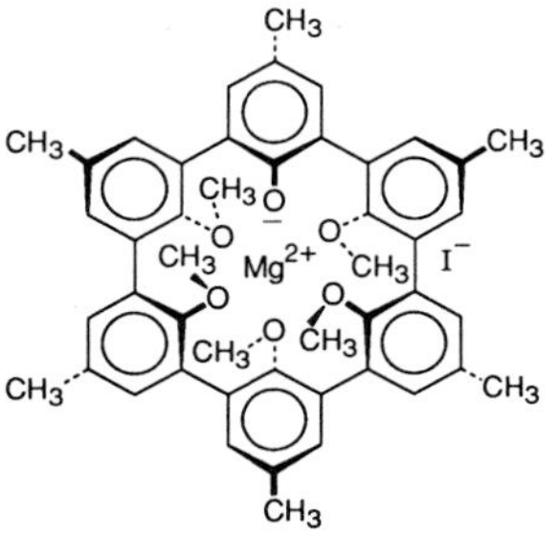

21 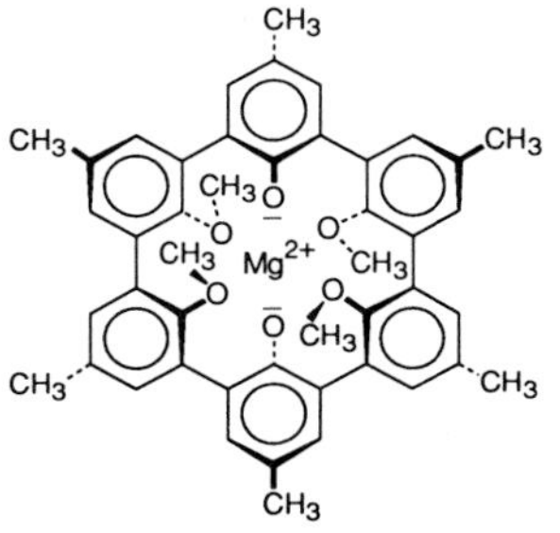 **22**

environment in the mixed salt **21** to make the ordinarily stable ArOCH$_3$ group very easily subject to nucleophilic substitution under mild conditions.

Although we were led to prepare **1** as a result of model examination, we were misled in the cases of **14** and **16**, *anti*-bridged stereoisomers of *syn*-bridged spherands **13** and **15**. Models of the *anti* hosts are easy to assemble, since the six aryl oxygens have an up–down-up–down-up–down arrangement as the eye 'circumnavigates' the structures. This alternation places the O—R—O bridges *anti* to one another in **14** and **16**. Experimentally stereoisomers **13** and **15**, whose two bridges are *syn* to one another, were obtained. Models of the two *syn* structures cannot be assembled without shaving substantial amounts of plastic from the four aryl oxygen atoms that terminate the bridges. The structures of our products in the syntheses were initially thought to be **14** and **16**, but crystal structure determinations proved them to be **13** and **15**. Thus models are excellent in predicting structures of complexes formed by mixing hosts and guests where the free energies of complexation are 5–20 kcal mol^{-1}, but are less reliable for predicting structures when complexes are prepared by high-energy aryl-radical coupling reactions. With variations in the reaction conditions for synthesis, small amounts of **14**·LiCl were prepared. This compound was found to decomplex with about the same degree of difficulty as the parent system, **1**·LiCl.[3] Lengthening the methylene bridge spanning the oxygens to four methylenes in dibromide **19** led upon ring closure to the *syn*-coupled spherand **20**, whose structure was established by crystal structure determination.[3]

2.3 Crystal Structures of Spherands and Spheraplexes

2.3.1 Non-bridged Hosts

The crystal structures of (OMe)$_6$-spherand **1**, **1·Li**$^+$, and **1·Na**$^+$ are depicted in **23**, **24**, and **25** respectively. Structure **23** has a beautiful snowflake-like appearance. Its oxygens have no conformational freedom, but exist in an approximately octahedral arrangement that defines an enforced cavity. As expected from CPK model examination, the hole in **23** is lined with 24 electrons, which are shielded from solvation by the six aryl and six methyl groups. The crystal structures of free host **1** (**23**) and of complexed host in **24** and **25** are very similar, so upon complexation the host undergoes very little reorganization. The structures exhibit D_{3d} (**23** and **24**) or close to D_{3d} (**25**) molecular symmetry.[1]

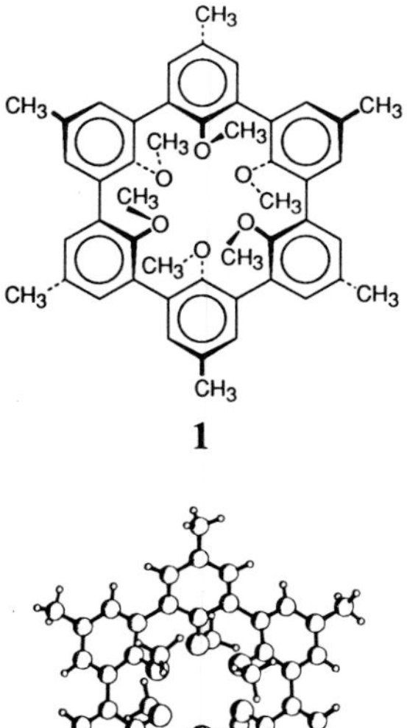

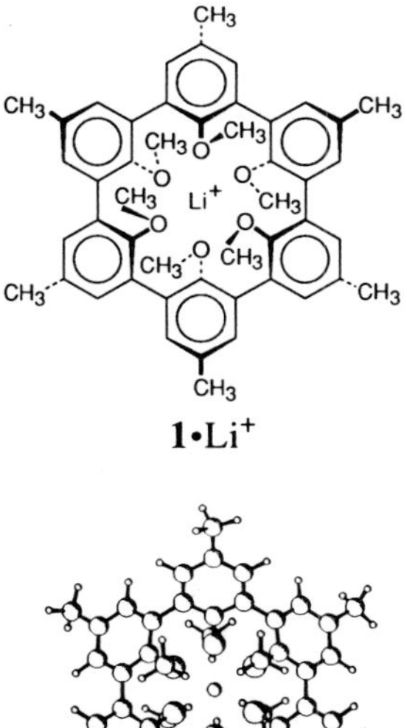

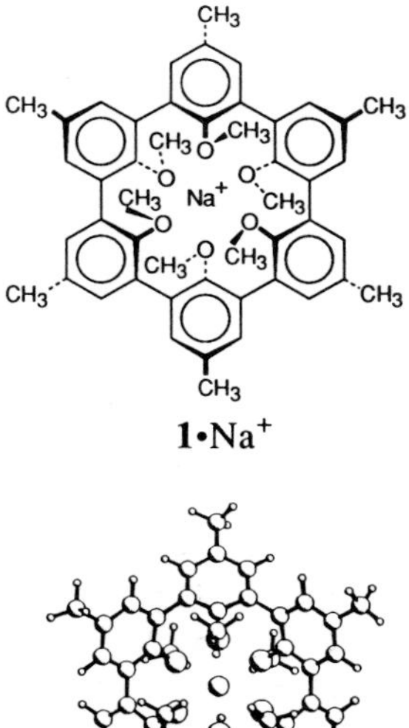

The hole diameters in **23**, **24**, and **25** are 1.62, 1.48, and 1.73, respectively, and the Li$^+$----O distances are all the same, as are the Na$^+$----O distances, thus the hole must shrink slightly to embrace Li$^+$, and expand slightly to accommodate Na$^+$. The change in hole diameter is accomplished through a combination of Ar—Ar dihedral-bond angle changes (52°–60° as limits), and small deformations of C—C and C—O bond angles throughout the structure that modify the distances between the oxygens and the best plane of their attached aryls. All of the six pseudo-*ortho* O-to-O distances in **24** (crystal structure of (OMe)$_6$-spheraplex, **1·Li**$^+$), at 2.78 Å, are close to normal, whereas these distances in the free host **1** are above normal at 2.92 Å.

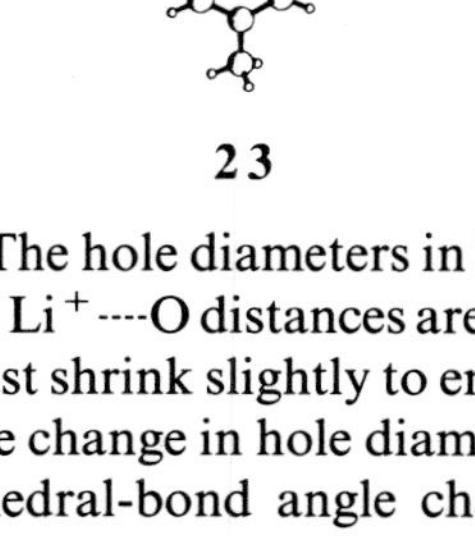

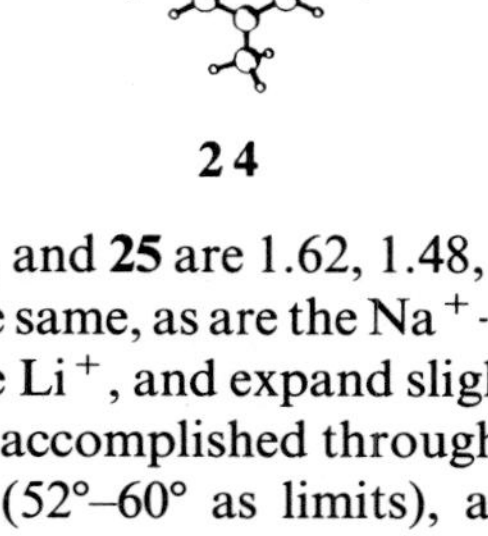

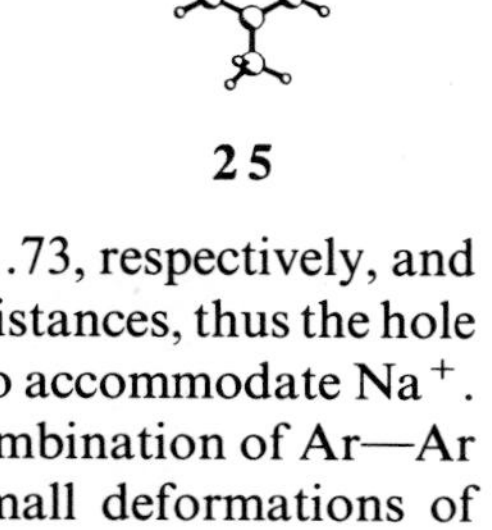

2.3.2 Bridged Hosts

The crystal structures of *syn*-bridged spheraplexes **26** and **28** are depicted in stereoviews **27** and **29**, respectively. Both structures have a near-mirror plane bisecting the lithium and CH_3O oxygens. The effect of the two bridges in each structure is to 'squeeze out' of ligating range *one* of the CH_3O oxygens to provide long Li^+----OCH_3 distances of 2.89 and 3.47 Å in **27** and **29**, respectively. The average Li^+----O ligating distance in **27** is 2.04 Å (range, 2.00–2.08 Å), and in **29** is 2.253 Å (range, 2.03–2.43 Å) compared with 2.14 Å in **1·Li^+**, **24**.[1]

Distances are measured atom center to atom center. The effective Li^+ diameters were calculated by averaging the short Li^+----O distances, subtracting the radius of oxygen (1.40 Å),[4] and multiplying by two. The diameters calculated this way vary with the number of ligands as follows: 1.28 Å with five oxygens in **27**; 1.48 Å with six oxygens in **24**; and 1.72 Å with seven oxygens in **29**. These diameters compare with those taken from complexes in which Li^+ organizes its ligands as follows: four-coordinate, tetrahedral, 1.12 Å; five-coordinate, 1.36 Å; six-coordinate, octahedral, 1.50 Å. No seven-coordinate distances were found.[1]

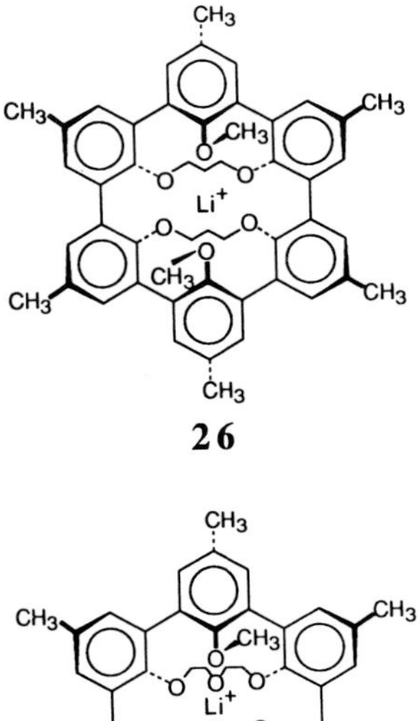

26 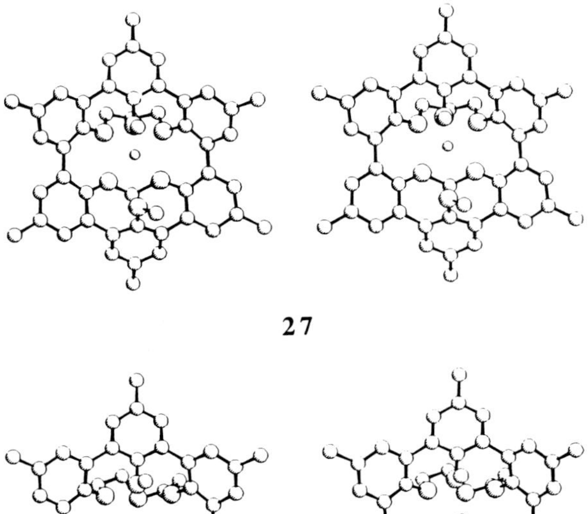 **27**

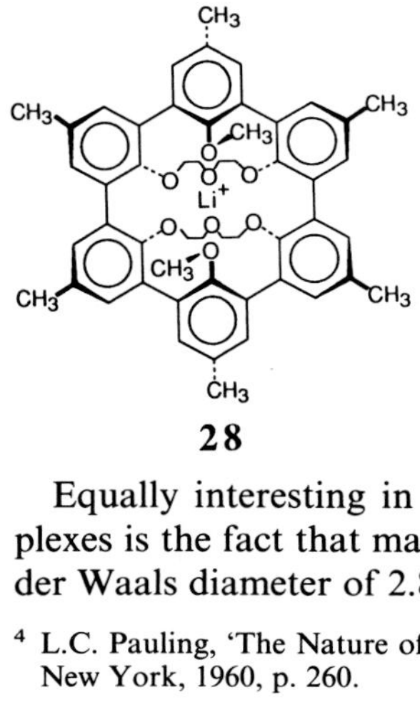

28 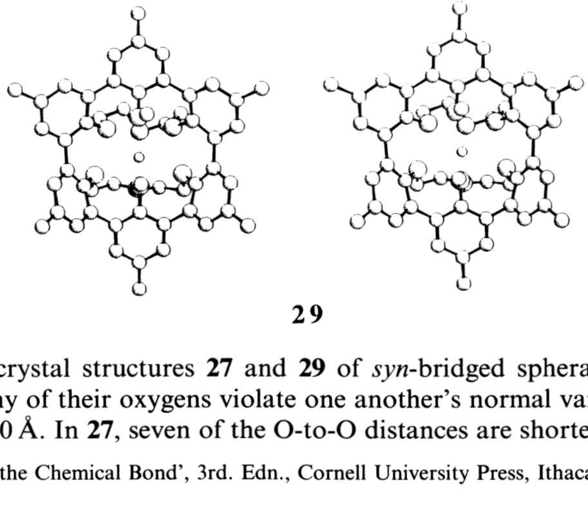 **29**

Equally interesting in crystal structures **27** and **29** of *syn*-bridged spheraplexes is the fact that many of their oxygens violate one another's normal van der Waals diameter of 2.80 Å. In **27**, seven of the O-to-O distances are shorter

---

[4] L.C. Pauling, 'The Nature of the Chemical Bond', 3rd. Edn., Cornell University Press, Ithaca, New York, 1960, p. 260.

than normal, all six of the pseudo-*ortho* and one of the pseudo-*meta* interoxygen distances averaging 2.64 Å, (range 2.50–2.73). In **29**, eight of the O-to-O distances are shorter than normal, the two bridge-terminating, pseudo-*ortho* O-to-O distances, the two shortest of the four CH$_2$O-to-pseudo-*ortho* ArOCH$_3$, and the four near O-to-O distances, in the two–O(CH$_2$CH$_2$O)$_2$ bridges average 2.67 Å (range 2.57–2.79 Å).

The crystal structure determinations of *anti*-bridged spheraplex **14·Li$^+$** provided two different, but very similar, spheraplexes, each having a center of symmetry (C_i point group). The Li$^+$ diameter of each complex (**30** is one of them) is 1.44 Å, and the six ligating oxygens are roughly octahedrally arranged about the Li$^+$, except that the four oxygens involved in the two–OCH$_2$CH$_2$CH$_2$O–bridges have shorter than usual Li$^+$----O ligating distances (1.94 and 1.95 Å), and the Li$^+$----OCH$_3$ ligating distances are longer than usual (2.46 and 2.39 Å). These contrast with the 2.14 Å distances in **24** (**1·Li$^+$**) and the average 2.04 Å distance (five ligating oxygens) in the *syn* isomer **13·Li$^+$**. Four of the pseudo-*ortho* O-to-O distances in **30** (both molecules) are less than the normal 2.80 Å, averaging 2.74 and 2.75 Å in the two molecules. The other two O-to-O pseudo-*ortho* distances are 2.83 and 2.83 Å, to provide average distances in each molecule of 2.77 and 2.78 Å, both lower than the normal 2.80 Å distance. The pseudo-*ortho* distances are much closer to normal than those in the *syn* isomer **27**, seven of whose O-to-O (six are pseudo-*ortho*) distances are less than 2.80 Å to give an average of 2.64 Å. Thus the *syn*-spheraplex is much more compressed than the *anti*-spheraplex.

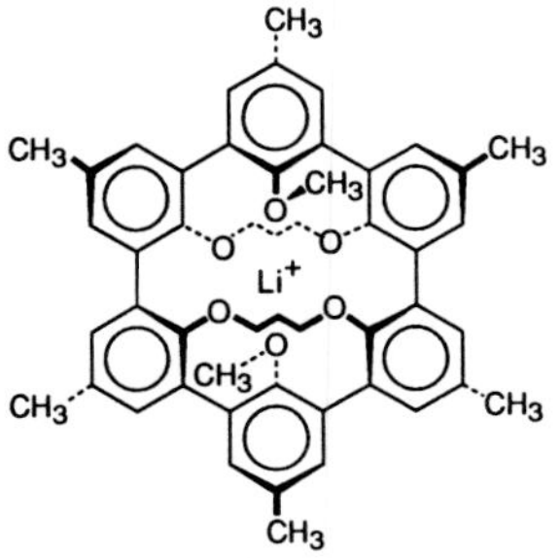

14·Li$^+$

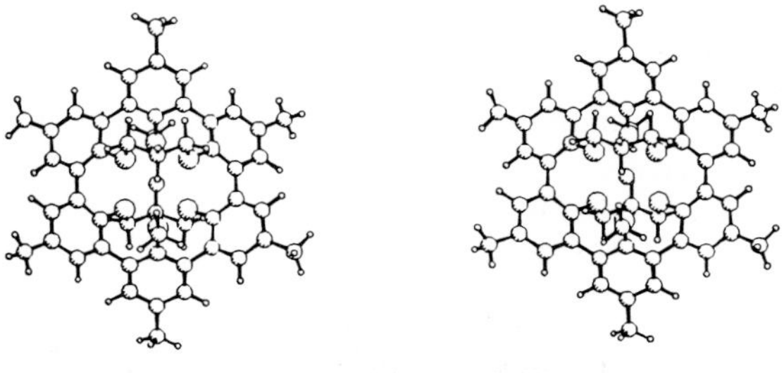

30

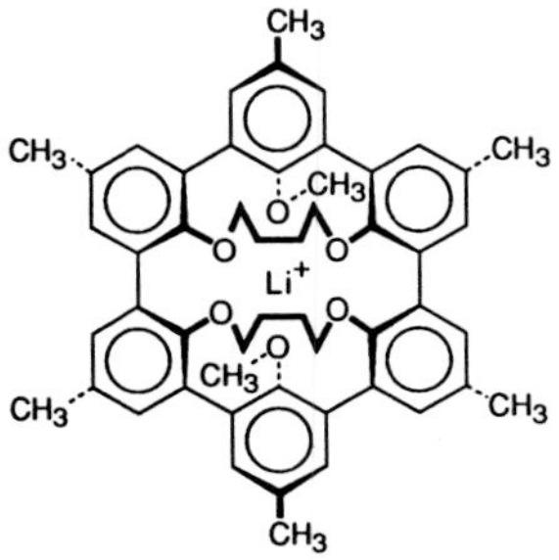

31

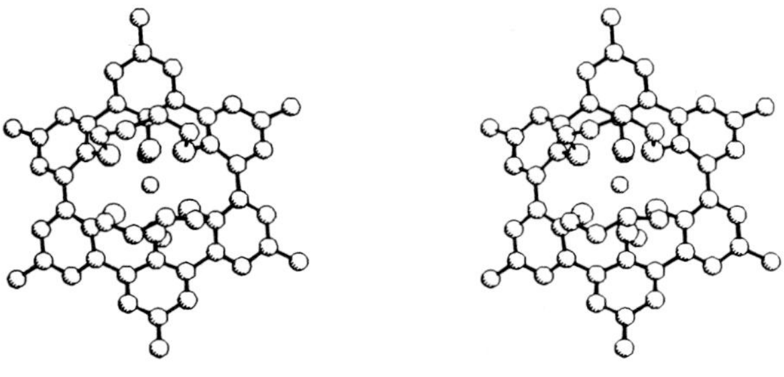

32

Other interesting comparisons are provided by the crystal structures of the *syn* isomers, **27** containing $O(CH_2)_3O$ and **32** with $O(CH_2)_4O$ bridges. In **32**, six oxygens ligate Li^+, the average $Li^+\cdots O$ distance being 2.17 Å (2.05–2.35 Å range), the effective diameter of Li^+ being 1.54 Å. In **27**, only five oxygens ligate Li^+, the average $Li^+\cdots O$ distance being 2.04 Å (2.00–2.08 Å). Thus the effect of the space occupation of the extra CH_2 in the bridges of **32** is to push the sixth oxygen into a position to ligate Li^+ which is too remote in **27** for contact. As a result, the effective diameter of Li^+ in **32** at 1.54 Å is closer to that in **24** (1.48 Å) than in **27** (1.28 Å). In **32**, the six pseudo-*ortho* O-to-O distances average 2.63 Å (2.53–2.71 Å) which is very close to the 2.64 Å (range 2.51–2.73 Å) of its *syn*-bridged propylene analogue **27**. Interestingly, the oxygens of **32** exhibit a more perfect octahedral arrangement than those of **24** or **27**. The free hosts of **14** and **20** were not obtained in adequate amounts to allow their study.[3] The fact that the more highly strained *syn*-bridged isomers **26**, **28**, and **31** dominated the products of ring closure in their syntheses remains a mystery that invites further study.

2.4 Hemispherand and Hemispheraplex Crystal Structures

2.4.1 Three Preorganized Ligands

Hemispherands are hosts at least half of whose ligating heteroatoms are preorganized for binding prior to complexation, the remaining structure being corand in character. A prototypical hemispherand is depicted in **33** and its crystal structure in **33a**. In CPK models of **33**, the three anisole oxygens are arranged in an up–down–up manner with their methyl groups diverging from the cavity. The cycle is completed with a $(CH_2OCH_2)_3$ conformationally flexible bridge, two of whose methylenes can turn inward to fill the cavity. The crystal structure **33a** shows that the cavity is filled with two methylene groups which must be displaced by any guest upon complexation.[5]

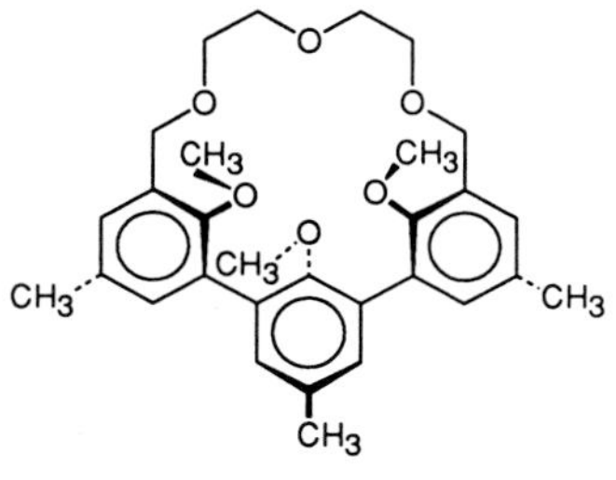

33 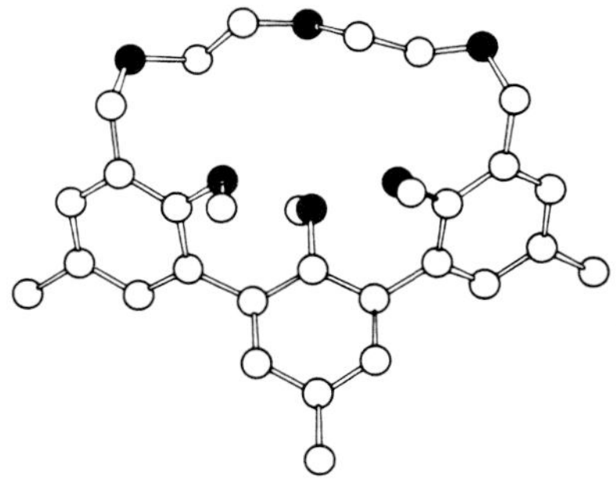   **33a**

[5] G.M. Lein and D.J. Cram, *J. Am. Chem. Soc.*, 1985, **107**, 448.

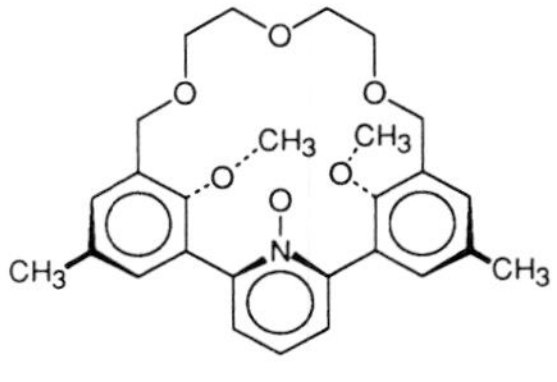

Spherands, Spheraplexes, and Their Relatives

34

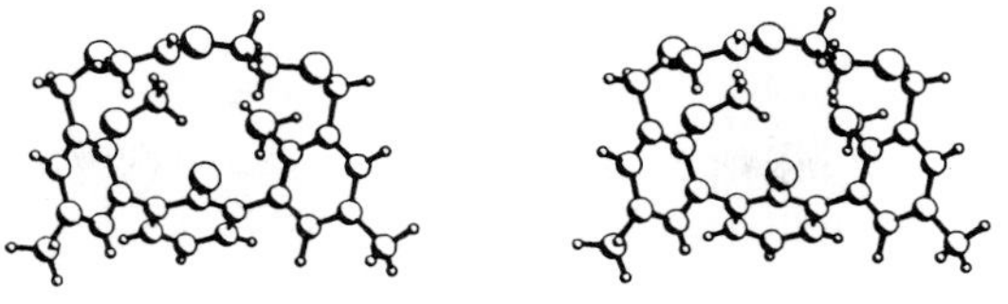

34a

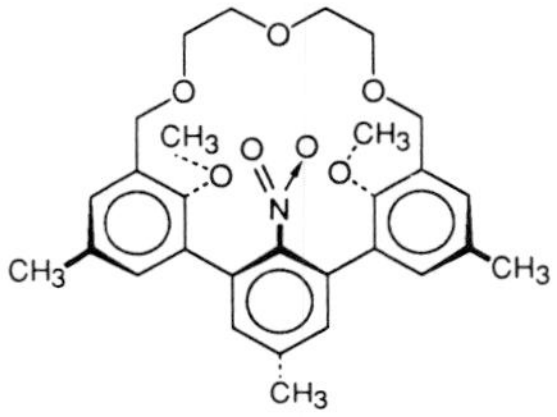

35

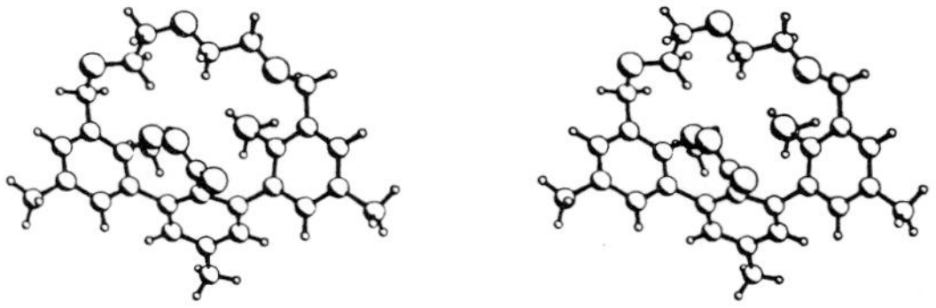

35a

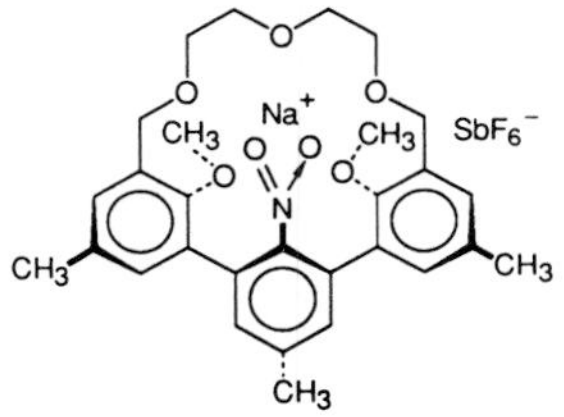

36

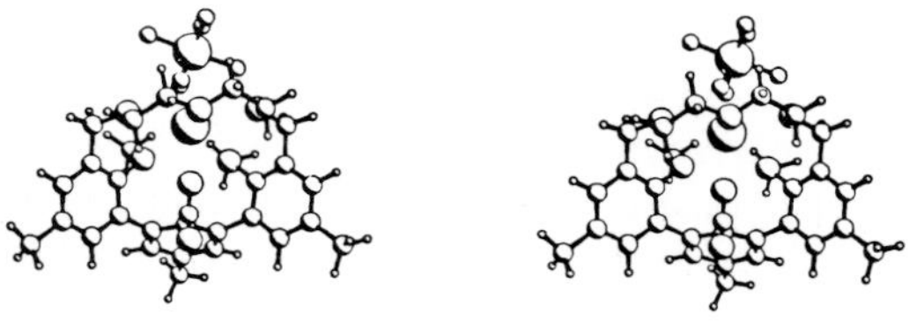

36a

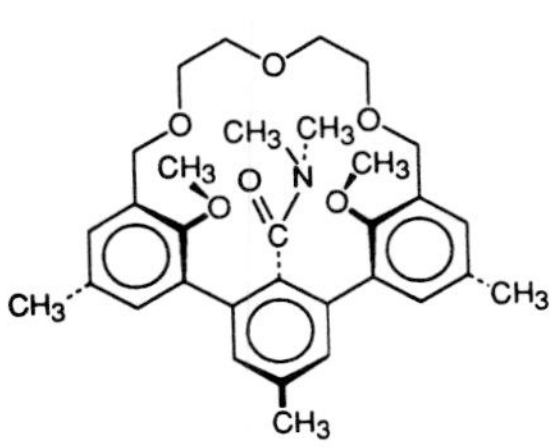

37

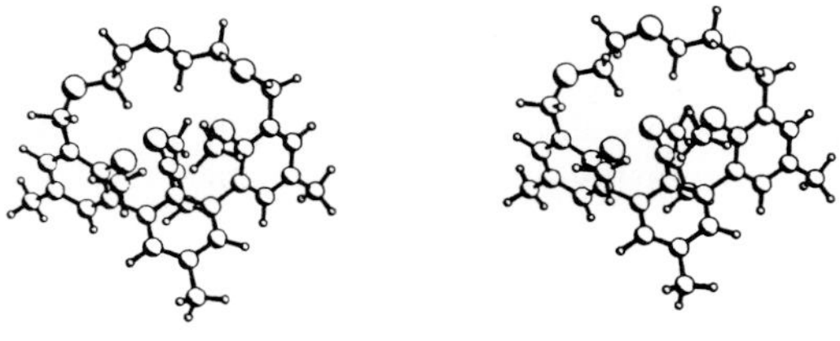

37a

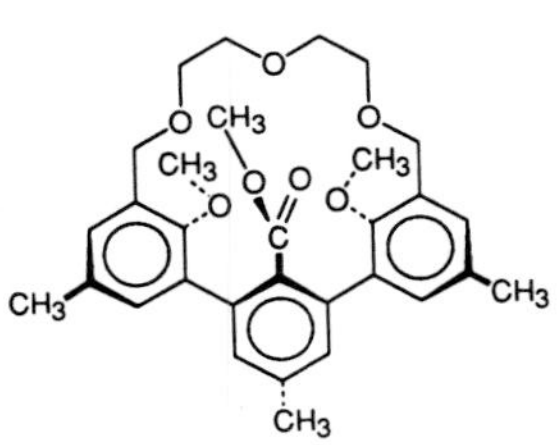

38

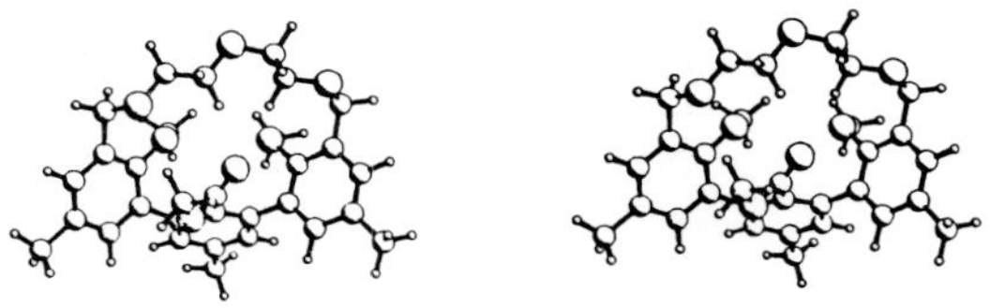

38a

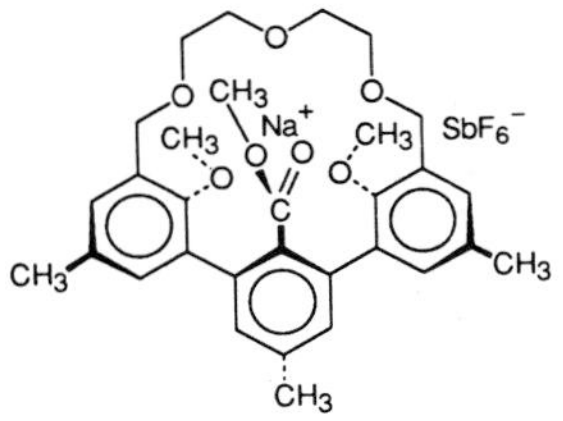

39 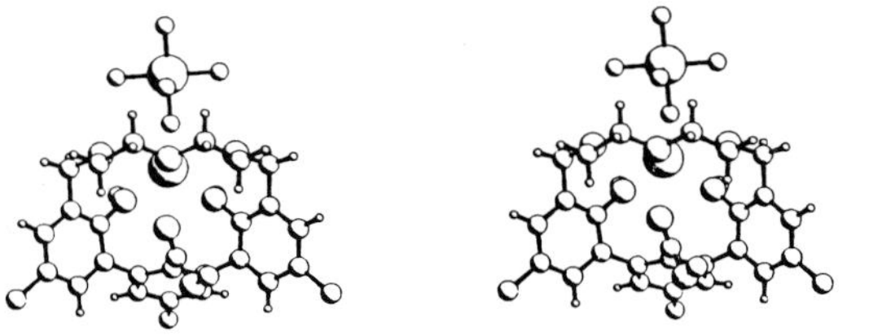 **39a**

A wide range of functional groups were preorganized for complexation through their attachment to aryls substituted in their 2,6-positions by anisyl groups, as in hemispherands **34**, **35**, **37**, and **38**, whose crystal structures are shown as stereoviews **34a**, **35a**, **37a**, and **38a**. The stereoviews **36a** and **39a** of the two hemispheraplexes **36** and **39** illustrate how the hosts **35** and **38** must reorganize upon complexation.[6]

2.4.2 Four Preorganized Ligands

Another family of hemispherands contains four preorganized anisyl units, of which **40** is the prototype. Crystal structures **40**·Li^+·H_2O, **40**·Na^+·H_2O, **40**·KSCN, and **40**·RbPicrate were obtained and are depicted in the stereoviews. In those cases where the counterion or water also ligated the alkali metal ion, these species are included. Notice that as the diameter of the guest ions increases, the host adapts by a variety of conformational adjustments. Complex **40**·Li^+·H_2O is capsular, **40**·Na^+·H_2O and **40**·K^+SCN^- are nesting, whereas **40**·Rb^+Pic^- is perching. The average Ar—Ar dihedral angles are as follows: **40**, 56.5°; **40**·Li^+·H_2O, 50.8°; **40**·Na^+·H_2O, 55.3°; **40**·K^+·SCN^-, 67.2°; and **40**·Rb^+Pic^-, 65.7°. The angle between the best plane of the six oxygens and the 1,2-$C_6H_4O_2$ group also changes: **40**, 77.9°; **40**·Li^+·H_2O, 48.2°; **40**·Na^+·H_2O, 35.8°; **40**·K^+·SCN^-, 38.7°; and **40**·Rb^+·Pic^-, 43.7°. Thus the system uses these (and more) available structural stratagems to minimize its energy by adjusting to the increasing diameter of the guest.[7]

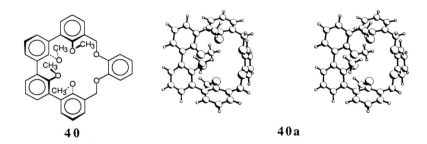

40 **40a**

---

[6] J.A. Bryant, R.C. Helgeson, C.B. Knobler, M.P. de Grandpre, and D.J. Cram, *J. Org. Chem.*, 1990, **55**, 4622.

[7] J.A. Tucker, C.B. Knobler, I. Goldberg, and D.J. Cram, *J. Am. Chem. Soc.*, 1989, **54**, 5460.

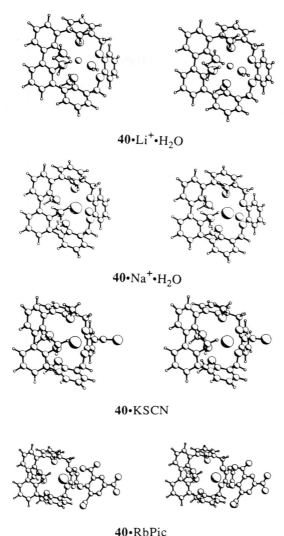

40·Li⁺·H₂O

40·Na⁺·H₂O

40·KSCN

40·RbPic

2.4.3 Cryptahemispherands

The conformational adaptivity of the cryptands[8] was merged with the preorganized spherands in the cryptahemispherands,[9] whose Na^+, K^+, and Cs^+ complexes provided good crystal structures (**41**·Na^+, **42**·KSCN, **43**·Na^+, **43**·K^+ H₂O, and **43**·Cs^+ H₂O). In all of these capsular structures, the three anisyl groups possess the usual up–down–up arrangement with the unshared electron pairs of the oxygens oriented inward toward the guest. The bridging hetero-

---

[8] B. Dietrich, J-M. Lehn, and J.P. Sauvage, *Tetrahedron Lett.*, 1969, 2885–2892.
[9] D.J.Cram, S.P. Ho, C.B. Knobler, E.F. Maverick, and K.N. Trueblood, *J. Am. Chem. Soc.*, 1986, **108**, 2989.

atoms' unshared electron pairs are also oriented inward in all of the complexes, even though some of the electron pairs are outside of ligating range. An impression of the variation of cavity size as the hosts and guests are changed is provided by the N-to-N distances (Å) which vary as follows: **41·Na$^+$**, 4.64; **42·KSCN**, 5.48; **43·Na$^+$**, 6.68; **43·K$^+$·H$_2$O**, 6.36; and **43·Cs$^+$·H$_2$O**, 6.67. Notice that K$^+$ shrinks the cavity in **43·K$^+$** more than does Na$^+$ in **43·Na$^+$**, which reflects the fact that Na$^+$ is just too small to ligate all nine heteroatoms at the same time, whereas K$^+$ is complementary enough to the cavity to draw the heteroatoms toward it.

The effective diameters of Na$^+$ and K$^+$ in these complexes vary with the number of ligating atoms. The Na$^+$ in **41·Na$^+$** exhibits strong interactions with its seven ligands, the Na$^+$----N and Na$^+$----O distances being close to the standard values of 2.66 and 2.35 Å, respectively.[4] The diameter of Na$^+$ ligated to seven heteroatoms is 2.14 Å, greater than the diameter of 1.75 Å for Na$^+$ bound to six ligands in **1·Na$^+$**. In **42·K$^+$SCN$^-$**, K$^+$ interacts strongly with seven of its eight ligands, the central methoxy oxygen being 3.21 Å from the K$^+$ as compared to the average K$^+$----O distances of 2.84 Å (range 2.73–3.21 Å). The resulting K$^+$ diameter is 2.96 Å. In **43·Na$^+$**, the two Na$^+$----N distances are 3.18 and 3.64 Å, well above the standard value of 2.45 Å.[4] Five of the Na$^+$----O distances are between 2.53 and 2.69 Å, while the other two are 2.78 and 2.85 Å, compared to the standard distance of 2.35 Å.[4] Because of spatial constraints of the ligands, the cavity of **43** in **43·Na$^+$** is notably dissimilar from that of **43** in **43·Cs$^+$**. In effect, Na$^+$ in **43·Na$^+$** ligates the five oxygens close to one of the nitrogens, and the cavity is elsewhere unfilled. The apparent Na$^+$ diameter is 2.56 Å, much greater than normal. In solution, the Na$^+$ in **43·Na$^+$** undoubtedly 'rattles' in the cavity.

The normal diameter of K$^+$ is large enough in **43·K$^+$** for all nine ligating heteroatoms to be within reach of the metal ion. The K$^+$----N distances are 3.22 and 3.25 Å, which compare to the standard values of 2.83 Å. The average M$^+$----O distance is 2.88 Å compared to the standard value of 2.73 Å.[4] The diameter of this K$^+$ with nine ligands is 3.08 Å, leaving out the water molecule whose K$^+$----O distance is 3.44 Å.

Host **43** appears to be a nearly ideal structure to bind Cs$^+$. Comparisons of the 3.38 Å average Cs$^+$----N and the 3.03 Å average Cs$^+$----O distances in **43·Cs$^+$** with the respective standard distances of 3.19 and 3.09 Å, show that the Cs$^+$ ion just about fills the cavity. The effective diameter of the 9-ligated Cs$^+$ is 3.36 Å.

41·Na$^+$ **41·Na$^+$**

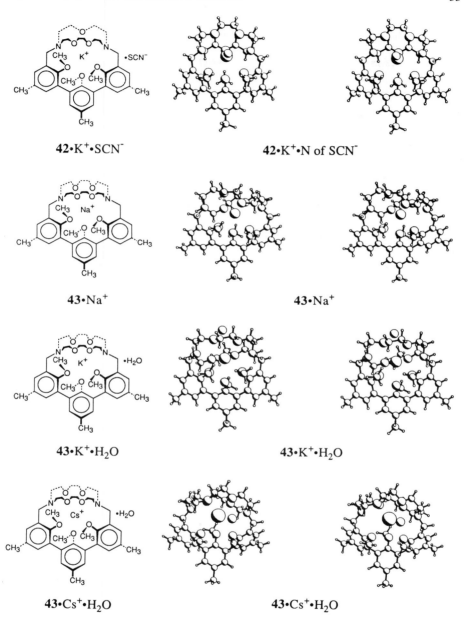

42·K⁺·SCN⁻   42·K⁺·N of SCN⁻

43·Na⁺   43·Na⁺

43·K⁺·H₂O   43·K⁺·H₂O

43·Cs⁺·H₂O   43·Cs⁺·H₂O

2.5 Correlation of Structure and Binding

2.5.1 Determination of Binding Power

To assess the binding power of our hosts, we developed a simple means of determining their binding free energies of complexation with the picrate salts of

Li^+, Na^+, K^+, Rb^+, Cs^+, NH_4^+, $CH_3NH_3^+$, and $t\text{-}BuNH_3^+$. The guest salts were distributed between $CDCl_3$ and D_2O at 25 °C in the presence and absence of host. The amounts of picrate distributed in each layer were measured spectrometrically. From the results, K_a (M^{-1}) and $-\Delta G^0$ values (kcal mol^{-1}) were calculated using Equation 2.1. This method was rapid and convenient for obtaining $-\Delta G^0$ values at 25 °C ranging from about 6–16 kcal mol^{-1} in $CDCl_3$ saturated with D_2O.[10] Higher values up to 22 kcal mol^{-1} were obtained by equilibration experiments between complexes of known and those of unknown $-\Delta G^0$ values. Others were determined from measured k_{-1} and k_1 values, all in the same medium at 25 °C.[11,12]

$$H + GPic \underset{k_{-1}}{\overset{k_1}{\rightleftarrows}} H \cdot G \cdot Pic \quad K_a = k_1/k_{-1} \quad \Delta G^0 = -RT \ln K_a \quad (2.1)$$

2.5.2 Corand and Anisyl Hemispherand Binding Comparison

Chart 2.1 summarizes the pattern of changes in binding of the simple hemispherands, **33**, **44**, **46**,[2] **47**,[13] and standard lipophilic corand **45**[2] in $CDCl_3$ at 25 °C. All of these hosts possess 18-membered macrorings and contain six oxygens. Compared to corand **45**, introduction of two anisyl units in **46** decreases, and three anisyls in **33** increases the general binding power of the systems. The presence of a stiffening additional bridge and another oxygen in **44** makes it the strongest general ion binder of this group, which peaks at K^+ with $-\Delta G^0$ at about 13.8 kcal mol^{-1}. Introduction of a fourth anisyl unit, as in **47**, results in maximum binding of Na^+ at 13.7 and K^+ at 10.8 kcal mol^{-1}. The hemispherands are generally better binders than the corands, because they are at least half preorganized.

Table 2.1 summarizes data for hosts of general structure **48** binding the eight cations in $CDCl_3$ at 25 °C.[6] These hosts **48A–48M** provide a range of $-\Delta G^0$ values from <6 to 15.1 kcal mol^{-1}, and are arranged generally in decreasing order of their binding power. Compound **48A** with a central $CON(CH_3)_2$ group is the best binder, being particularly good for Li^+, Na^+, and K^+ with $-\Delta G^0$ values (kcal mol^{-1}) of 11.7, 15.1, and 12.9, respectively. All of the other systems show a much lower affinity for Li^+ (<6 to 7.7 kcal mol^{-1}). The systems **48B–48F** are similar in their binding of the alkali metal and NH_4^+ ions, maximizing with Na^+ at 11.0–12.4 kcal mol^{-1}, and dropping to Cs^+ at 6.9–9.0 kcal mol^{-1}. These systems include central functional groups CO_2CH_3, OCH_3, $N\rightarrow O$, $(R_2N)_2C{=}O$, and $ArNO_2$. Systems **48G–48K** containing pyridyl, anilino, furanyl, phenolic, and CH_3S groups are relatively poor and nondiscriminating binders. Noteworthy about **48L** and **48M**, containing $SOCH_3$ and SO_2CH_3 groups, is their high ability to discriminate in binding Na^+ over Li^+ and K^+. Thus, **48L** and **48M** bind Na^+ 3.7 and 2.8 kcal mol^{-1} better than

---

[10] R.C. Helgeson, G.R. Weisman, J.L. Toner, T.L. Tarnowski, Y. Chao, J.M. Mayer, and D.J. Cram, *J. Am. Chem. Soc.*, 1979, **101**, 4928.
[11] D.J. Cram and G.M. Lein, *J. Am. Chem. Soc.*, 1985, **107**, 3657.
[12] D.J. Cram and S.P. Ho, *J. Am. Chem. Soc.*, 1985, **107**, 2998.
[13] S.P. Artz and D.J. Cram, *J. Am. Chem. Soc.*, 1984, **106**, 2160.

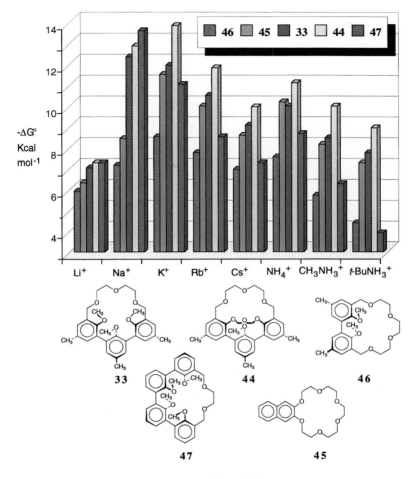

Chart 2.1

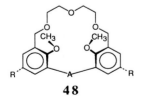

R = CH$_3$ or (CH$_3$)$_3$C, and A = a variety of ligating groups preorganized by two flanking anisyl moieties.

Li$^+$ and 3.8 and 3.2 kcal mol^{-1} better than K$^+$, respectively. Generally, Rb$^+$ and NH$_4^+$ have $-\Delta G^0$ values less than 1 kcal mol^{-1} from one another, although the absolute values range from < 6 to > 10 kcal mol^{-1}. The $-\Delta G^0$ values for host complexing CH$_3$NH$_3^+$ and (CH$_3$)$_3$CNH$_3^+$ are highest with pyridine as the central group, **48G**, (11.1 and 11.7 kcal mol^{-1}, respectively).

Table 2.1 Binding free energies ($-\Delta G^0$, kcal mol^{-1}) of hosts **48** for picrate salt guests at 25°C in CDCl$_3$ saturated with D$_2$O

Compound	Host Central A group	Guest cation							
		Li$^+$	Na$^+$	K$^+$	Rb$^+$	Cs$^+$	NH$_4^+$	MeNH$_3^+$	t-BuNH$_3^+$
48A	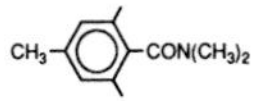 CH$_3$–⌬–CON(CH$_3$)$_2$	11.7	15.1	12.9	10.8	9.1	9.8	8.6	8.9
48B	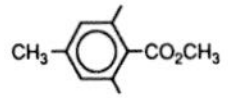 CH$_3$–⌬–CO$_2$CH$_3$	7.2	12.4	10.9	8.4	6.9	7.8	6.4	<6
48C	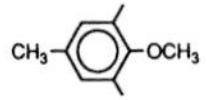 CH$_3$–⌬–OCH$_3$	7.0	12.2	11.8	10.5	9.0	9.9	8.2	7.7
48D	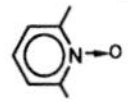 N-oxide pyridine	6.8	12.2	11.9	10.1	8.8	9.5	9.1	9.9
48E	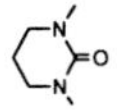 cyclic urea	6.7	12.0	11.3	9.9	8.7	9.3	9.1	9.5
48F	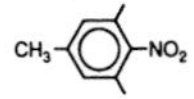 CH$_3$–⌬–NO$_2$	7.1	11.0	10.6	9.1	7.8	8.3	7.0	<6
48G	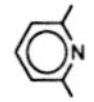 pyridine	7.2	10.8	10.9	10.1	9.7	10.8	11.1	11.7
48H	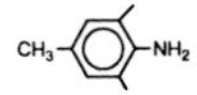 CH$_3$–⌬–NH$_2$	6.9	9.3	9.0	7.8	7.2	7.4	6.6	
48I	furan	<6	8.9	9.2	8.2	7.4	7.5	6.8	6.8
48J	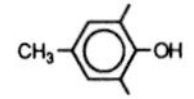 CH$_3$–⌬–OH	6.8	7.9	8.0	6.7	6.7	6.4	<6	<6
48K	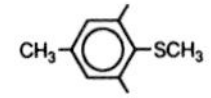 CH$_3$–⌬–SCH$_3$	<6	10.8	10.5	8.8	7.8	8.3	7.3	6.0
48L	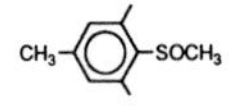 CH$_3$–⌬–SOCH$_3$	7.7	11.4	7.6	7.1	<6	6.6	<6	<6
48M	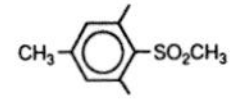 CH$_3$–⌬–SO$_2$CH$_3$	6.7	9.5	6.3	6.6	<6	<6	<6	<6

Table 2.2 Binding free energies ($-\Delta G^0$, kcal mol^{-1}) of cryptahemispherands with picrate salts in CDCl$_3$ at 25 °C

	Li^+	Na^+	K^+	Rb^+	Cs^+	NH_4^+
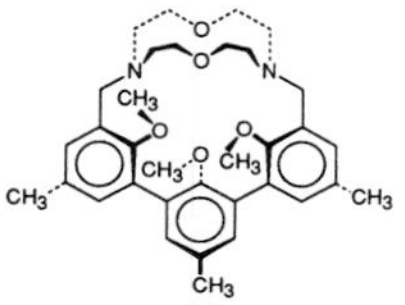 41	18.8	20.6	15.0	12.3	10.4	
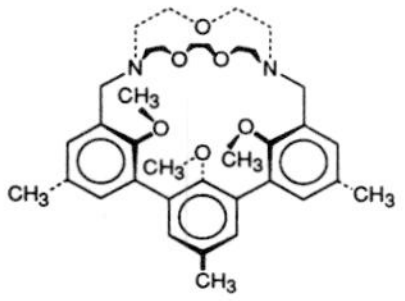 42	13.3	21.0	>19.9	20.4	16.4	18.6
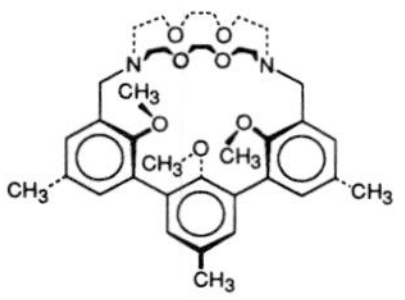 43	9.9	13.5	19.0	20.3	21.7	20.7

2.5.3 Cryptahemispherand Binding and Specificity

Cryptahemispherands **41–43** possess the following desirable properties: they are very strong binders of alkali metal ions; **41** and **43** exhibit high specificity toward the physiologically important ions Na$^+$ and K$^+$; the hosts and their complexes equilibrate rapidly on the human time scale. The $-\Delta G^0$ values of the three hosts are found in Table 2.2. Notice that **41**, with the shorter cryptand bridges, exhibits 5.6 kcal mol^{-1} higher binding toward Na$^+$ than K$^+$, and **43**, with longer cryptand bridges exhibits 5.5 kcal mol^{-1} higher binding toward K$^+$ than Na$^+$.

2.5.4 Spherand Binding and Specificity

The spherands showed the highest specificity and strongest binding towards their most complementary guests (see Chart 2.2). Thus O(Me)$_6$-spherand, **1**, the most powerfully binding host yet encountered[1] binds LiPicrate in CDCl$_3$ at 25 °C with > 23 kcal mol^{-1}, and NaPicrate with 19.2 kcal mol^{-1}. Replacement of one OCH$_3$ group of **1** with H, as in **2**, retains the specificity to provide 10.4 and 6.6 kcal mol^{-1}, respectively, for complexing the two guests, but the general level of binding drops by about 10 units with this substitution. Potential host **6**, the open-chain counterpart of **1** in which one Ar—Ar bond is replaced with two H's, drops the binding of all alkali metal picrates to <6 kcal mol^{-1}. This

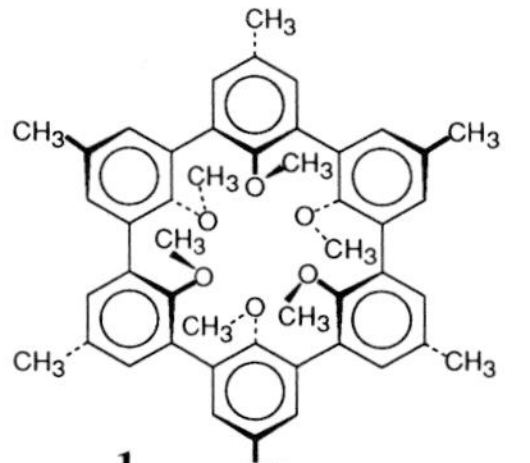

	1	**2**	**6**
Li⁺	>23	10.4	<6
Na⁺	19.2	6.6	<6

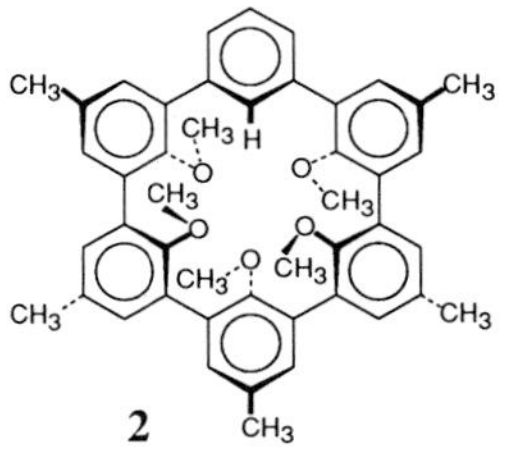

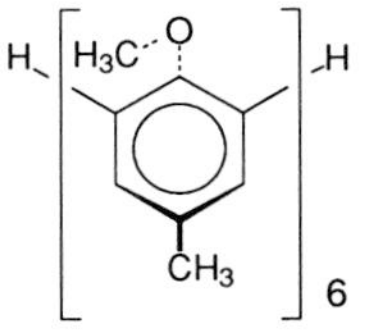

	49	**4**	**5**
Li⁺	<6	16.8	15.9
Na⁺	<6	13.3	18.7

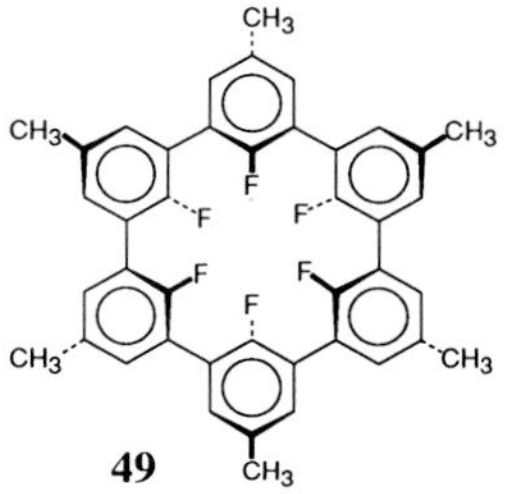

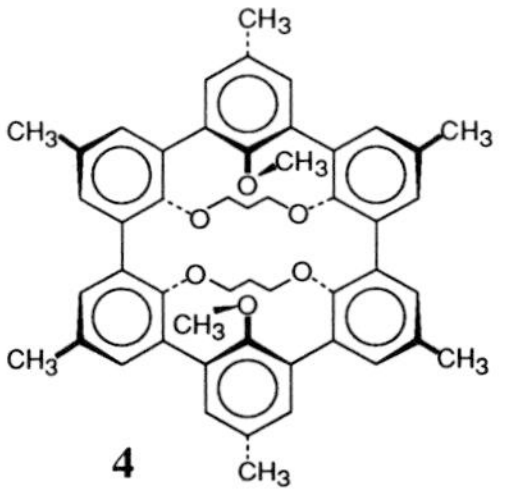

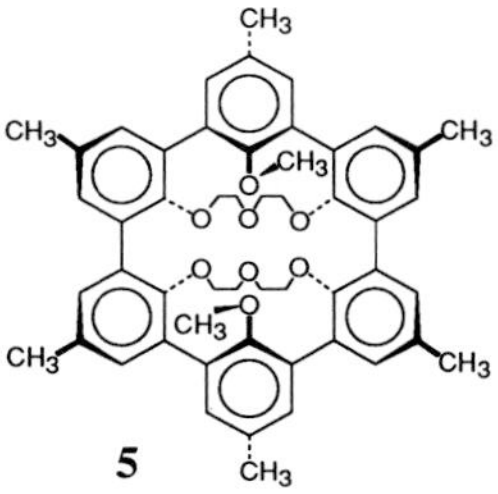

Chart 2.2 *Binding free energies ($-\Delta G°$, kcal mol^{-1}) in CDCl$_3$ at 25°C*

dramatic drop[1] is attributed to the over 500 conformations that must be frozen out in **6** to approximate the shape of **1**. In potential host **49**, six F atoms are substituted for the six OCH$_3$ groups of **1** without disturbing the alternating (up–down)$_3$ geometry. This substitution of F for O completely destroyed the binding power of **49**.[14] Although well preorganized for binding, **49** failed to complex detectably any of the alkaline earth metal ions. The introduction of two *syn*-CH$_2$CH$_2$CH$_2$ bridges into **1**, as in **4**, reduced its binding by about 6 kcal mol^{-1} for both LiPicrate and NaPicrate to provide $-\Delta G^0$ values of 16.8 and 13.3 respectively. When the two *syn*-OCH$_2$CH$_2$OCH$_2$CH$_2$O bridges in **5** replaced four OCH$_3$ groups of **1**, the binding of LiPicrate was reduced by >7 to 15.9 kcal mol^{-1}, whereas the binding of NaPicrate was essentially unchanged at 18.7. All six of the compounds of Chart 2.2 showed no detectable binding of any of the other alkali metal picrate salts. These data dramatically illustrate the importance of both stereoelectronic complementarity and preorganization to both ion specificity and strong complexation.[11]

When the numbers of anisyl groups were increased to eight as in **50**, the two CH$_3$ groups of the OCH$_3$ moieties at 12 and 6 o'clock turned inward to fill the cavity, as shown by the crystal structure **50a**.[15] When complexed with Cs$^+$, as in **51**, these methyl groups turned outward, as shown by its crystal structure **52**. The binding power of **50** peaks at Cs$^+$ as expected, as shown by the following values (CDCl$_3$, 25°C, picrate salts): Li$^+$, 8.3; Na$^+$, 10.1; K$^+$, 8.9; Rb$^+$, 10.4;

[15] D.J. Cram, R.A. Carmack, M.P. de Grandpre, G.M. Lein, I. Goldberg, C.B. Knobler, E.F. Maverick, and K.N. Trueblood, *J. Am. Chem. Soc.*, 1987, **109**, 7068.

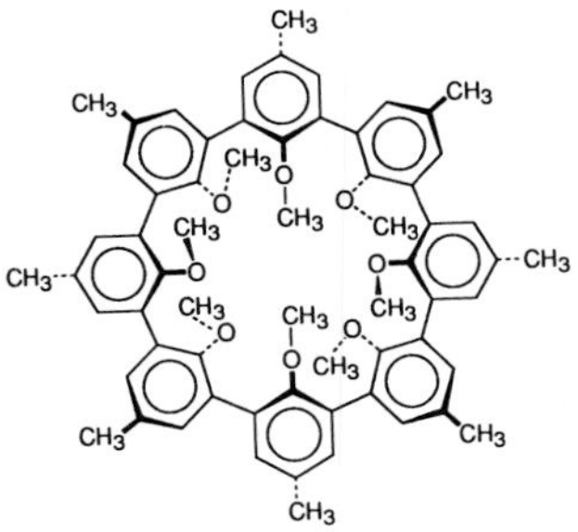

50

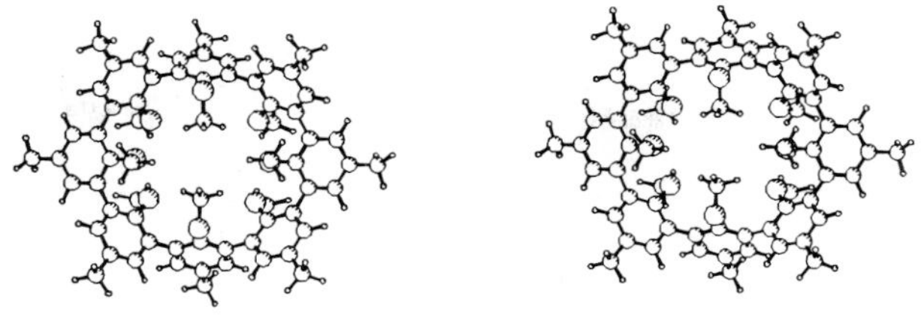

50a

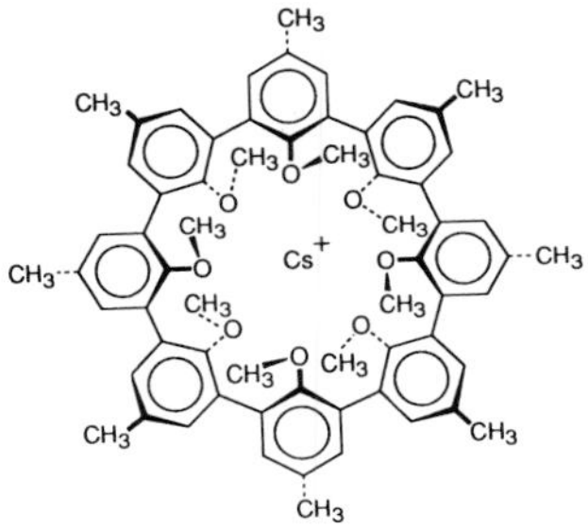

51

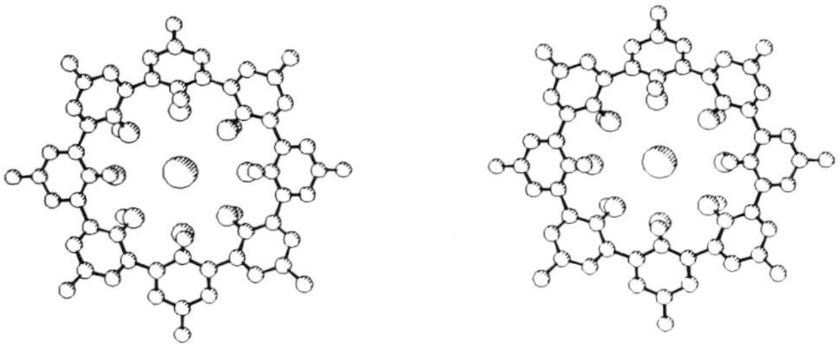

52

Cs^+, 13.9; NH_4^+, 9.1; $CH_3NH_3^+$, 9.0; $(CH_3)_3CNH_3^+$, 4.0 kcal mol^{-1}. Of these ions only Cs^+ is large enough to contact all of the oxygens at the same time, and it shows the strongest binding. However, CPK model examination shows the cavity of **50** to be large enough to accommodate Li^+----OH_2 and Na^+----OH_2, which may explain the relatively high values of 8.3 and 10.1 kcal mol^{-1}, respectively, for **50** complexing with these ions.[15]

2.6 Principles of Complementarity and Preorganization

Solvation of hosts and guests makes use of the same forces as complexation, such as hydrogen-bonding, ion-pairing, π-acid to π-base attractions, metal-ion ligation, induced dipole–dipole effects, and van der Waals attractions. Usually each contact involves <1 kcal mol^{-1} of free energy, two orders of magnitude less than the energies of covalent bonds. Complexation can compete with solvation only when many contact sites are collected in the same molecule and *can act simultaneously*, so that binding forces at each site are additive. This idea is implicit in all host–guest studies, but is so important that it deserves to be put in explicit form and given a name. We have called it the 'principle of complementarity', and have stated it as follows: *to complex, hosts must have binding sites which can simultaneously contact and attract the binding sites of guests without generating internal strains or strong non-bonded repulsions.* Thus complementarity is the central determinant of structural recognition.[16]

In Chart 2.3 are depicted a variety of host structures, and listed beneath are

---

[16] D.J. Cram, 'From Design to Discovery', American Chemical Society, Washington, DC, 1990, p. 91.

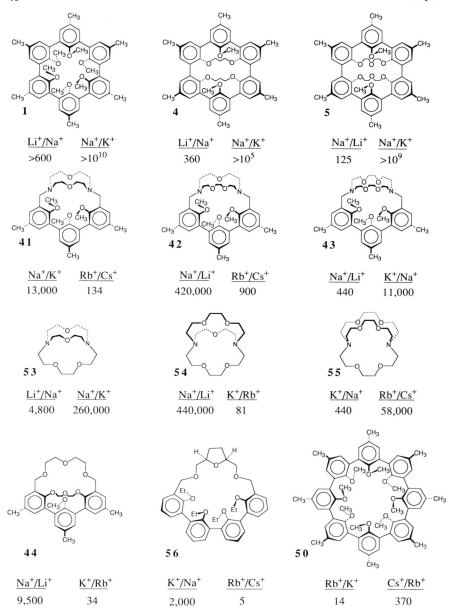

Chart 2.3 *Structural recognition measured by K_a/K_a' values for alkali metal picrates at 25°C in CDCl₃ saturated with D₂O*

K_a (M^{-1}) ratios for the alkali metal ion guests to which the hosts are the most strongly bound (CDCl$_3$ at 25 °C). These ratios provide a quantitative measure of structural recognition of one guest over another. In each ratio, the two metal ions involved are those closest to one another in the periodic table. Notice that spherands provide ratios ranging from 125 to $> 10^{10}$, that cryptahemispherands give ratios that vary from 134 to 420 000, Lehn's cryptands[8] from 81 to 440 000, and the hemispherands from 5 to 9500. The guest ions most important to physiology are Li^+, Na^+, and K^+, and fortunately ratios involving these ions have the largest values. The lowest ratios involve $K^+:Rb^+$, cryptand **54** providing a value of 81, and hemispherands **44** and **56**,[13] values of 34 and 5, respectively. Generally, the arrangement of the classes of hosts in decreasing order of their ability to exhibit structural recognition in complexation of the alkali metal ion is as follows: spherands > cryptahemispherands ~ cryptands > hemispherands > corands > podands. Of course, exceptions exist.

Just as complementarity governs structural recognition, *preorganization governs binding power*. This generalization has been named the 'principle of preorganization', and has been explicitly stated as follows: *the more highly hosts and guests are organized for binding and low solvation prior to their complexation, the more stable will be their complexes*. Both enthalpic and entropic components are involved in preorganization, since solvation involves the complexing partners.[16]

The most dramatic example of the importance of preorganization to binding power involves a comparison of the abilities of $(OMe)_6$-spherand, **1**, and its podand **6** to bind Li^+ and Na^+ in $CDCl_3$ at 25 °C. The two compounds differ constitutionally only in the sense that **6** has two Ar—H bonds replacing one Ar—Ar bond in **1**. The ratio of K_a values of **1** and **6** binding LiPicrate are $> 10^{12}$, and for binding NaPicrate are $> 10^{10}$ (see Chart 2.2). Whereas **1** exists in a single enforced conformation, **6** has over a thousand different conformations, only two of which can ligate these ions octahedrally. The organization and desolvation costs for **1** complexing these ions were paid during its synthesis, whereas these costs would have to come from the complexation free energies were **6** to complex these ions. These costs were too high, and as a result, complexation of **6** was not detectable.[11]

1   **6**

The families of hosts listed in the order of decreasing free energies of binding of their most complementary guest ions provide this sequence: spherands > cryptahemispherands > cryptands > hemispherands > corands > podands. This order is also that of their degree of preorganization. Thus preorganization appears to be a centrally important determinant of binding power.

2.7 Illustration of the Effects of Preorganization on Binding

The synthesis, crystal structure determination, and binding properties of **57** completed the study of a series of 18-membered ring compounds in which the CH_2OCH_2 units of 18-crown-6 were systematically replaced by 2,6-disubstituted anisyl units.[17] The stereoview of **57**, depicted in **57a**, shows the compound to be fully preorganized for binding with the top side of the cavity shielded from solvation by three methyl groups, and the under side by only *two* methyls. This disposition of the five methyls allows three of the six oxygens that line the cavity to be solvated by small molecules such as H_2O from the less shielded under face.

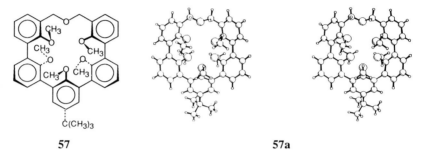

 57   57a

Table 2.3 lists the free energies of complexation ($-\Delta G^0$, kcal mol^{-1} at 25 °C in $CDCl_3$ saturated with D_2O) by **57** and its relatives of the alkali metal and ammonium ions. Corand **58** was used in place of 18-crown-6 because its lipophilicity keeps it from being soluble in the aqueous layer during the determination of its $-\Delta G^0$ values. In proceeding from corand **58** at the top of Table 2.3 to spherand **1** at the bottom, the trends provide the following conclusions: (1) the introduction of one or two anisyl groups in place of CH_2OCH_2 groups as in **59** and **46** depresses the binding and specificity by lowering the degree of preorganization intrinsic to corand **58** without disturbing K^+ as the most strongly bound guest; (2) when three anisyl groups are introduced as in **33**, their degree of preorganization of ligands and inhibition of solvation increase the binding of almost all ions, but favor Na^+ binding over that of the others; (3) in passing from **33** to **60** with four anisyl groups, the binding strength and specificity for Na^+ increase and for all of the larger ions decrease; (4) the trends are accentuated in going from **60** to **57** with five anisyl groups, but the binding of Li^+ increases dramatically by $-\Delta\Delta G^0 = 5.6$ kcal mol^{-1}; and (5) in going from **57** to **1**, which is fully preorganized for binding Li^+ and Na^+, the binding of Li^+ jumps by > 10 kcal mol^{-1} and that of Na^+ by 4.8 kcal mol^{-1}, making **1** more specific for Li^+ than for Na^+ by ~ 4 kcal mol^{-1}. Unlike **57**, **1** binds no other ions.

These results taken as a whole show that once the number of anisyls reaches the self-organizing state at three units, the greater the number of additional

---

[17] R.C. Helgeson, B.J. Selle, I. Goldberg, C.B. Knobler, and D.J. Cram, *J. Am. Chem. Soc.*, 1993, **115**, 11506.

Table 2.3 *Free energies of complexation ($-\Delta G^0$, kcal mol^{-1} at 25°C in CDCl$_3$ saturated with D$_2$O) by increasingly preorganized hosts of alkali metal and ammonium picrates*

Host	Li$^+$	Na$^+$	K$^+$	Rb$^+$	Cs$^+$	NH$_4^+$
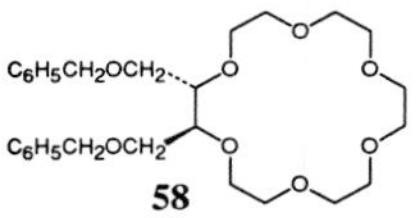 58	6.3	8.4	11.4	9.9	8.5	10.1
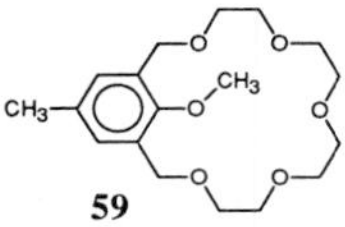 59	5.5	6.4	8.5	7.5	6.9	7.6
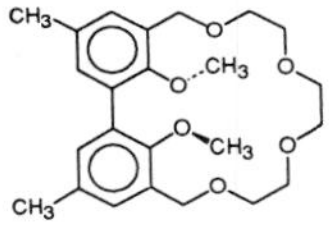 46	6.5	8.7	9.8	8.6	7.8	7.9
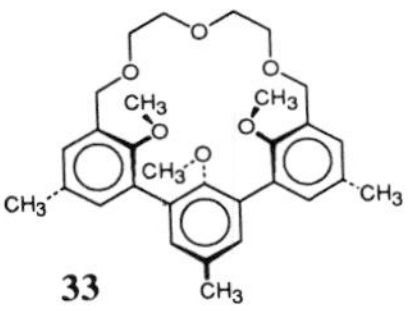 33	7.2	12.2	11.7	10.5	9.0	9.8
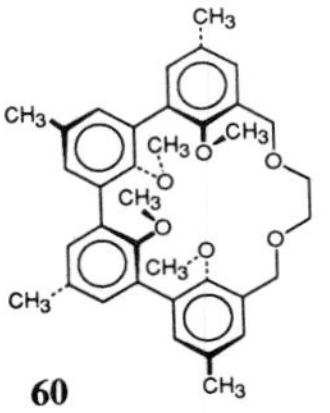 60	7.2	13.5	10.7	8.4	7.1	8.7
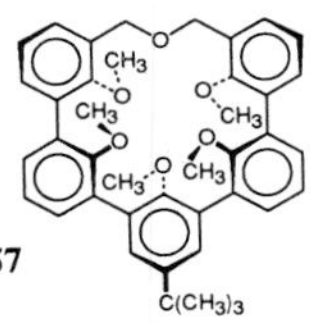 57	12.8	14.4	10.4	7.9	6.7	8.7
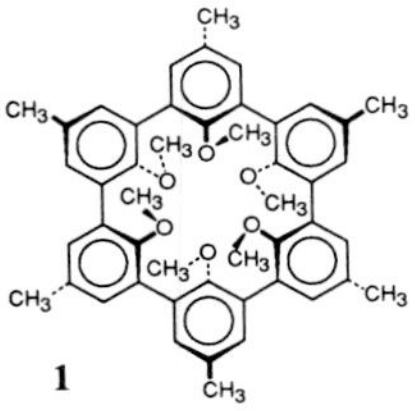 1	>23	19.2				

anisyl units, the smaller the cavity becomes, the greater becomes the specificity for the small ions, and the larger becomes the $-\Delta G^0$ values for binding Li$^+$ and Na$^+$ ions.[17]

2.8 Rates of Complexation and Decomplexation of Spherands and Hemispherands

The second-order rate constants for spherands **1**, **4**, and **5** (Chart 2.3) complexing LiPicrate and NaPicrate (CDCl$_3$ saturated with D$_2$O, 25 °C) fell between about 10^5 and 10^6 M^{-1} s^{-1}. The first-order decomplexation rate constants ranged from about 10^{-4} to $<10^{-12}$ s^{-1}, or by a factor $>10^8$. The K_a values for the three spherands ranged from $>10^{16}$ to 10^9 M^{-1}, or by a factor of $>10^7$. Thus the changes in binding power of the three spherands are mainly determined by the changes in the dissociation, rather than the association constants.[11] The same was true for the cryptands binding the alkali metal ions in a variety of solvents, the association rate constants varying between 10^4 to 10^7 M^{-1} s^{-1}.[18]

Values for the association and dissociation rate constants for a number of hemispherands binding KPicrate in CDCl$_3$ saturated with D$_2$O (25 °C) have been estimated, *e.g.* that of **40**. Those for association ranged from 10^7 to 10^9 M^{-1} s^{-1}, whereas those for dissociation fell between 27 and 4 s^{-1}. For these partially preorganized and solvated hosts binding less strongly solvated K$^+$ ion (as compared with Li$^+$ and Na$^+$), the association rate constant played the larger role in determining the K_a and $-\Delta G^0$ values, in contrast to the spherands and cryptands.[16]

2.9 Cyanospherands

In an attempt to preorganize cyano groups as ion binders, the octacyanospherand **61** was designed with the help of CPK models. The compound was synthesized, and its crystal structure determined, a face stereoview of which is shown in **62**. As anticipated from model examination, the molecule possesses approximate D_{4d} symmetry, with an (up–down)$_4$ arrangement of the eight cyano groups, four on each side of the best plane of the molecule. In the crystal structure, one water molecule on each side of the best plane beautifully bridges the cyano groups at 1 and 7 o'clock on the near face, and at 5 and 11 o'clock on the remote face. The molecule contains an empty, nearly spherical cavity of

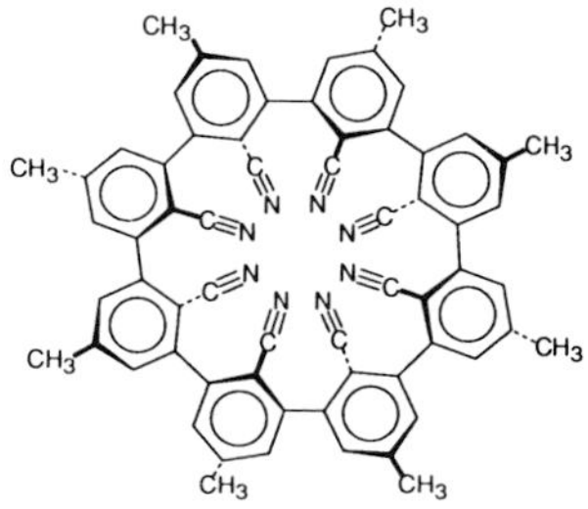

61 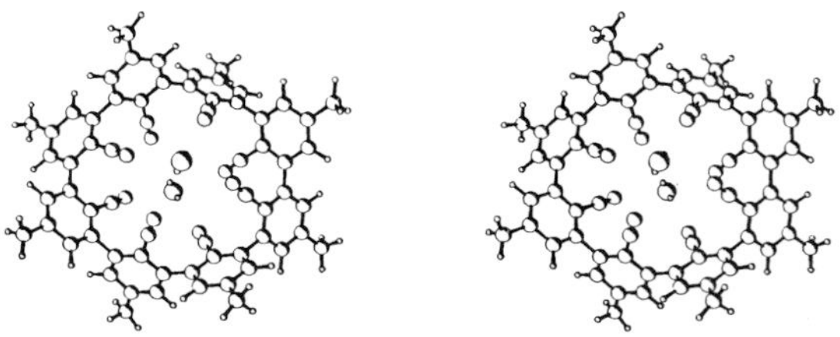   **62**

[18] B.G. Cox, J. Garcia-Ross, and H. Schneider, *J. Am. Chem. Soc.*, 1981, **103**, 1054.

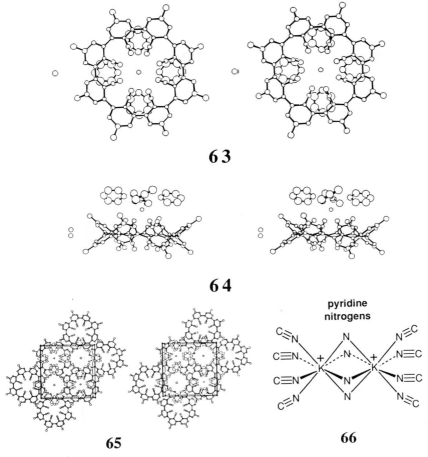

63

64

65   **66**

about 3.2 Å diameter defined by the eight C-C≡N groups in a square-antiprismatic arrangement.[19]

A crystal structure was also obtained of **61**·2KBr·4pyridine complex, face and side stereoviews of which are depicted in **63** and **64**, respectively. In both views one KBr is omitted. As in **62**, the host of **63** possesses approximately D_{4d} symmetry and contains an empty, nearly spherical cavity of about 2.4 Å diameter defined by the eight C-C≡N groups in a square-antiprismatic arrangement. A K$^+$ perches on each face of the macrocycle ligated by the *sp*-orbital electron pairs of four converging cyano groups. Each K$^+$ is also ligated by the four nitrogens of four pyridines arranged like a four-bladed propeller above and below each face of the spherand. Each pyridine bridges two K$^+$ ions to form the stacked, polymeric system shown in stereoview **65**, and abbreviated in **66**. The bromide ions are arranged around the equator of the macrocycle as shown in **64** and **65**. Thus the K$^+$ and Br$^-$ ions each lie on different, but parallel axes, which are 9.3 Å apart. Our expectation that the bromide ions would occupy the cavity of the cyanospherand was not fulfilled.

[19] K. Paek, C.B. Knobler, E.F. Maverick, and D.J. Cram, *J. Am. Chem. Soc.*, 1989, **111**, 8662.

The $-\Delta G^0$ values for octacyanospherand **61** binding Li^+, Na^+, K^+, Rb^+, Cs^+, NH_4^+, $CH_3NH_3^+$, and $(CH_3)_3CNH_3^+$ picrates were determined ($CDCl_3$, 25 °C), and found to be respectively 10.6, 13.5, 14.1, 12.6, 11.8, 11.8, 11.2, and 10.0 kcal mol^{-1}. At the concentrations employed (10^{-3} M for host and guest), vapor-pressure osmometry (46 °C) demonstrated the complex to be composed of one host and one guest molecule. The two striking features of the $-\Delta G^0$ values are their high numbers and lack of variation with guest change, all eight values of $-\Delta G^0$ lying between 10.0 and 14.1 kcal mol^{-1}. These values proved to be higher by >4 to >8 kcal mol^{-1} than those found for the open-chain counterpart $Br(4-CH_3C_6H_2CN)_8Br$, whose $-\Delta G^0$ values were all below the detectable level (< 6 kcal mol^{-1}). Again the power of preorganization of the host is demonstrated by these results.[19]

2.10 Chromogenic Ionophores

Others demonstrated that when chromophores were covalently attached to corands[20,21] and to a spherand[22] color changes occurred when these hosts came in contact with Li^+, Na^+, or K^+. Inspired by these early results, we utilized the high ion-specificity of our spherands and cryptahemispherands to design and synthesize chromogenic ionophores that might be used in solution for detecting and analyzing colorimetrically the amounts of Li^+, Na^+, and K^+ in the presence of one another.

Phenolic compound **67** was oxidized to quinone **68** which was condensed with $2,4-(NO_2)_2C_6H_3NHNH_2$ to give **71**. The binding properties of **67** and **68** toward Li^+, Na^+, and K^+ were determined by our standard picrate salt extractive technique[10] in $CDCl_3$ saturated with D_2O (25 °C). The respective $-\Delta G^0$ values obtained were as follows: for formation of **67·**Li^+, **67·**Na^+, and **67·**K^+, 12.3, 8.4, and <6 kcal mol^{-1}; for **68·**Li^+, **68·**Na^+, and **68·**K^+, 12.5, 9.8, and 6.5 kcal mol^{-1}. Thus the binding power and specificities of **67** and **68** were greatly reduced from that of parent spherand **1**, but were still large enough for our purposes. Intramolecular hydrogen bonding of the ArOH····OCH_3 variety is undoubtedly responsible for the reduced binding power of **67**. The orientation of the axes of the sp^2 orbitals containing the unshared electron pairs of the quinone oxygen of **68** is coplanar with the quinone ring and the unshared electron pairs are, therefore, unavailable to ligate metal ions.

The pK_a values for **71** and model compounds **69** and **70** were determined in 80% dioxane–20% water (v:v) at 30 °C both in the presence and absence of added potential complexing agents and of the bases diazobicyclo[4.3.0]non-5-ene (DBN) or K_2CO_3. The pK_a values obtained were as follows: **71**, DBN, 13.0; **71**, K_2CO_3, 12.7; **71**, $CaCl_2$, 12.8; **71**, $MgCl_2$, 13.2; **71**, $LiClO_4$, 5.9; **71**, $NaClO_4$, 6.9; **69**, DBN, 10.8; **70**, DBN, 11.2. Deionized water and very pure additives were used throughout. These results demonstrate that **71** complexes only Li^+

---

[20] M. Tagaki, H. Nakamura, and K. Ueno, *Anal. Lett.*, 1977, **10**, 1115.
[21] J.P. Dix and F. Vögtle, *Angew. Chem., Int. Ed. Engl.*, 1978, **17**, 857; H-G. Lohr and F. Vögtle, *Acc. Chem. Res.*, 1985, **18**, 65.
[22] T. Kaneda, K. Sugihara, H. Kamiya, and S. Misumi, *Tetrahedron Lett.*, 1981, **22**, 4407.

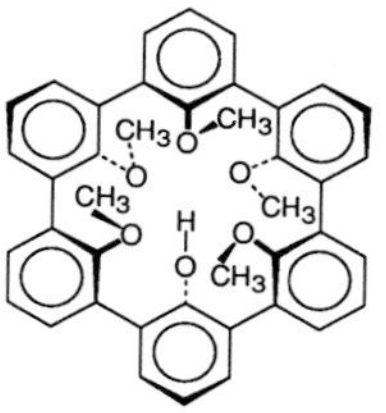

and Na$^+$ to form **72** and **73**, respectively. The pK_a values of **71**, **72**, and **73** are 13, 6, and 7, respectively. The presence of the Li$^+$ in **72** increases the acidity of the phenolic proton by 7 powers of 10, and the Na$^+$ in **73** by 6 powers of 10.

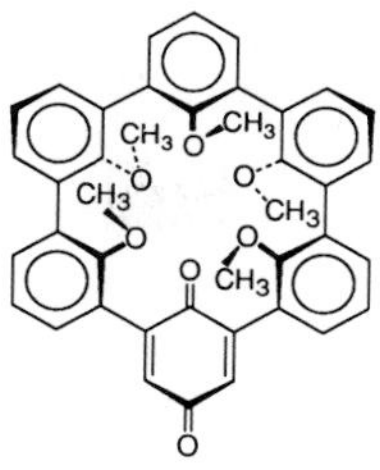

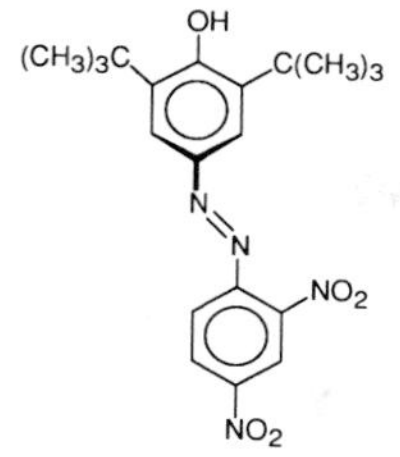

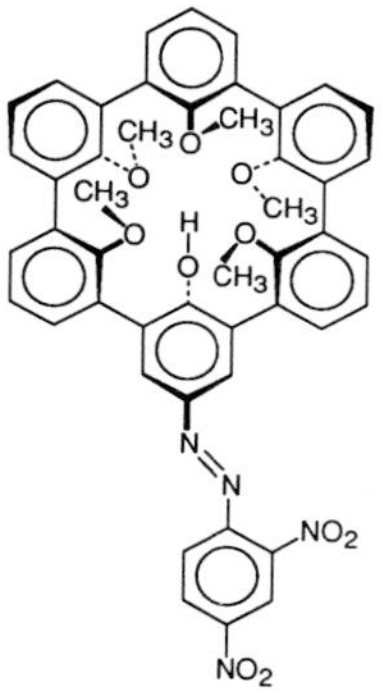

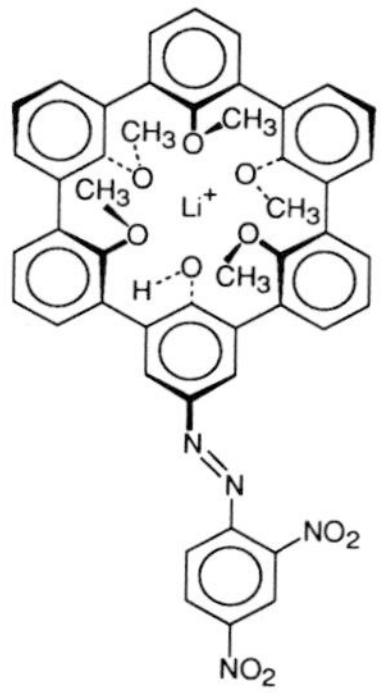

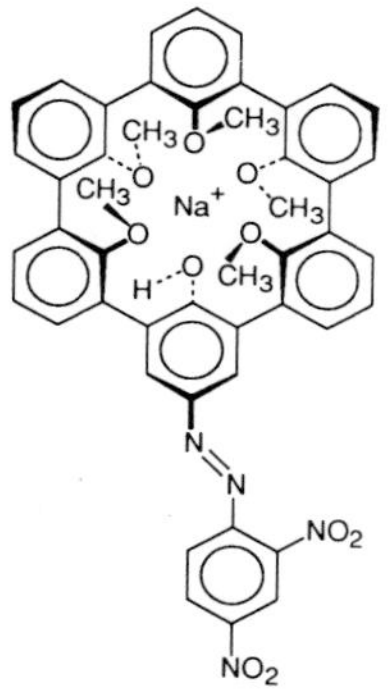

This dramatic effect was used to demonstrate visually the high specificity and sensitivity of **71** to the presence of Na$^+$ and Li$^+$ in solutions. Dilute solutions of **71**·LiClO$_4$ were deep violet in color showing complete deprotonation in the absence of base. Similar solutions of **71**·NaClO$_4$ were green showing a mixture of **71**$^-$·Na$^+$ and **71**·Na$^+$ClO$_4{}^-$, whereas solutions of **71** were pale yellow. Careful electronic spectral measurements demonstrated the longest wavelength bands of various chromogenic species in various solvents gave the respective λ_{max} (nm) and ε_{max} (l mol^{-1} cm^{-1}) values: in CHCl$_3$, **71**, 408, 26 000; **71**·LiClO$_4$, 483, 36 000; **71**$^-$·Li$^+$, 600, 47 000; **71**·NaClO$_4$, 476, 26 000; **71**$^-$·Na$^+$, 606, 54 000. In EtOH these values were as follows: **71**, 388, 26 000; **71**·LiClO$_4$, 382,

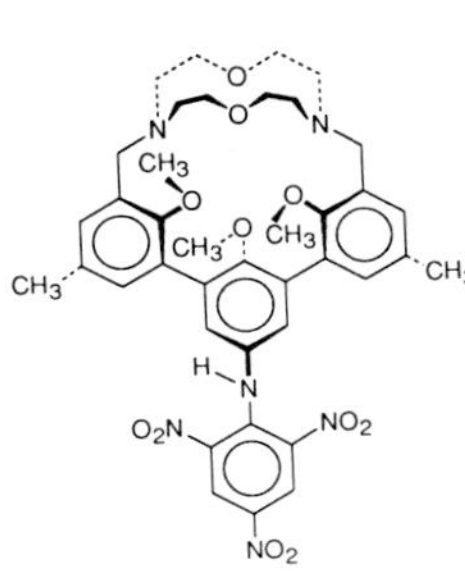

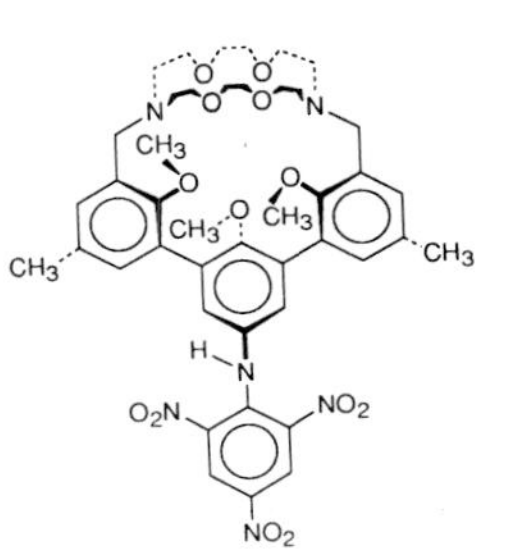

14 000; $71^-\cdot Li^+$, 564, 20 000; $71\cdot NaClO_4$, 470, 12 000; $71^-\cdot Na^+$, 576, 27 000. The differences make **71** and its analogues potentially useful compounds for analytical determinations of Li^+ and Na^+ in the presence of other ions, or even of each other. The sensitivity of **71** to Na^+ required that the compound be handled in quartz or teflon equipment, since both pyrex and soft glass visibly released Na^+ to organic solutions of **71**. One drop of Los Angeles tap water added to 100 ml of conductivity water gave high enough Na^+ concentration to produce a purple color when a drop of the diluted water was added to a dioxane–water solution of **71** containing a trace of pyridine.[23]

Two chromogenic cryptahemispherands **74** and **75** were designed and synthesized which now enjoy use by Technicon Instruments Corp. for Na^+ and K^+ analyses of serum. The pK_a values of **74**·LiBr, **74**·NaBr, and **74**·KBr in 1% $EtO(CH_2CH_2O)_2H$ in water were respectively 7.9, 7.0, and 7.9, which showed that **74**·NaBr is almost one power of 10 more stable than **74**·LiBr or **74**·KBr. The pK_a values of **75**·NaBr and **75**·KBr were found to be 7.75 and 7.05 in the same medium, the latter value being insensitive to the presence of as many as 100 equivalents of NaBr. In appropriately buffered systems, the selectivity factor for **74** complexing Na^+ is about 1000, and for **75** complexing K^+ is estimated to be about 1500. These factors were large enough to allow colorimetric determinations to be made routinely and directly on serum.[24]

---

[23] D.J. Cram, R.A. Carmack, and R.C. Helgeson, *J. Am. Chem. Soc.*, 1988, **110**, 571.
[24] R.C. Helgeson, B.P. Czech, E. Chapoteau, C.R. Gebauer, A. Kumar, and D.J. Cram, *J. Am. Chem. Soc.*, 1989, **111**, 6339.

CHAPTER 3

Chiral Recognition in Complexation

> 'Had the aphorism 'it takes one to recognize one' not been born prior to the discovery of enantiomerism among organic compounds, this discovery might have served as midwife for the expression'.

Chiral recognition in the crystalline state is as old as the discovery that racemic amines or acids could be enantiomerically resolved by differential crystallization of diastereomeric salts. Our early survey of chiral corands containing 2,2'-disubstituted-1,1'-binaphthyl units for use in resolving primary amine salts by extraction (Section 1.2) encouraged us to design and study systems useful in resolving amino acid and ester salts, and in preparing chiral catalysts for asymmetric induction investigations. This chapter provides a survey of the results.

3.1 Hosts Containing One Chiral Element

Hosts ideal for enantiomer separations by extraction should possess the following properties: (1) they should be strong complexing agents that discriminate between enantiomers by large factors; (2) both host enantiomers should be easily preparable to facilitate obtaining either guest enantiomer in pure form; (3) the compounds should be stable, recoverable, and reusable; (4) their absolute configurations should be known, so that configuration-binding free energy correlations can be made; (5) the chiral element should contain C_2 axes so that the same perching complex is produced by attachment of amine salts to either face of macrorings; (6) the systems should be simple and of the lowest possible molecular weight; and (7) a crystal structure of a coraplex composed of an alkylammonium salt and the host should be available.

System **1**, in which A groups were varied, was chosen for study since it came closest to fulfilling the above conditions.[1] In particular, a crystal structure of complex (S)-**2** was available,[2] which is drawn in **2a**.

Enantiomer distribution constants (EDC) were determined by equilibrating $CDCl_3$ solutions of enantiomerically pure host with racemic amino acid or ester salts in D_2O at 0 °C. The EDC values were calculated from Equation 3.1 in

---
[1] D.S. Lingenfelter, R.C. Helgeson, and D.J. Cram, *J. Org. Chem.*, 1981, **46**, 393.
[2] I. Goldberg, *J. Am. Chem. Soc.*, 1980, **102**, 4106.

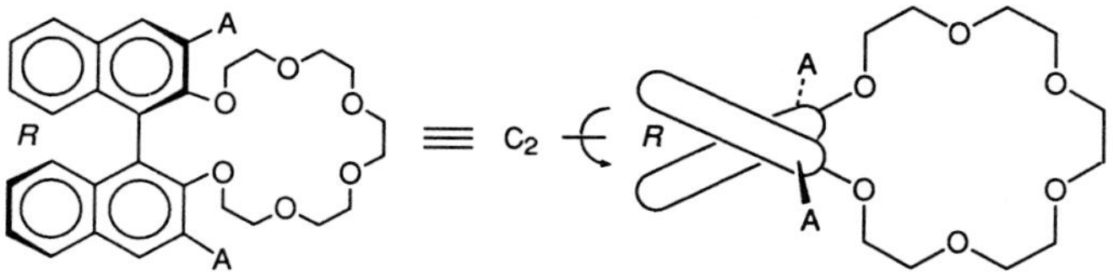

1

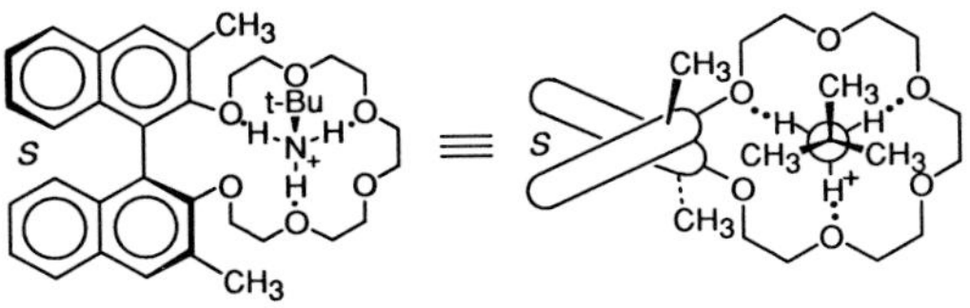

(S)-2 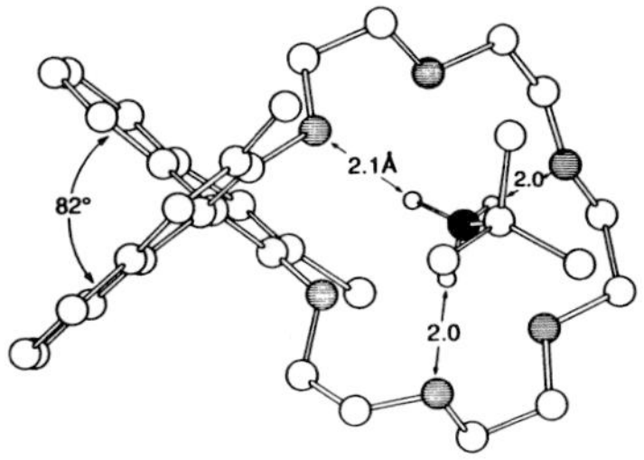   (S)-2a

which [H·G$_A$X] is the concentration at equilibrium of the complexed salt of the dominant guest G$_A$ in CDCl$_3$, [H·G$_B$X] is that of the subordinate complexed salt G$_B$ in CDCl$_3$, [G$_A$X] is that of guest A in D$_2$O, and [G$_B$X] is that of guest B in D$_2$O. The difference in the free energies of binding of the two diastereomeric salts is related to EDC by the expression, $-\Delta\Delta G^0 = RT\ln(\text{EDC})$. To adjust the concentration of the complexes in the CDCl$_3$ layer to the desired amount, the CDCl$_3$ layer was diluted to a 0.55 mole fraction with CD$_3$CN (0.45 mole fraction), and/or the aqueous layer was made 2 M in LiClO$_4$. The former tactic made the organic layer more polar, and the latter salted out the organic guest, driving it somewhat more into the organic layer. The amino acids required both tactics, whereas the least lipophilic amino ester (methyl alanine perchlorate salt) required only the salting-out technique.

$$\text{EDC} = ([\text{H·G}_A\text{X}][\text{G}_B\text{X}]/([\text{H·G}_B\text{X}][\text{G}_A\text{X}])) \qquad (3.1)$$

Of nine hosts of general structure **1** which were examined, compound **3**, in which A = C$_6$H$_5$, provided the generally highest EDC and $-\Delta\Delta G^0$ values in complexing the α-amino ester and acid salts. The values obtained are recorded in Table 3.1 along with the configuration of the dominant diastereomeric complex in the organic phase. The respective EDC and $-\Delta\Delta G^0$ values for the amino ester salts range from a high of 19.5 and 1.6 kcal mol^{-1} for phenylglycine

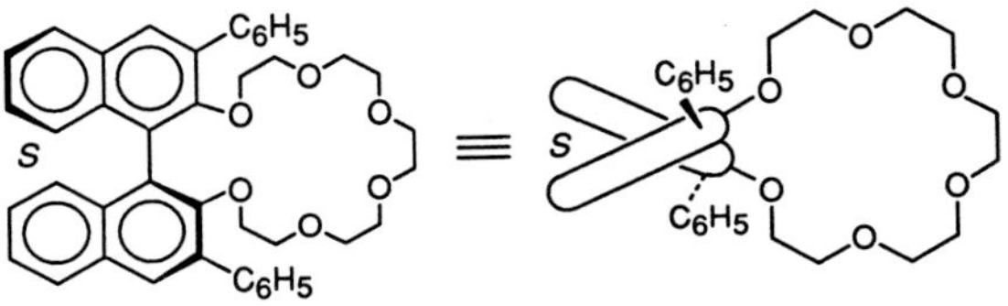

(S)-3

Table 3.1 EDC and $-\Delta\Delta G^0$ (kcal mol^{-1}) values for host (S)-3 complexing $RCH(CO_2CH_3)NH_3ClO_4$ and $RCH(CO_2H)NH_3ClO_4$ enantiomers at 0°C, respectively, in $CDCl_3$ and 0.45 mol fraction CD_3CN in $CDCl_3$

R group of guest	Ester		Acid		Configuration of dominant diastereomer
	EDC	$-\Delta\Delta G^0$ (kcal mol^{-1})	EDC	$-\Delta\Delta G^0$ (kcal mol^{-1})	
C_6H_5	19.5	1.6	19.2	1.6	(S)(L)-H·G
$(CH_3)_2CH$	7.7	1.1	8.1	1.1	(S)(L)-H·G
$C_8H_6NCH_2$[a]	7.9	1.1	5.1	0.9	(S)(L)-H·G
$C_6H_5CH_2$	4.4	0.8	6.4	1.0	(S)(L)-H·G
$CH_3SCH_2CH_2$	6.0	1.0	13.5	1.4	(S)(L)-H·G
CH_3	3.9	0.7	6.4	1.0	(S)(L)-H·G

[a]3-Indolyl group of tryptophane.

ester (R = C_6H_5) to a low of 3.9 and 0.7 kcal mol^{-1} for alanine ester (R = CH_3), and for the acid salts from a high of 19.2 and 1.6 kcal mol^{-1} for phenylglycine to a low of 4.4 and 0.8 kcal mol^{-1} for phenylalanine (R = $C_6H_5CH_2$). This host, **3**, exhibited slightly higher chiral recognition toward the amino acid salts than the ester salts, but generally the values for each type of guest were close to one another.

The direction of the chiral bias in all of the examples of Table 3.1 favored the (S)(L)- or (R)(D)-diastereomeric complexes over the corresponding (S)(D)- or (R)(L)-complexes. The same was true for the binding of standard ester salts of phenylglycine and valine by hosts of general structure **1** in which the A group was varied as follows: A = CH_3, CH_2OCH_3, $CH_2OC_6H_5$, $CH_2OC_6H_4OCH_3$-p, $CH_2SC_6H_5$, $CH_2SCH_2C_6H_5$, and $C_6(CH_3)_5$. The total number of different host–guest combinations was 28. Structures (S)(L)-**4** and (S)(D)-**4** provide a rationale for the common direction of this configurational bias. Of the various host–guest close contacts in these two complexes, the diastereomers differ only in one's A-to-CO_2R' and naphthalene-to-H interactions vs. the other's A-to-H and naphthalene-to-CO_2R' interactions. Of these two sets of close contacts, the former appears to be the more stabilizing. The conformation pictured for each diastereomeric complex places the bulky R group of the guest in a position remote from the walls of the naphthalene. These structures were selected over alternatives on the basis of analogies with crystal structure **2**, and conclusions

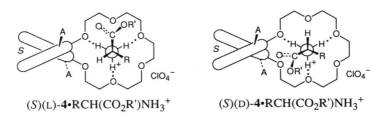

(S)(L)-**4**·RCH(CO_2R')NH_3^+ (S)(D)-**4**·RCH(CO_2R')NH_3^+

derived from CPK model examination. The ^{1}H NMR spectra of the diastereomeric amino ester salt complexes $(S)(L)$-**4**·$C_6H_5CH(CO_2CH_3)NH_3ClO_4$ and $(S)(D)$-**4**·$C_6H_5CH(CO_2CH_3)NH_3ClO_4$ supported the above explanations.[1]

Molecular model examinations of amino acid complexes of **3** indicate that the phenyl groups substituted in the 3,3'-positions of the host play two roles. The phenyl proximate to the perching guest extends the chiral barrier of its attached naphthalene ring to a position alongside the macroring to about 12 o'clock of formula **3**. The phenyl on the side opposite the perching guest inhibits the macroring from bending away from the Ar—C_6H_5 on the guest side to relieve steric compression. When the A substituent of **1** is CH_3, these steric effects are too small to shape the host, and when the A substituent of **1** is $C_6(CH_3)_5$, the steric effects are too large to allow tripodal binding of the guest. When A = C_6H_5, as in **3**, the steric effects are well enough balanced to just allow tripodal binding of the guest, and maximum EDC and $-\Delta\Delta G^0$ values are obtained.[1]

3.2 A Chiral Breeding Cycle

The greater stability of dissolved $(S)(L)$-**3**·$C_6H_5CH(CO_2H)NH_3ClO_4$ over $(S)(D)$-**3**·$C_6H_5CH(CO_2H)NH_3ClO_4$ was found to carry over to the crystalline state by the success of the following experiment. Racemic $(R)(S)$-**3** was prepared in 31% yield from readily available starting materials. This racemate was enantiomerically resolved by *crystallizations* of the $(R)(D)$- and $(S)(L)$-complexes with commercially available (D)- and (L)-$C_6H_5CH(CO_2H)NH_3ClO_4$ as guests. The diastereomerically pure complexes were readily decomposed by pH changes and extraction to give pure (R)-**3** and (S)-**3**. These enantiomeric hosts were then used to resolve racemic $C_6H_5CH(CO_2H)NH_3ClO_4$ into its enantiomers by either crystallization or extractive procedures as shown in Equations 3.2 and 3.3. Alternatively, ester $C_6H_5CH(CO_2CH_3)NH_3ClO_4$ could be substituted for the acid in this scheme. In the crystallizations, 0.5 mol. of pure enantiomer and 1 mol. of racemate were always used, and in all cases only $(R)(D)$- or $(S)(L)$-complexes crystallized. The direction of the chiral bias in the crystallization and extraction experiments was the same. The experiments taken in sum complete a chiral-breeding cycle.

$$(R)(S)\text{-host} \xrightarrow[\text{with D- or L-guest and decomplexation}]{\text{crystalline}} (R)\text{- or }(S)\text{-host} \quad (3.2)$$

$$(D)(L)\text{-guest} \xrightarrow[\text{(R)- or (S)-host and decomplexation}]{\text{countercurrent extraction or crystallization}} (D)\text{- or }(L)\text{-guest} \quad (3.3)$$

3.3 Hosts Containing Two Chiral Elements

Hosts (S,S)-**5** and (S,S)-**6** contain two binaphthyl and two bitetralyl chiral elements, respectively, incorporated into a corand macroring. The first of these systems gave the highest chiral recognition in complexing amino acid and ester salts of several examined.[3,4] Of the $RCH(CO_2R')NH_3X$ salts studied, the highest chiral recognition was observed when $R = C_6H_5$, $R' = CH_3$, and $X = PF_6^-$. The $CDCl_3$-D_2O extractive technique at 0 °C applied to (S,S)-**5** and $(D)(L)$-$C_6H_5CH(CO_2CH_3)NH_3PF_6$ gave an EDC value of 31 and a $-\Delta\Delta G^0$ value of 1.9 kcal mol^{-1}. Similarly, in $CDCl_3$–CD_3CN, (S,S)-**5** applied to $(D)(L)$-$C_6H_5CH(CO_2H)NH_3ClO_4$ gave an EDC value of 13.9 and a $-\Delta\Delta G^0$ value of 1.4 kcal mol^{-1}. The $(S,S)(L)$- and $(R,R)(D)$-complexes were found to be the more stable of the diastereomer pairs for all of the 13 amino ester salts[3] and the seven amino acid salts[4] examined. Although HPF_6 salts gave higher chiral recognition than $HClO_4$ salts,[3] the latter were more convenient to use. In similar experiments, (S,S)-**6** applied to $(D)(L)$-$C_6H_5CH(CO_2CH_3)NH_3ClO_4$ (see **7**) at 0 °C in $CDCl_3$–CD_3CN gave EDC = 13.6 and $-\Delta\Delta G^0 = 1.4$ kcal mol^{-1}, with the $(S,S)(L)$-diastereomeric complex dominating.

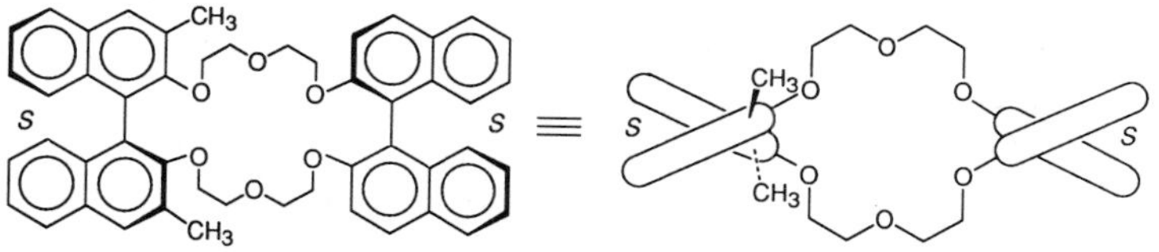

(S,S)-**5**

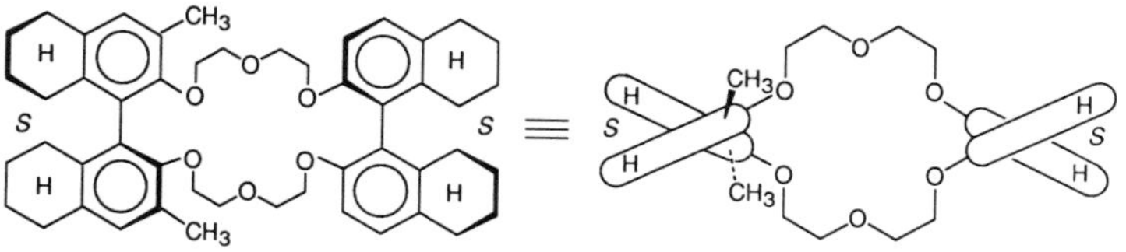

(S,S)-**6**

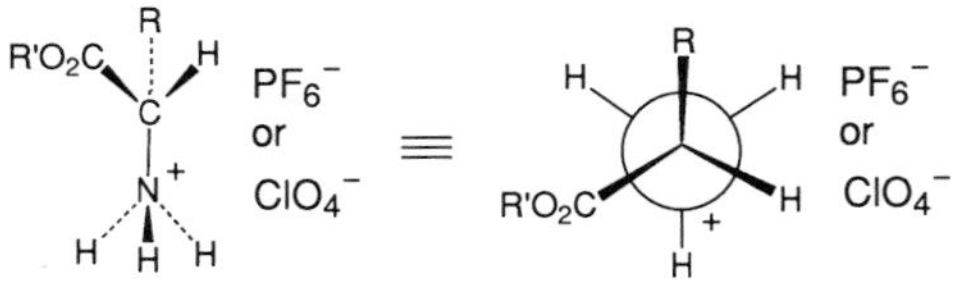

7

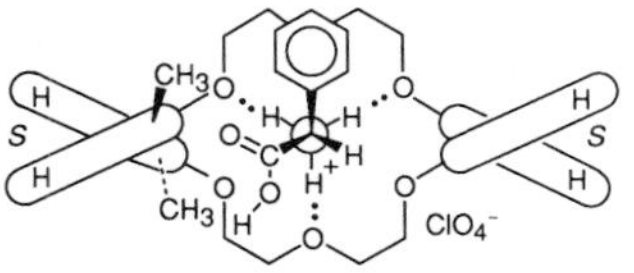

(S,S)-(L)-**6**·$C_6H_5CH(CO_2H)NH_3ClO_4$

Although many of the diastereomeric complexes were crystalline, only one proved to be amenable to crystal structure determination. In 1977 suitable crystals of $(S,S)(L)$-**6**·$C_6H_5CH(CO_2H)NH_3ClO_4$ were obtained, but the unit cell

---

[3] S.C. Peacock, L.A. Domeier, F.C.A. Gaeta, R.C. Helgeson, J.M. Timko, and D.J. Cram, *J. Am. Chem. Soc.*, 1978, **100**, 8190.

[4] S.C. Peacock, D.M. Walba, F.C.A. Gaeta, R.C. Helgeson, and D.J. Cram, *J. Am. Chem. Soc.*, 1980, **102**, 2043.

contained two independent complexes plus one molecule of EtOAc to give a total of 298 atoms. This structure was solved after ten years of intermittent effort involving a variety of approaches to give the structures **8a** and **8b**, each depicted in side and top stereoviews. The two crystal structures are remarkably similar, which suggests that host–guest intracomplex interactions determine their structures rather than intercomplex lattice interactions. Both crystal structures involve tripodal binding, both exhibit the guest conformation shown in **7**, and the disposition of C_6H_5, H, and CO_2H groups in cavities shown in $(S,S)(L)$-**6**·$C_6H_5CH(CO_2H)NH_3ClO_4$. The large C_6H_5 group occupies the cavity centered at 12 o'clock, the H is located at 4 o'clock in the least spacious part of the cavity centered at 6 o'clock, and the CO_2H is found in the more spacious part of the same cavity at 8 o'clock.[5] These structures are consistent with ^{1}H NMR spectra of solutions of similar complexes dissolved in $CDCl_3$–CD_3CN,[3] and are close to those whose CPK molecular models inspired the study of the systems.[3,4]

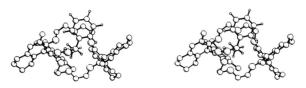

8a, stereoview with eye looking down C-N axis of molecule a.

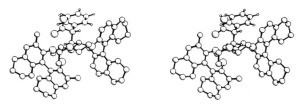

8a, stereoview with C-N axis in the plane of the page.

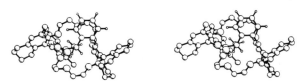

8b, stereoview with eye looking down C-N axis of molecule b.

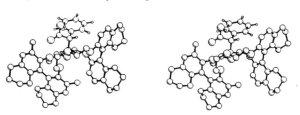

8b, stereoview with C-N axis in plane of the page.

[5] C.B. Knobler, F.C.A. Gaeta, and D.J. Cram, *J. Chem. Soc., Chem. Commun.*, 1988, 330.

Racemic host $(R,R)(S,S)$-**5** was resolved into its two enantiomers by treating an EtOAc solution of the racemate with 0.5 mol. of (D)-$C_6H_5CH(CO_2H)NH_3$ ClO_4. The (R,R)(D)-diastereomeric complex crystallized leaving in solution the (S,S)-host, which was purified through its complex with (L)-C_6H_5 $CH(CO_2H)NH_3ClO_4$. Thus chiral recognition in solution and in the crystalline state involves the same configurational bias. Attempts to crystallize the (S,S)(D)-diastereomer failed, even when enantiomerically pure components were mixed. These experiments taken in sum complete the chiral-breeding cycle formulated in Equations 3.4 and 3.5.[6]

$$(R,R)(S,S)\text{-host} \xrightarrow[\text{with D and L guests}]{\text{crystallization}} (R,R)\text{-host} + (S,S)\text{-host} \quad (3.4)$$

$$(D)(L)\text{-guest} \xrightarrow[\text{extraction with} \ (R,R)\text{- or }(S,S)\text{-hosts}]{\text{countercurrent}} (D)\text{-guest} + (L)\text{-guest} \quad (3.5)$$

3.4 An Amino Ester Resolving Machine

An amino acid and ester resolving machine was designed, built, and tested, which made use of the ability of the enantiomers of **5** to complex and transport through bulk $CHCl_3$ the enantiomers of α-amino acid or ester salts. Figure 3.1 shows a cross section of the machine consisting of a W-tube and stirred compartments. The lower left compartment contained a $CHCl_3$ solution of (S,S)-**5**, and the lower right a $CHCl_3$ solution of (R,R)-**5**. The central upper compartment contained an aqueous solution of racemic amino acid or ester HPF_6 salt floating on the heavier $CHCl_3$ solutions. The two flanking upper compartments contained water floating on the $CHCl_3$ solutions. When the five stirrers were activated, (S,S)-**5** in the left hand lower pool complexed (L)-amino acid or ester at the H_2O–$CHCl_3$ interface, transported it through the left $CHCl_3$ pool, and delivered it to the left hand upper aqueous pool. Simultaneously, (R,R)-**5** in the right hand lower pool transported and delivered (D)-amino acid or ester to the right hand aqueous pool. The thermodynamic driving force for the machine's operation involved exchange of an energy-lowering entropy of dilution of each enantiomer for an energy-lowering entropy of mixing. To maintain the concentration gradients down which the enantiomers traveled in each arm of the W-tube, fresh racemic host was continuously added to the central aqueous pool, and the separated (L)- and (D)-salts were continuously removed from the left and right pools.[6]

With $C_6H_5CH(CO_2CH_3)NH_3PF_6$ as the guest transported in the W-tube at 0 °C, (L)- and (D)-salts of 86–90% enantiomeric excess were removed from the left and right aqueous pools. Those values correspond to an (L):(D) ratio of thirteen for salts delivered to the left arm, and a (D):(L) ratio of nineteen for salts

[6] M. Newcomb, J.L. Toner, R.C. Helgeson, and D.J. Cram, *J. Am. Chem. Soc.*, 1979, **101**, 4941.

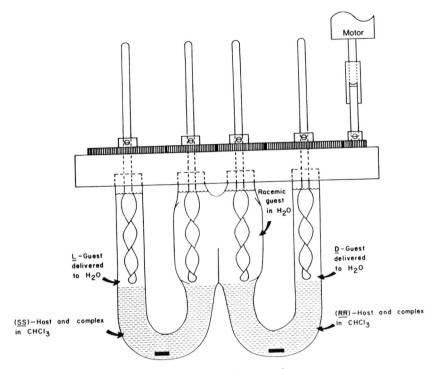

Figure 3.1 *Resolving machine*

delivered to the right arm. The differential transport depends on lipophilization of the enantiomeric guest salts by complexation with a chiral host, and on the amount of transport without complexation being very low.[6]

A good correlation between (L):(D) or (D):(L) values obtained in transport experiments and EDC values obtained in one plate extraction experiments suggested the following conclusions about the transport: (1) the mixing in the organic layer was the slow step; (2) the enantiomeric organic salts rapidly equilibrated between the two phases at the H_2O–$CHCl_3$ interfaces; and (3) the stirring in the two aqueous layers was good enough to provide uniform concentrations of all salts throughout that phase.[6]

3.5 Chromatographic Resolution of Racemic Amine Salts

The working part of (R,R)-**9**, the enantiomer of (S,S)-**5**, was attached at its remote 6′-position to a macroreticular resin (polystyrene–divinylbenzene) to give immobilized host of ~18 000 mass units per average active site. The resulting material was used to give complete enantiomeric resolution by chromatography of ten different racemic α-amino acid HClO$_4$ salts with CHCl$_3$–CH$_3$CN or CHCl$_3$–EtOAc as the mobile phase. Separation factors ranged from a high of 24 ($-\Delta\Delta G^0 = 1.7$ kcal mol^{-1}) with C$_6$H$_5$CH(CO$_2$H)NH$_3$ClO$_4$ to a low of 1.5 for CH$_3$CH(CO$_2$H)NH$_3$ClO$_4$ ($-\Delta\Delta G^0 = 0.2$ kcal mol^{-1}). Similarly, five different racemic α-amino ester HClO$_4$ or HPF$_6$ salts were resolved into their enantiomers to give separation factors that ranged from a high of 26 ($-\Delta\Delta G^0 = 1.8$ kcal mol^{-1}) for 4-HOC$_6$H$_4$CH(CO$_2$CH$_3$)NH$_3$ClO$_4$ to a low of 4.7 ($-\Delta\Delta G^0 = 0.8$ kcal mol^{-1}) for 4-HOC$_6$H$_4$CH$_2$CH(CO$_2$CH$_3$)NH$_3$ClO$_4$. Formula **10** represents a generalized structure of the more stable complex, where the large group **L** is the R group of the guest, **M** is the medium sized CO$_2$H or CO$_2$CH$_3$ group, and **S** is the small group H attached to the chiral center of the guest. The behavior of the working part of this host was semiquantitatively similar in the chromatographic, the transport, and the simple extraction experiments.[7]

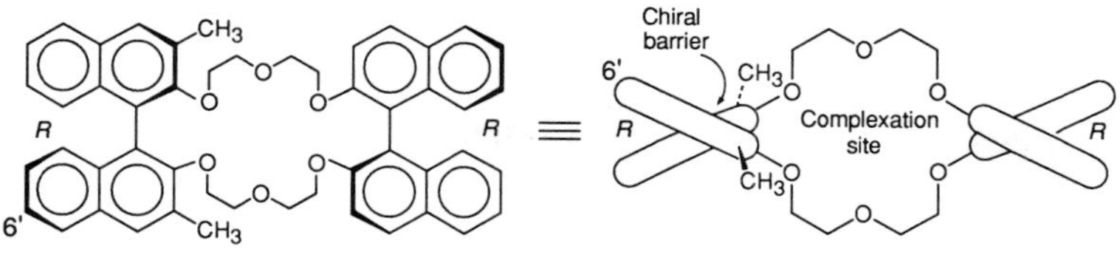

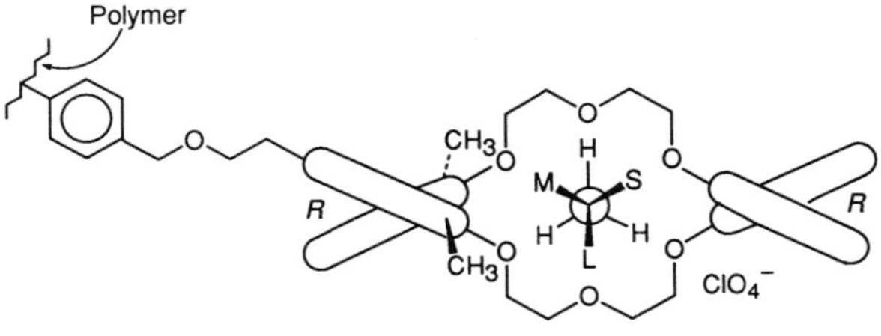

3.6 Failure of a Magnificent Idea Guides Research

Compounds **11–13** and **15** are more noteworthy for their decorative shapes than for their chiral recognition properties. Host (S,S)-**12** was found to bind alkylammonium salts stronger than (S,S)-**11**, but provided lower chiral recognition than (S,S)-**11**, and much lower than its dimethyl analogue (S,S)-**5** (Section 3.3). Notice (S,S)-**11** possesses three C_2 axes of symmetry to give it D_2 overall symmetry, whereas (S,S)-**12** possesses only C_2 overall symmetry.[8] Compound **13** contains three binaphthyl units, two of (S)- and one of (R)-configuration, giving the system C_2^* symmetry. The area around the central binding site is divided into three subcavities, one large (L'), one medium (M'), and one small (S') in size, which are complementary in shape to **LMSC*NH$_3^+$** salts, as

---

[7] G.D.Y. Sogah and D.J. Cram, *J. Am. Chem. Soc.*, 1979, **101**, 3035.
[8] E.P. Kyba, J.M. Timko, L.J. Kaplan, F. de Jong, G.W. Gokel, and D.J. Cram, *J. Am. Chem. Soc.*, 1978, **100**, 4555.

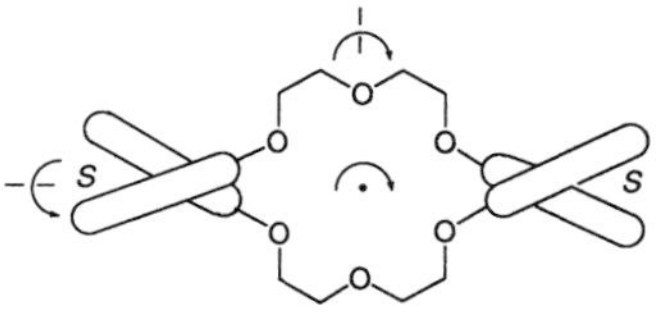

(S,S)-**11**, D_2 symmetry 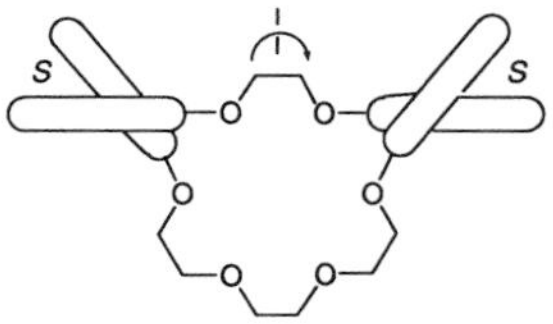   **12**, C_2 symmetry 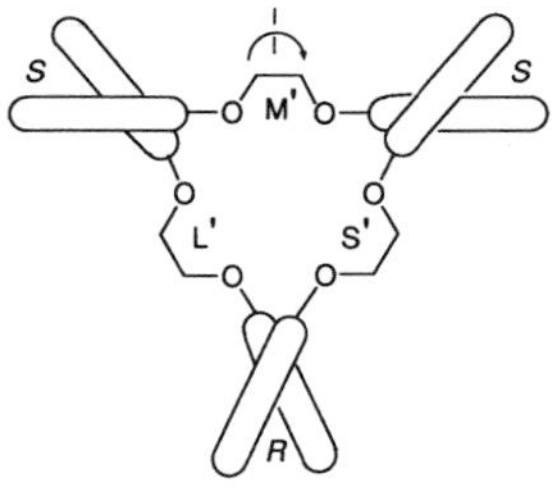   **13**, C_2 symmetry

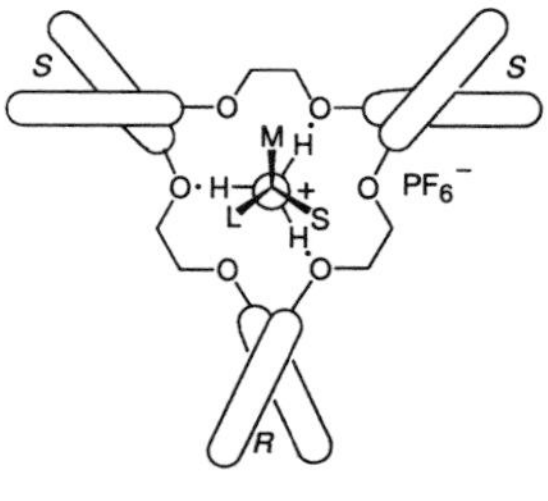

14 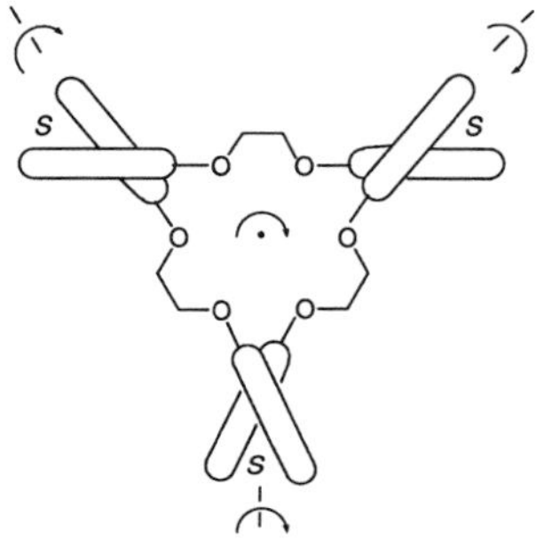   **15**, D_3 symmetry 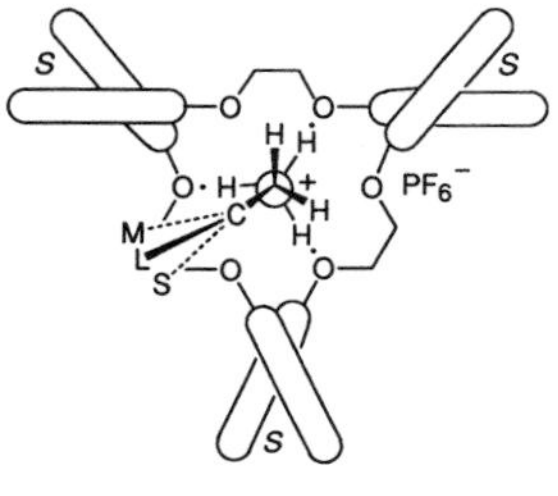   **16**

envisioned in potential complex **14**. Unfortunately the naphthalene groups attached to the oxygens reduced their basicity to the point that no complexation could be detected.[9] Potential host **15** contains three (S)-binaphthyl units to give it overall D_3 symmetry and a group of identical subcavities, each of which is complementary to potential guests of structure $LMSC^*CH_2NH_3^+$ salts, as is suggested by the envisioned complex **16**. Disappointingly, **15** also failed to complex primary ammonium salts.[9]

Although our attempts failed to find extractive conditions that would bring compounds **13** and **15** on scale for chiral recognition studies, we found that chiral recognition by (S,S)-**11** in $CDCl_3$ of $C_6H_5CH(CH_3)NH_3X$ in D_2O varied dramatically with the character of the counterion X^- present. Extractions of racemic amine salt solutions with $CDCl_3$ at 0 °C provided EDC values in the $CDCl_3$ layer. With no added inorganic salt present, RNH_3X failed to be extracted when $X =$ F, Cl, Br, CF_3CO_2, Cl_3CCO_2, or picrate, but with $X = PF_6$, AsF_6, or SbF_6, extraction occurred to give EDC values of 1.9–2.1. The structures of the complexes formed seemed to be independent of the counterion with these large anions, whose charge is highly dispersed. When $X =$ I or SCN, complexes were formed but were not sufficiently ordered to provide chiral recognition. Extractions of RNH_3PF_6 could be performed in the presence of Li^+, Na^+, F^-, Cl^-, or Br^- ions used for salting out without interfering with chiral recognition. The ability of I^- and SCN^- to destroy chiral recognition, we believe, is due to their ability to enter the organic medium by hydrogen bonding

[9] F. de Jong, M.G. Siegel, and D.J. Cram, *J. Chem. Soc., Chem. Commun.*, 1975, 551.

one of the N^+—H protons, thus destroying the tripodal binding on which chiral recognition depends.[8]

3.7 Chiral Catalysis in Michael Addition Reactions

Bergson[10] reported the first example of chiral catalysis of the Michael addition reaction, and Wynberg[11] reported optical purities of products as high as 76% enantiomeric excess in the cinchona alkaloid-catalyzed additions of β-ketoesters to methyl vinyl ketone. We found that (R)-**2** and (R,R)-**9** complexed to t-BuOK or KNH_2 catalyzed the addition of carbon acids to CH_2=$CHCOCH_3$ or CH_2=$CHCO_2CH_3$. High asymmetric induction and catalytic turnover were observed in toluene, the catalysis being dependent on the solubilization of the base through complexation with (R)-**2** or (R,R)-**9** in the non-polar medium. The reactions probably proceed by the catalytic chain reaction mechanism formulated in Equations 3.6–3.8.

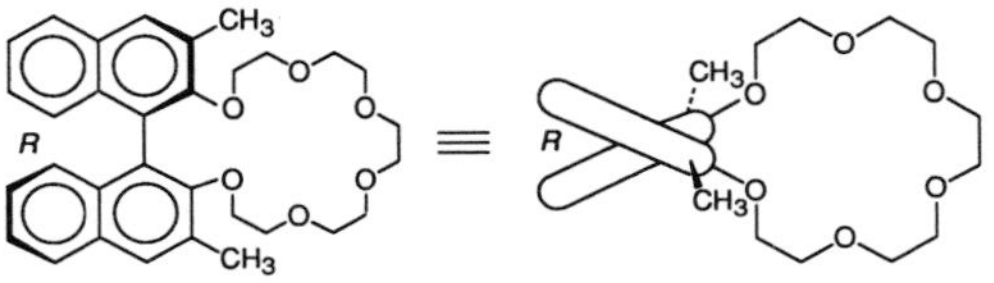

(R)-**2**

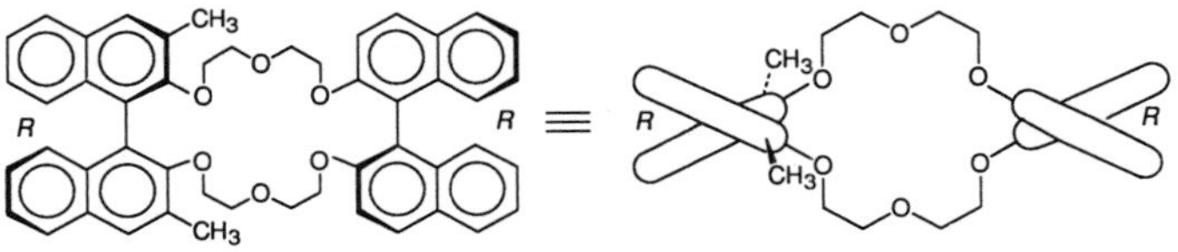

(R,R)-**9**

Initiation:

$$\text{Host} \cdot K^+B^- + H\text{-}R \longrightarrow \text{Host} \cdot K^+R^- + BH \tag{3.6}$$

Addition:

$$\text{Host} \cdot K^+R^- + {}^-C{=}C{-}C{=}O \longrightarrow \overset{*}{R}{-}C{-}C{=}C{-}O^-K^+ \cdot \text{Host} \tag{3.7}$$

Chain transfer:

$$\overset{*}{R}{-}C{-}C{=}C{-}O^-K^+ \cdot \text{Host} + H\text{-}R \longrightarrow \overset{*}{R}{-}C{-}\overset{H}{C}{-}C{=}O + \text{Host} \cdot K^+R^- \tag{3.8}$$

The most stereospecific reaction involved the addition of **17** to **18** to give (S)-**19** in the presence of (R,R)-**9**·$(CH_3)_3$COK as catalyst. At −78 °C in toluene, the reaction went in 48% stoichiometric yield to give product of ca. 99%

---

[10] B. Langstrom and G. Bergson, *Acta Chem. Scand.*, 1973, **27**, 3118.
[11] H. Wynberg and R. Helder, *Tetrahedron Lett.*, 1975, 4059; K. Hermann and H. Wynberg, *J. Org. Chem.*, 1979, **44**, 2738.

enantiomeric excess (e.e.) with catalytic turnover number of 10. In this reaction, the difference in free energies of the two transition states ($\Delta\Delta G\ddagger$) was in excess of 2 kcal mol^{-1}. When the temperature was raised to 25 °C, the stoichiometric yield increased to 67%, the catalytic turnover number increased to 15, but the optical yield decreased to 67% e.e. ($\Delta\Delta G\ddagger = 1.0$ kcal mol^{-1}).

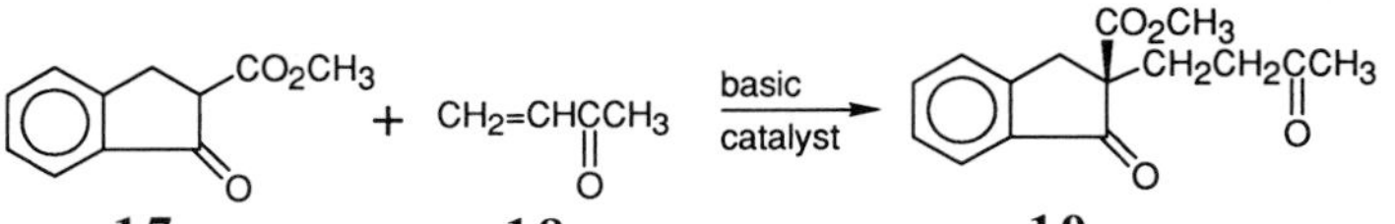

The configuration-determining step for the production of **19** is formulated in **20**, whereas **21** rationalizes on steric grounds the configurational direction of the chiral tilt. The rectangle embracing K$^+$ in **21** symbolizes the plane of the β-keto ester ring in the complex. The C$^-$ occupies the position in the complex sterically most amenable to protonation by the carbon acid **17** in the chain transfer step, Equation 3.8. The face of the delocalized carbanion is *sided* because of its location between the chiral naphthalene barriers.[12,13]

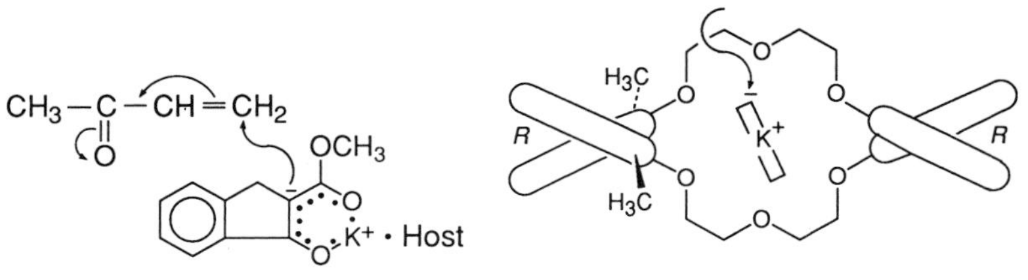

The addition of weaker carbon acid **22** to **23** catalyzed by (R)-**2**·KNH$_2$ gave (S)-**24** in 75–100% stoichiometric yields, optical yields of 83–60% e.e. ($\Delta\Delta G\ddagger = 0.9$–0.5 kcal mol^{-1}), and catalytic turnover numbers of 7–55. The configuration-determining step is formulated in **25**. Structures **26** and **27** rationalize on steric grounds how the configuration of host (R)-**2** leads to (S)-**24** as the dominant enantiomer. The rectangle in **26** and **27** embracing K$^+$ symbolizes the plane of the four-membered ring of the ion pair which in the complex is perpendicular to the best plane of the host's oxygens. The relative positions of the CH$_3$ and C$_6$H$_5$ groups are determined by the low steric requirements of the CH$_3$ and the large steric requirements of the naphthalene, against whose face the CH$_3$ group rests. The proton-donating carbon acid **22** approaches the carbanion face from the side opposite the K$^+$ (the top) as is indicated in **27**.

When (R,R)-**9**·KOC(CH$_3$)$_3$[13] was employed as catalyst, (R)-**24** was obtained in 80% stoichiometric yield, 65% e.e. optical yield, with 19 as the catalytic turnover number. Formula **28** rationalizes how the configuration of the catalyst

---

[12] D.J. Cram and G.D.Y. Sogah, *J. Chem. Soc., Chem. Commun.*, 1981, 625.
[13] In **9** and **21**, the (R,R)-catalyst is formulated to simplify comparisons of the drawings. Experimentally, (S,S)-catalyst produced (R)-**19** and (S)-**24**, respectively.

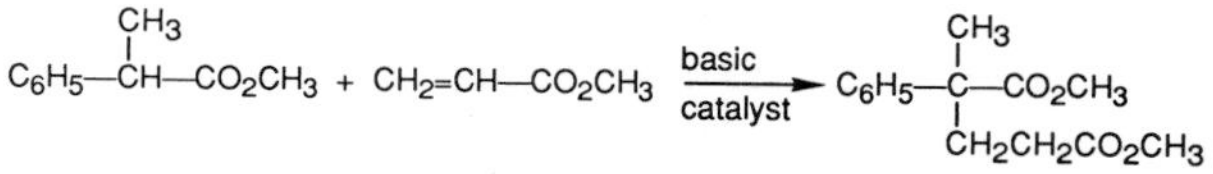

Chiral Recognition in Complexation 61

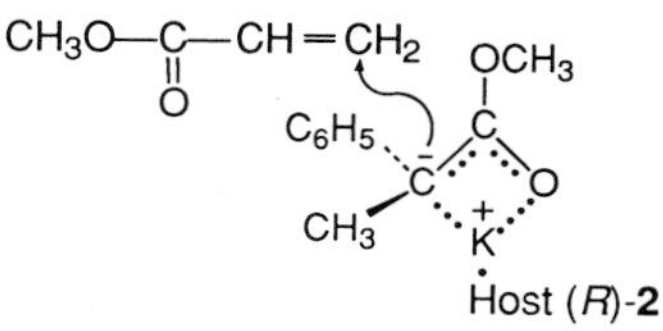

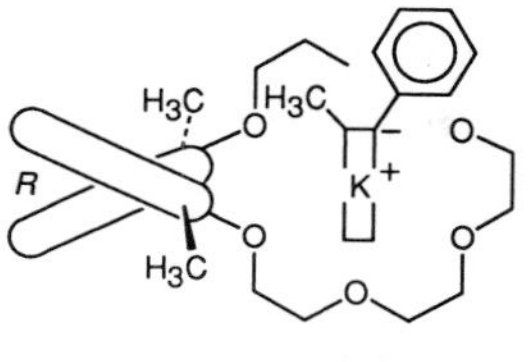

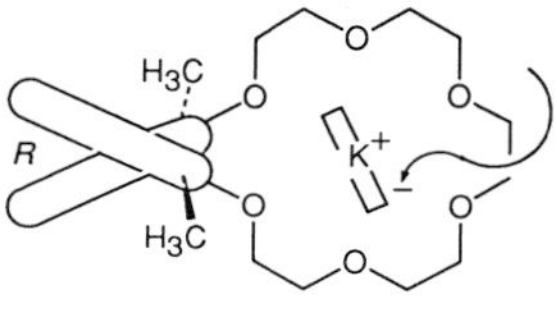

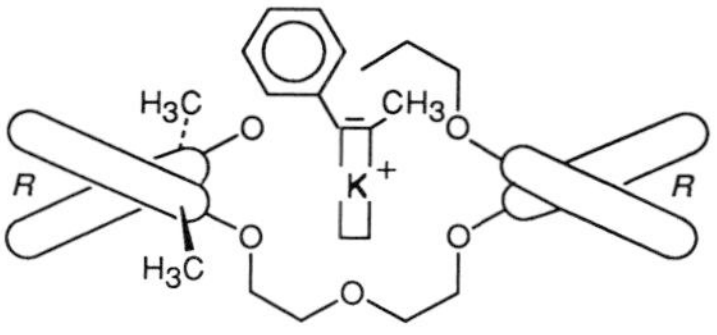

controls the configuration of the dominant product. In **28** the C_6H_5 and CH_3 groups are oriented to produce the least steric strains.

3.8 Chiral Catalysis in Methacrylate Ester Polymerization

Since anionic polymerization of methacrylate esters is a special example of the Michael addition reaction, we anticipated that (*R*)-**2** and (*R,R*)-**9** complexed with metal bases might serve as chiral catalysts to give optically active isotactic polymers (see **29**). The carbanion ion-paired to the host-bound cation is present at the growing end of the chain, in effect 'stamping out' asymmetric units. Unlike ordinary polymerization, any configurational 'mistake' would tend to be corrected rather than perpetuated in the successive generations of asymmetric centers.[14,15]

After we had observed high optical rotations for isotactic polymer,[15] Yuki *et al.* reported that (−)-spartaine-BuLi-initiated polymerization of trityl methacrylate gave optically stable isotactic polymer whose high optical rotation was attributed to helicity.[16] At that time, the only other known synthetic polymers that were helically chiral were the polyalkyl isocyanides, which showed high configurational stability.[17]

Methyl, *tert*-butyl, and benzyl methacrylates polymerized with (*S*)-**2**·$(CH_3)_3COK$ or (*S,S*)-**9**·$(CH_3)_3COK$ as catalysts (ester:catalyst ratios,

---

[14] D.J. Cram and K.R. Kopecky, *J. Am. Chem. Soc.*, 1959, **81**, 2748; D.J. Cram and D.R. Wilson, *J. Am. Chem. Soc.*, 1963, **85**, 1249.
[15] D.J. Cram and D.Y. Sogah, *J. Am. Chem. Soc.*, 1985, **107**, 8301.
[16] Y. Okamoto, K. Suzuki, K. Ohta, K. Hatada, and H. Yuki, *J. Am. Chem. Soc.*, 1979, **101**, 4763.
[17] W. Drenth and R.J.M. Nolte, *Acc. Chem. Res.*, 1979, **12**, 30.

10–100), in 95% toluene–5% tetrahydrofuran (THF) at $-78\,°C$, to give 80–90% isotactic polymers of 1600–2100 mean molecular weight and initial specific rotation that varied from $[\alpha]_{578}^{25°} = +350°$ for benzyl ester with (S)-**2**·$(CH_3)_3COK$ to $-180°$ for *tert*-butyl ester with (S,S)-**9**·$(CH_3)_3COK$. The methyl and *tert*-butyl ester polymers mutarotated at ambient temperature over a period of hours to give material with rotations close to zero. When *tert*-butyl polyester, with $[\alpha]_{578}^{25°} = +117°$, was cleaved with methanolic HCl under mild conditions and the polyacid produced was treated with CH_2N_2, the methyl polyester obtained gave $[\alpha]_{578}^{25°} = +2°$. We conclude that the high rotations are due to helicity of the polymers induced by the chiral cavities of the catalyst in which each unit is added to the growing chain. The helices of the polymeric esters appear to be thermodynamically unstable kinetic products of the polymerization, whose conformations randomize over time. The small residual rotation after randomization is associated with the meso-like interior and different end groups of the isotactic polymer. The greater kinetic stability of the benzyl ester we attribute to its higher mean average molecular weight, and possibly to partial crystallinity.

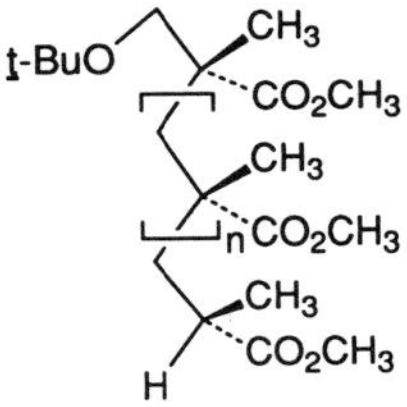

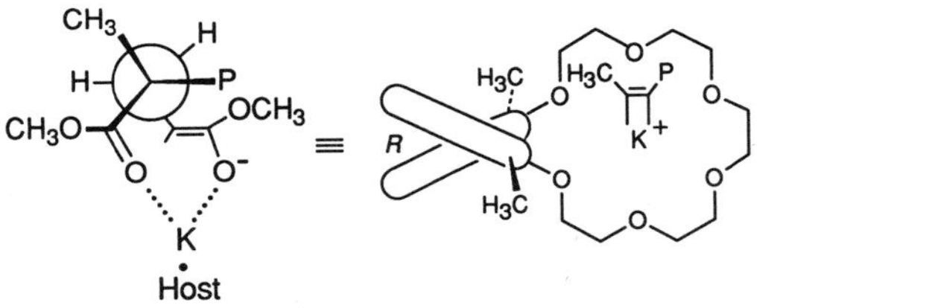

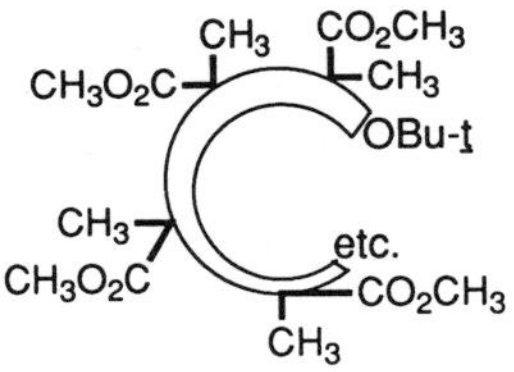

The patterns of relationships between the configurations of the catalysts employed and the signs of rotations of the helical polymers produced are coherent with the patterns observed for the simple Michael reactions. Thus (S)-**2**·$(CH_3)_3COK$ always gave (+)-polymer and (S,S)-**9**·$(CH_3)_3COK$ gave (−)-polymer. As in the simple Michael additions of Section 3.8, these two catalysts gave opposite chiral biases. Structures **30** and **32** are models of suggested conformations of the dominant transition states for the carbon–carbon bond-forming steps of the propagation reactions. Thus **30** predicts helical polymer **31** with a counterclockwise helix, whereas **32** predicts an enantiomeric polymer with a clockwise configuration.

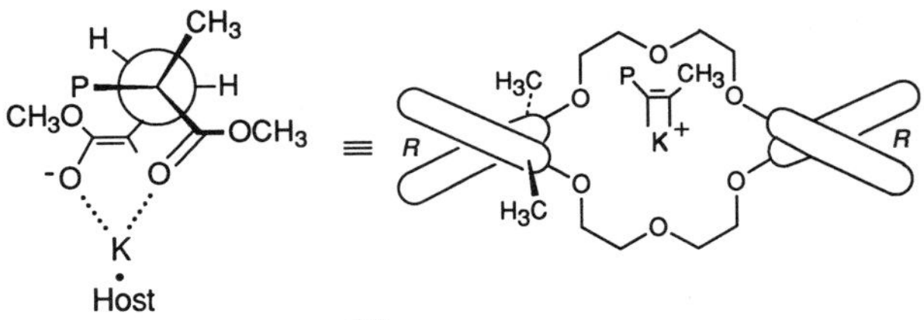

3.9 Chiral Catalysis of Additions of Alkyllithiums to Aldehydes

Tetramethylethylenediamine activates organolithium reagents through complexation, a fact which suggested that hosts **33** and **34**, complexed to such reagents, should induce stereoselectivity in their addition reactions to aldehydes. In CPK models of their alkyllithium complexes, the rigid naphthalene rings coupled with the spirane structures provide a high degree of 'sidedness' to carbonyl groups ligated to the lithium of the complexes, as is envisioned in **35** and **36**. Both hosts **33** and **34** contain C_2 axes, which reduces the number of possible conformations for diastereomeric transition states for addition reactions, and therefore gives less averaging of reactions providing opposite chiral biases.[18]

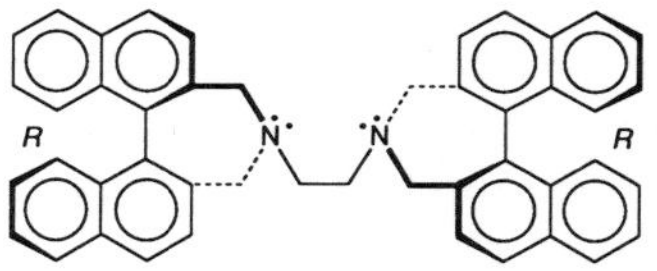

(R,R)-**33** 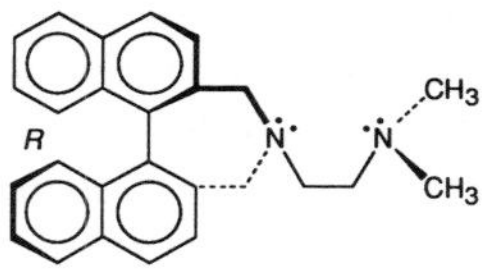   (R)-**34**

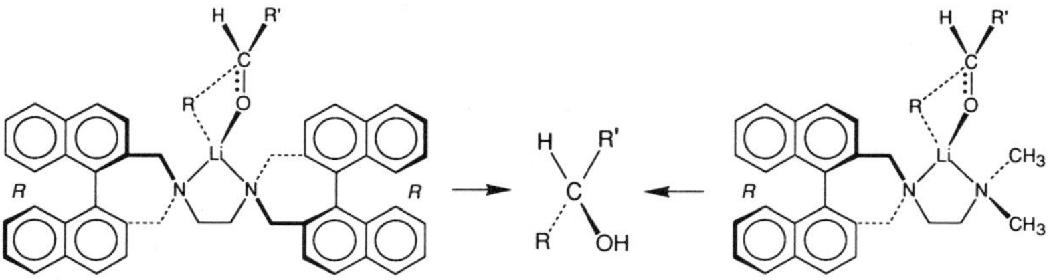

35   **37**   **36**

Preliminary experiments established that $CH_3(CH_2)_3Li$ added to C_6H_5CHO in Et_2O at $-120\,°C$ gave an 81% yield of product, $CH_3(CH_2)_3CHOHC_6H_5$. When the reaction was carried out in the presence of a molar ratio of catalyst to RLi of less than unity, the asymmetric induction was less than when that ratio exceeded unity. Thus the catalyzed addition rate exceeded the non-catalyzed rate, but by a factor too small to provide useful catalyst turnover. Accordingly, ratios of 1.2 ± 0.2 were used in subsequent *stoichiometric catalysis* experiments. The best experimental conditions for the reactions were found to involve Et_2O as solvent at $-120\,°C$.

Chiral catalysis in 9 different reactions of the type **35** → **37** or **36** → **37** gave optical yields of secondary alcohols of 92–22% e.e., and stoichiometric yields of 35–75%. In all nine reactions, the configurations of the dominant enantiomers produced conformed to that of **37**. Thus models **35** and **36** based on steric factors rationalize the results. The highest stereoselectivity was observed when **35**·$Li(CH_2)_3CH_3$ was added to C_6H_5CHO to give (R)-$C_6H_5CHOH(CH_2)_3CH_3$ with 92% e.e. As predicted by general model **35**→**37**, when **35**·LiC_6H_5 was

---

[18] J-P. Mazaleyrat and D.J. Cram, *J. Am Chem. Soc.*, 1981, **103**, 4585.

added to $CH_3(CH_2)_3CHO$, (S)-$C_6H_5CHOH(CH_2)_3CH_3$ of 45% e.e. was produced. Thus an inversion in the order of attachment of groups to the forming asymmetric center inverts the configuration of that center.

Because of the rigid polyspirane structures and freedom from conformational ambiguity of the presumed transition states in **35** and **36**, the only structural choice involves the placement of the H and R' groups. That diastereomer is formulated in **35** and **36** in which the H is directed toward the face of the upper left naphthalene, and in **35**, R is directed toward the cavity alongside the face of the lower right naphthalene. Models of **35** and **36** appear less compressed than those of their diastereomers in which the positions of the H and R' groups are interchanged.

The correlations implicit in **35** and **36** giving **37** show the following similarities to the *rule of steric control of asymmetric induction* for additions to carbonyl groups adjacent to chiral centers.[19] In both correlations, $\Delta\Delta G\ddag$ values for the diastereomeric transition states differ by about 1.0–0.1 kcal mol^{-1}. Both correlations depend on steric effects for asymmetric induction. In both correlations, an inversion in the order of attachments of groups to the forming asymmetric center inverts the configuration of that center.

At the time of publication of this study,[18] both the Seebach[20] and Mukaiyama[21] research groups had reported that organometallics complexed to chiral amines added to aldehydes to give optical yields ranging from 92 to 15% e.e.

A chirophile is a chemist devoted to the symmetry properties of objects, particularly those containing chiral elements. Such objects may be molecules, transition states, short-lived reaction intermediates, complexes, or molecular conformations. Historically, stereochemistry was developed largely by carbohydrate chemists, but with the elucidation of the structures of antibiotics, steroids, alkaloids, peptides, proteins, nucleosides, and nucleotides, every chemist had to be concerned with symmetry properties. With the invention of propylene and methacrylate polymers and their many chiral elements, some polymer chemists became stereochemists, since the physical properties of polymers depend on the patterns of configurations of the many chiral elements in a linear polymer chain. Finally the symmetry properties of nuclear magnetic resonance spectra correlate frequently with molecular symmetry properties, forcing all organic chemists to become stereochemists. Chirophiles are those delighted by these developments.

---

[19] D.J. Cram and F.A. Abd Elhafez, *J. Am. Chem. Soc.*, 1952, **74**, 5828.
[20] D. Seebach, G. Crass, E.M. Wilka, D. Hilvert, and E. Brunner, *Helv. Chim. Acta*, 1979, **62**, 2695.
[21] T. Mukaiyama, K. Soai, T. Sato, H. Shimizu, and K. Suzuki, *J. Am. Chem. Soc.*, 1979, **101**, 1455.

CHAPTER 4
Partial Enzyme Mimics

'Enzyme catalysts collect and orient chemical reactants through complexation, a relationship which converts potentially reacting groups into neighboring groups, thereby lowering the energies of their transition states. Why shouldn't this idea be applicable to designable organic catalysts?'

An exciting challenge to organic chemists is the design, synthesis, and study of organic catalysts of lowest possible molecular weight that possess some of the properties of enzyme systems. The properties most worth mimicking are as follows: (1) enzymes induce fast reaction rates under mild conditions; (2) they show high structural and chiral recognition of the substrates whose reactions they catalyze; (3) each enzyme molecule catalyzes the reactions of enormous numbers of substrate molecules without itself being destroyed. In other words, enzymes have high *turnover* values; (4) enzymes are subject to competitive inhibition by compounds that bind to the enzyme but do not undergo reaction. Properties not possessed by most enzymes, but which are desirable in their mimics, are molecular weights not higher than 2000–3000, the fewest number possible of chiral elements, stability to handling even at elevated temperature, and solubilities in a variety of solvents. We chose the transacylases as enzymes to model because the mechanisms of catalysis have been the most studied and understood.

Other investigators, particularly Cramer, Bender, Kaiser, Breslow, Tabushi, and van Hooidonk, have used the inner lipophilic surface of the naturally-occurring cyclodextrins as binding sites for organic guest compounds in hydroxylic media. The hydroxyl groups attached to the rim provided nucleophiles for reactions of the bound guest.[1] Proton transfer catalysts have also been attached to the rim.[2] Low stereospecificity was observed in acylations[3] of the cyclodextrin systems. However, use of a completely synthetic complexing carbomacrocyclic hydroxamic acid as host and nucleophile gave dramatic rate enhancements in a transacylation.[4]

---

[1] [a] W.D. Griffiths and M.L. Bender, *Adv. Catal.*, 1973, **23**, 209-261; [b] M.L. Bender, 'Mechanisms of Homogeneous Catalysis from Protons to Proteins', Wiley-Interscience, New York, 1971, Chapter 11; [c] M.L. Bender and M. Komiyama, 'Cyclodextrin Chemistry', Springer-Verlag, West Berlin, 1978.
[2] Y. Iwakura, K. Uno, T. Toda, S. Onozuka, K. Hattori, and M.L. Bender, *J. Am. Chem. Soc.*, 1975, **97**, 4432.
[3] K. Flohr, R.M. Paton, and E.T. Kaiser, *J. Chem. Soc., Chem. Commun.*, 1971, 1621.
[4] R. Hershfield and M.L. Bender, *J. Am. Chem. Soc.*, 1972, **94**, 1376.

4.1 A Partial Transacylase Mimic Based on a Corand

We designed and synthesized (S)-**1**, a binaphthyl corand, which provides a binding site for RNH_3^+, and which, like papain, possesses a sulfhydryl group to act as nucleophile. Open-chain compound (S)-**2** was also prepared for comparison. System **1** has the following virtues. In CPK models, **1** is complementary to $RCH(NH_3^+)CO_2Ar$ in the sense that in the complex **1**·$RCH(NH_3^+)CO_2Ar$ the CH_2SH of the host is proximal to the CO_2Ar group. Similarly, in the crystal structure of the related dimethyl host complex **3**·$(CH_3)_3CNH_3^+$, one methyl of the host contacts and lies between two methyls of the $(CH_3)_3CNH_3^+$ guest. In $CDCl_3$ saturated with D_2O at 25 °C, cycle **3** complexed $(CH_3)_3CNH_3^+$ picrate with $\Delta G^0 = -6.4$ kcal mol^{-1}, which is 3 kcal more negative than the binding of the same guest by the dimethyl open-chain compound **4**. A CPK molecular model of the diastereomeric complex between (S)-**1** and (L)-$RCH(NH_3^+)CO_2Ar$ is visibly less compressed and presumably more stable than that between (S)-**1** and (D)-$RCH(NH_3^+)CO_2Ar$. Finally, (S)-**1** possesses a C_2 axis which means that the same complex is formed irrespective of which face of the macroring is complexed.[5,6]

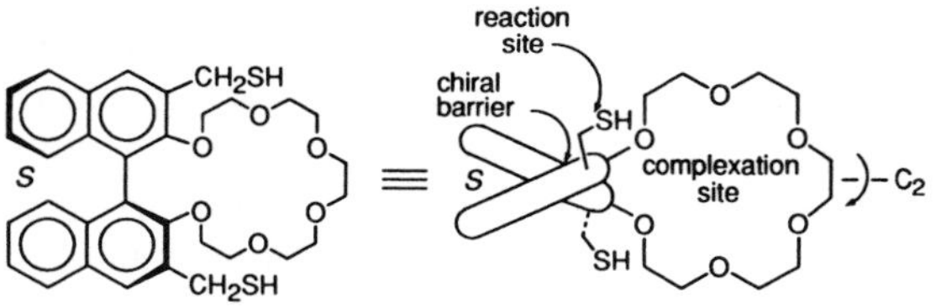

(S)-**1** 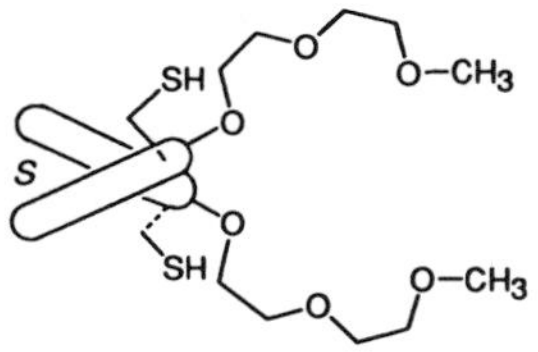   (S)-**2**

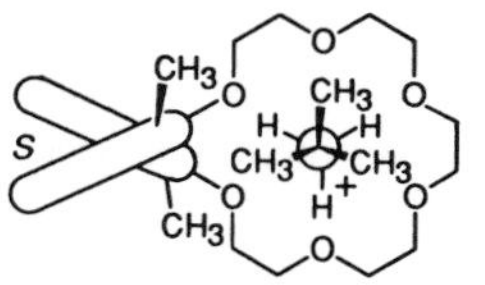

(S)-**3**·$(CH_3)_3CNH_3^+$ 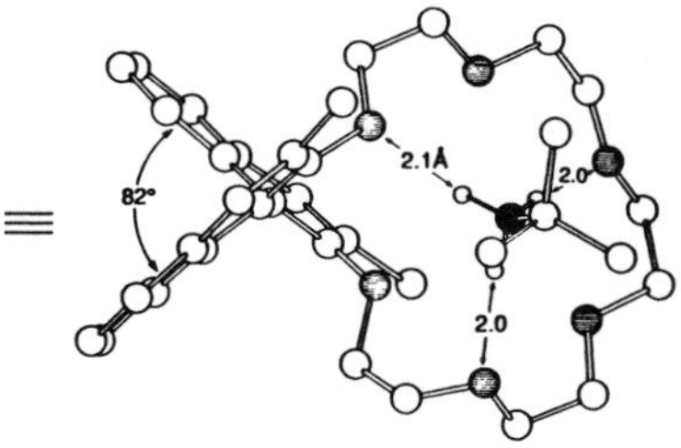   (S)-**3**·$(CH_3)_3CNH_3^+$ crystal structure 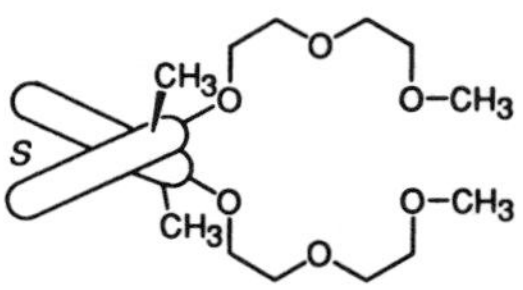   (S)-**4**

4.1.1 Kinetic Acceleration

The kinetics of the appearance of p-$NO_2C_6H_4OH$ in the reactions between p-$NO_2C_6H_4O_2CCH(NH_3^+)RClO_4^-$ and **1** or **2** were measured in solvents, buffered to a pH of 4.3–7.0 (referenced to water), in which the RS^- was the active nucleophile. The ratio of host to guest concentration was a factor of 50 or more, the reactions followed pseudo-first-order kinetics, and were not subject

---

[5] Y. Chao and D.J. Cram, *J. Am. Chem. Soc.*, 1976, **98**, 1015.
[6] Y. Chao, G.R. Weisman, G.D.Y. Sogah, and D.J. Cram, *J. Am. Chem. Soc.*, 1979, **101**, 4948.

Partial Enzyme Mimics

Table 4.1 Rate acceleration factors and free energy differences for thiolysis by complexing (S)-**1** vs. noncomplexing (S)-**2** of (R)-aminoester salts at 25 °C in 20% EtOH–80% CH_2Cl_2(v)[a]

Aminoester salt (Br^-)[a]	k_1/k_2	$\Delta(\Delta G^{\ddagger})$ (kcal mol^{-1})
$(CH_3)_2CHCH_2CH(NH_3^+)CO_2Ar$	>1170	>4.2
$C_6H_5CH_2CH(NH_3^+)CO_2Ar$	490	3.7
$(CH_3)_2CHCH(NH_3^+)CO_2Ar$	160	3.0
$CH_3CH(NH_3^+)CO_2Ar$	>130	>2.9
$CH_2(NH_3^+)CO_2Ar$	>130	>2.9
$C_4H_9N^+CO_2Ar$[b]	0.8	−0.1

[a] The thiols all possessed the (L)- and the aminoester salts, the (D)-configurations.
[b] Proline ester salt.

to general acid or base catalysis. Since **1** binds and orients the amino ester salts and **2** does not, the k_1:k_2 ratios and the derived $-\Delta(\Delta G^{\ddagger})$ values provide measures of the effect of the complexation of the transition states for thiolysis on the reaction rates. Table 4.1 records these values measured in 20% EtOH–CH_2Cl_2(v) at 25 °C for various amino ester salts.[5,6]

The results of Table 4.1 show that the leucine, phenylalanine, valine, alanine, and glycine ester salts acylated macrocycle **1** from >10^2 to >10^3 times faster than they acylated non-cycle **2**, as shown by the k_2:k_1 values of Table 4.1. The free energies of the rate-limiting transition states relative to their equilibrium ground state are lower for cycle **1** than for the non-cycle **2** by the following values, kcal mol^{-1}, for the various amino ester salts: leucine, >4.2; phenylalanine, 3.7; valine, 3.0; alanine, >2.9; glycine, >2.9. All of these ester salts are capable of tripodal binding with cyclic host **1**, but not with non-cyclic host **2**. Thus tripodal binding stabilizes the rate-limiting transition state by several kcal mol^{-1}.[5,6]

In contrast to the primary amino ester salts, proline is a secondary amino ester salt whose NH_2^+ group is incapable of tripodal binding. The k_2:k_1 value for this salt was 0.8 and the $-\Delta\Delta G^{\ddagger}$ value was −0.1 kcal mol^{-1}. This result confirms the conclusion that tripodal binding is required for the large rate accelerations observed for the other amino ester salts. For further comparison, the rate constants were determined for acylation by $CH_2(NH_3^+)CO_2Ar$ of the medium, of **2**, and of the medium by $CH_2(NH_3^+)CO_2Ar \cdot 18$-crown-6. These rate constants were all within a factor of three of one another. Thus, only when the thiol group was part of a strong complexing agent were the large rate increases observed.[5,6]

4.1.2 Competitive Inhibition

The transacylation of (S)-**1**·(L)-$C_6H_5CH_2CH(NH_3^+)CO_2Ar$ was subject to competitive inhibition by K^+ ion. In 20% EtOH–$CHCl_3$ (v) at 25 °C, k_1 for this ester salt reacting with (S)-**1** in the absence of KOAc was 560 times that in the presence of K^+ ion at a concentration ratio of $[K^+]:[RNH_3^+] \sim 120$. Thus K^+ acted as a competitive and dominant binder of **1** and eliminated the acceleration caused by structured complexation.[6]

The effect of solvent on $k_1:k_2$ ratios was determined at 25 °C for the thiolation of (L)-$C_6H_5CH_2CH(NH_3^+)CO_2Ar$ with cycle **1** and non-cycle **2**. The ratio decreased as the solvent became more hydroxylic as follows: 1600 in 0.04% $EtOH$–3% CH_3CN–$CHCl_3$ (v); 960, in 20% $MeOH$–$CHCl_3$ (v); 490 in 20% $EtOH$–$CHCl_3$ (v); 8.2 in 40% H_2O–CH_3CN (v). Thus the hydroxylic solvents appear to be weak competitive inhibitors for complexation of both host and guest, and therefore are rate inhibiting.[5,6]

The rate constant k_1 for (S)-**1** reacting with (L)-$RCH(NH_3^+)CO_2Ar$ in 20% $EtOH$–CH_2Cl_2 (v) at 25 °C decreased with decreasing size for the R groups at their points of attachment to the amino ester salt. The rate constants when normalized to that for the ester salt with R = $(CH_3)_2CH$ gave the following rate factors: CH_3, 290; $(CH_3)_2CHCH_2$, 46; $C_6H_5CH_2$, 17; $(CH_3)_2CH$, 1. Thus about 3.4 kcal mol^{-1} of transition state stabilization is represented by the spread in rate factors, which value acts as a measure of the range of structural recognition in the rate-limiting transition states for these acylations.[5,6]

4.1.3 Chiral Recognition

The chiral recognition involved in these transacylations is measured by rate factors and $-\Delta\Delta G\ddagger$ values. In 20% $EtOH$–CH_2Cl_2, the k_1 values for the reactions of (S)-**1** and (R)-**1** with (L)-$RCH(NH_3^+)CO_2Ar$ were determined. Values of $k_{S,L}:k_{R,L}$, rate constants for the formation of diastereomeric thiol ester products, correlated with the nature of the R group of the amino ester salts as follows: CH_3, 1; $(CH_3)_2CHCH_2$, 6.4; $C_6H_5CH_2$, 8.2; $(CH_3)_2CH$, 9.2. Both the direction of the chiral bias and the relative values of these chiral recognition factors for complexation in the rate-limiting transition states correlate with expectations based on CPK molecular model examination of the diastereomeric *ortho* ester intermediate for transacylation, as formulated in **5** and **6**. Thus the transition state resembling **5** is 1.3 kcal mol^{-1} more stable than

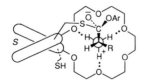
5, (S) to (L) relationship, more stable

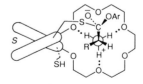
6, (S) to (D) relationship, less stable

that resembling **6** when R = $(CH_3)_2CH$. In **5**, the R group is distant from the chiral barrier, whereas in **6**, R is proximal to the chiral barrier, particularly, when R branches close to its point of attachment to the amino ester moiety.[5,6]

The thiolation of the amino ester salts by **1** proceeds by a mechanism in which a rapidly established equilibrium is followed by a slow acyl-transfer step.

The question arises as to whether the chiral recognition is associated with just the pre-equilibrium, with the acyl-transfer step only, or with both. In a control experiment, the dimethylbinaphthyl host, (S)-**3**, in $CDCl_3$ at 0 °C, was found to extract (L)-$(CH_3)_2CHCH(NH_3^+)CO_2CH_3$ preferentially from racemic material dissolved in D_2O, but the chiral recognition was only 0.6 kcal mol^{-1}. This extraction was carried out under solvent and temperature conditions which were more ideal for chiral recognition than those for the kinetics experiment, yet the latter gave the higher chiral recognition. Both experiments provided the same configurational bias. The facts suggest that more of the 1.3 kcal mol^{-1} of chiral recognition observed in the kinetics experiment was associated with the organization of the rate-limiting transition state than with the complex formed in the pre-equilibrium. The transition state with its covalent bond acting as an additional binding site should provide a much more structured species than a complex organized only by hydrogen bonding. As expected, no chiral recognition was observed in the acylation of open-chain model host (S)-**2** by (D)(L)-$C_6H_5CH_2CH(NH_3^+)CO_2Ar$.[6]

The solvolysis of the initially formed thiol ester of (S)-**1** and (L)-$C_6H_5CH_2CH(NH_3^+)CO_2Ar$ was followed polarimetrically in 20% EtOH–CH_2Cl_2. The first-order rate constant for this reaction was 10^4 smaller than the pseudo-first-order rate constant for acylation of the amino ester salt by (S)-**1**. Thus catalytic turnover was not observed in these systems.[6]

4.1.4 Abiotic and Biotic Comparisons

The similarities and differences between our abiotic and the biotic transacylases of nature are interesting to consider. In both systems, the rate for the first transacylation is greatly accelerated by complexation, which lowers the free energy of the transition states. The catalysis of the first acylation by complexation in both abiotic and biotic systems is subject to competitive inhibition by substances that complex reversibly, but do not react. In both the abiotic and biotic systems, the reacting partners show structural recognition of one another by discriminating in their reaction rates between amino ester salts with different side chains. Both systems show chiral recognition favoring amino ester salts of the (L)-configuration, and the degree of chiral recognition depends on the side chain of the amino ester salt.[7] The biotic system catalyzes both the acylation and deacylation by complexation to give turnover, which the abiotic host does not provide. The rate factors and discrimination between potential substrates are greater for the biotic than the abiotic systems.

Others subsequently reported that corands attached to thiol groups catalyze by complexation the transacylation of the thiol groups by *p*-nitrobenzoate esters of primary amino acid salts.[8,9] One communication reported an example of very high chiral recognition in this acylation.[9]

---

[7] W.D. Griffiths and M.L. Bender, *Adv. Catal.*, 1973, **23**, 637.
[8] T. Matsui and K. Koga, *Tetrahedron Lett.*, 1978, 1115.
[9] J-M. Lehn and C. Sirlin, *J. Chem. Soc., Chem Commun.*, 1978, 949.

4.2 Hosts Containing Cyclic Urea Units

The results described in Section 4.1 demonstrated that corand complexation, which gathers and orients reacting species, enhances reaction rates by several orders of magnitude, but not nearly as well as enzyme systems. The more highly preorganized systems of spherands and hemispherands of Chapter 2 were shown to bind alkali metal and ammonium ions more strongly than do the corands. We, therefore, thought it possible to apply the principles of preorganization and complementarity to the design and preparation of systems that might complex alkylammonium ions much more strongly than the corands. The results are described in this section. Such systems in turn might serve as a starting point for the design of abiotic transacylases that provide greater stabilization of transition states by complexation, and thus produce higher rate factors for transacylation (see Section 4.4).

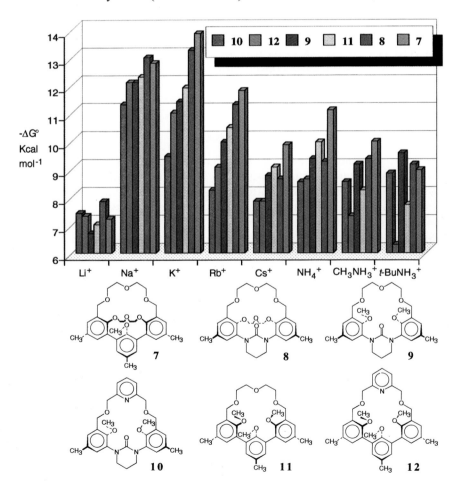

Chart 4.1

4.2.1 Preorganization of Cyclic Urea Units by Anisyl Unit Attachment

A survey of complexation of hemispherands with alkylammonium and alkali metal ions demonstrated that the cyclic urea unit flanked by two anisyl units incorporated into an 18-membered ring, as in **8** and **9**, provides strong binding toward $(CH_3)_3CNH_3^+$ picrate in $CDCl_3$ at 25 °C. The results are shown in Chart 4.1. Notice that of the six hemispherands examined, **7–12**, **8** and **9** are the strongest binders of $(CH_3)_3CNH_3^+$ at 9.5 and 9.1 kcal mol^{-1} respectively. The crystal structure of free host **9** depicted in **13** shows that two CH_2 groups of the $(OCH_2CH_2)_2O$ portion of the ring are turned inward to fill the cavity. The crystal structure of the complex **9**·$(CH_3)_3CNH_3^+$ pictured in **14** shows the hoped-for tripodal binding of the NH_3^+ containing a hydrogen bond to the urea oxygen. Comparison of crystal structures **13** and **14** shows that the flexible $(OCH_2CH_2)_2O$ portion of the ring reorganizes upon complexation, as was observed with complexes of **11** (see Section 2.4).[10]

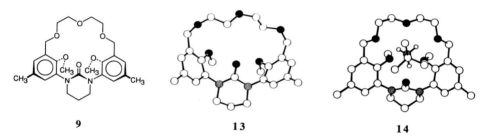

4.2.2 Binding Properties of Hosts Containing Multiple Cyclic Urea Units

These results coupled with CPK model examination and viable synthetic strategies led to the design and preparation of hosts **15** and **16**, with three cyclic urea units, and **17** and **18**, with two units. The binding free energies of hosts **15–18** toward Li^+, Na^+, K^+, Rb^+, Cs^+, NH_4^+, $CH_3NH_3^+$, and $(CH_3)_3CNH_3^+$ picrates in $CDCl_3$ at 25 °C are shown in Chart 4.2. Lipophilic corand **19** is included for comparison. Chart 4.2 provides several interesting conclusions. (1) The maximum structural recognition in complexation of the eight guests by the five hosts is shown by **17**, which binds Li^+ better than $(CH_3)_3CNH_3^+$ by ~9 kcal mol^{-1}. (2) The four urea-unit-containing hosts **15–18** are better binders toward almost all ions than is corand host **19**, by as much as ~12 kcal mol^{-1} in the case of **15** binding Li^+, even though **19** contains six and **15** only five potential binding sites. (3) The systems containing three cyclic urea units, **15** and **16**, are the strongest general binders, particularly of NH_4^+, $CH_3NH_3^+$, and $(CH_3)_3CNH_3^+$. Thus all the $-\Delta G^0$ values for these two hosts binding all eight ions range from about 11.3 to 18.2 kcal mol^{-1}. The less rigid of these two hosts

---

[10] D.J. Cram, I.B. Dicker, G.M. Lein, C.B. Knobler, and K.N. Trueblood, *J. Am. Chem. Soc.*, 1982, **104**, 6827.

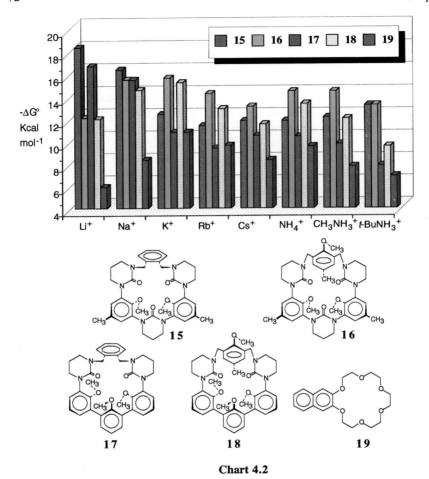

Chart 4.2

is **16**, which contains a 20-membered macroring, whereas the ring of **15** is 19-membered. (4) Of the five hosts, **16** is the best general binder of the eight guests, the range being 12.1 for Li$^+$ to 15.6 kcal mol^{-1} for K$^+$; NH$_4^+$, CH$_3$NH$_3^+$, and (CH$_3$)$_3$CNH$_3^+$ are bound by 14.4, 14.4, and 13.2 kcal mol^{-1}, respectively. (5) The higher the number of urea units in the host, the greater is the binding by the hosts of CH$_3$NH$_3^+$ and (CH$_3$)$_3$CNH$_3^+$. Thus **16** > **18** > **19** in binding CH$_3$NH$_3^+$, a spread of 6.8 kcal mol^{-1}, and in binding (CH$_3$)$_3$CNH$_3^+$, a spread of 6.3 kcal mol^{-1}.[11]

4.2.3 Crystal Structures

Examination of CPK models suggested that complexes **16·(CH$_3$)$_3$CNH$_3^+$** and **18·Na$^+$·H$_2$O** would have good binding properties. Chart 4.3 shows these

[11] D.J. Cram, I.B. Dicker, M. Lauer, C.B. Knobler, and K.N. Trueblood, *J. Am. Chem. Soc.*, 1984, **106**, 7150.

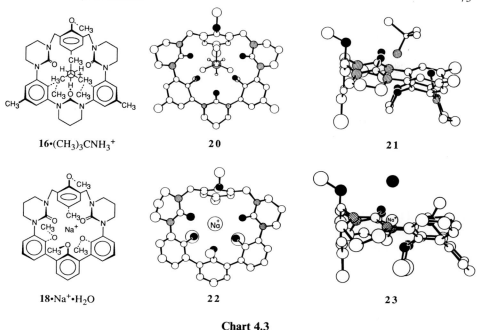

Chart 4.3

complexes along with the face and side views **20** and **21** of the crystal structure of the former, and similar views **22** and **23** of the latter. Notice the good correspondence between expectation and experiment. In **16**·$(CH_3)_3CNH_3^+$, the guest is hydrogen-bonded to the three upturned urea oxygens to form a tripod-perching complex, while the two flanking anisyl oxygens at 4 and 8 o'clock contact the N^+ from its underside. In **18**·Na^+·H_2O, the Na^+ lies in a nest of five oxygens of the host, with a water molecule also ligating the Na^+ from the top. In both crystal structures, the plane of the anisyl group at 12 o'clock is tilted away from the top binding face at an angle slightly less than 90 to the best plane of the macroring. This arrangement brings the aryl methyl of this group into contact with the methyls of the CH_3O groups of the two anisyls at 4 and 8 o'clock to form a closed surface of carbon–hydrogen or carbon–carbon bonds on the face of the macrocycle remote from the carbonyl oxygens.[11]

4.2.4 Binding Power Dependence on Structure

Of the hosts containing three cyclic urea units, **16** was the strongest binder of $CH_3NH_3^+$ and $(CH_3)_3CNH_3^+$ ions. Potential hosts **24**, **25**, and **26**, which also contain three cyclic urea units, were prepared to determine the effect of different degrees of preorganization on the binding free energies of RNH_3^+ guests. The $-\Delta G^0$ values, in kcal mol^{-1}, are listed below the formulas.

Host **24** contains the same ring system as **16**, except that the OCH_3 and aryl CH_3 groups are absent. This change reduced the $-\Delta G^0$ value from 14.4 to 7.4

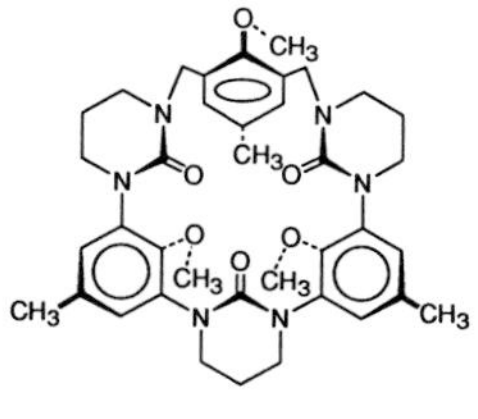

	16	24	25	26
		$-\Delta G^0$ (kcal mol^{-1})		
$CH_3NH_3^+$	14.4	7.4	14.9	7.3
$(CH_3)_3CNH_3^+$	13.2	8.6	14.3	7.7

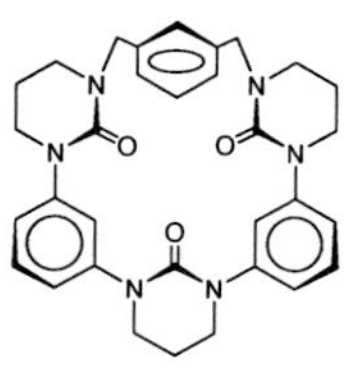

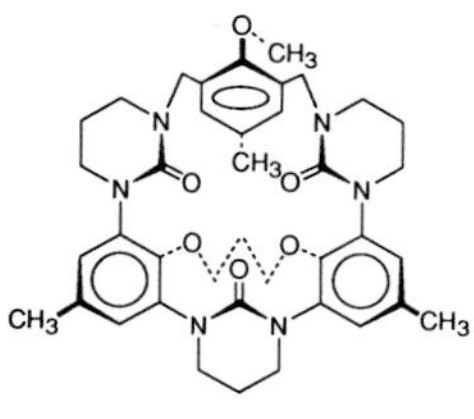

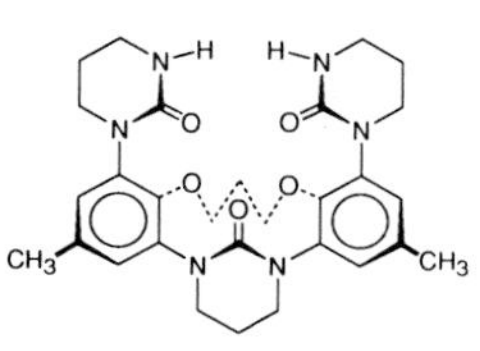

kcal mol^{-1} for the binding of $CH_3NH_3^+$, and from 13.2 to 8.6 kcal mol^{-1} for that of $(CH_3)_3CNH_3^+$. Molecular model examination of the four complexes suggests that much of this reduction in binding is due to the large contribution the CH$_3$O groups at 4 and 6 o'clock in **16** make to the preorganization of the three urea units.[12]

In models, the additional (CH$_2$)$_3$ bridge of **25** further preorganizes the three cyclic urea units for binding, and in experiment increases the $-\Delta G^0$ values by 0.5 for $CH_3NH_3^+$ and by 1.0 kcal mol^{-1} for $(CH_3)_3CNH_3^+$ over those for **16**. When the anisyl unit at 12 o'clock was omitted while the (CH$_2$)$_3$ bridge of **25** was retained, as in **26**, the $-\Delta G^0$ of binding dropped from 14.9 to 7.3 kcal mol^{-1} for $CH_3NH_3^+$, and from 14.3 to 7.7 kcal mol^{-1} for $(CH_3)_3CNH_3^+$. This reduction reflects the lack of preorganization of the two cyclic urea units in **26** at 2 and 10 o'clock. The $-\Delta G^0$ average for **16** binding Li$^+$, Na$^+$, K$^+$, Rb$^+$, Cs$^+$, NH$_4^+$, $CH_3NH_3^+$, and $(CH_3)_3CNH_3^+$ is 14.1 kcal mol^{-1} compared to 15.6 kcal mol^{-1} for **25**, which makes **25** the best general complexing agent for these ions yet prepared.[13]

The crystal structure of **25**·$(CH_3)_3CNH_3^+$ is drawn as a face view in **27**. The crystal was orderly enough, unlike that of **16**·$(CH_3)_3CNH_3^+$ (see **20**), to provide interesting structural details. The C—N bond of the guest is essentially normal,

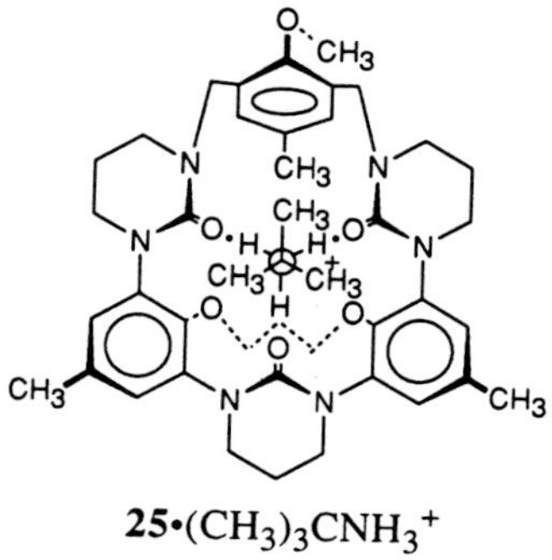

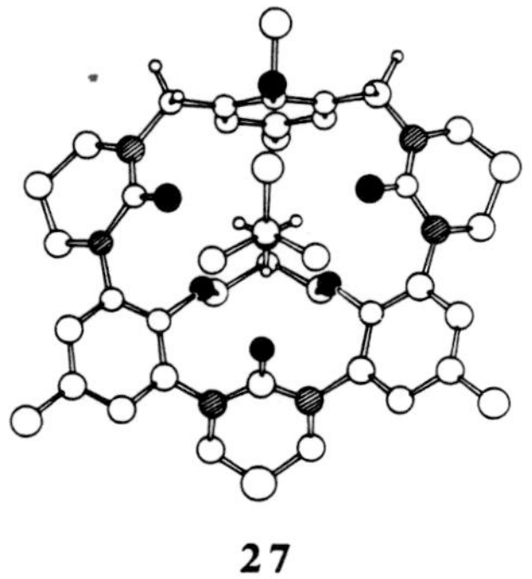

[12] R.J.M. Nolte and D.J. Cram, *J. Am. Chem. Soc.*, 1984, **106**, 1416.
[13] K.M. Doxsee, M. Feigel, K.D. Stewart, J.W. Canary, C.B. Knobler, and D.J. Cram, *J. Am. Chem. Soc.*, 1987, **109**, 3098.

Partial Enzyme Mimics 75

88.3°, to the plane defined by the three carbonyl oxygens, and the CH_3—C—N—H dihedral angles are 62°, close to the 60° anticipated. The three NH····O bond lengths are all about 2.75 Å, indicating strong hydrogen bonds. The angle of intersection of the plane of the *m*-xylyl unit at 12 o'clock and the best plane of the macroring is 84°. The $OCH_2CH_2CH_2O$ bridge is *anti* to the CH_3O group of the *m*-xylyl unit, and the central CH_2 of the bridge is oriented toward the *m*-xylyl unit. The overall structures of **20** and **27** are very similar to one another.[13]

Analysis of much crystallographic and neutron diffraction data indicates there to be a distinct preference for C=O····H-N hydrogen bonds to form in, or near to, the directions of the sp^2 unshared electron pairs on carbonyl oxygens.[14] In **27**, the O····H bond length at 6 o'clock is 1.9 Å, the C=O····H bond angle is 133°, and the O····H-N bond angle is 167°; the O····H bond lengths at 2 and 10 o'clock are both 2.0 Å, the C=O····H bond angles are 145°, and the O····H-N bond angles are 174°. These locations of the NH_3^+ hydrogens suggest that the hybridization at oxygen might be close to sp^3 with high negative charge density on oxygen. However, the average of eighteen O····H-O bond angles taken from four crystal structures of hydrates of tris-cyclic urea systems is 163°. These facts taken in sum suggest the cyclic urea oxygens are *soft ligands* whose hybridization at oxygen is flexible and responds to the demands of the guest.[13]

Hosts containing the cyclic urea units have many attractive properties. They are very soluble in organic solvents, they extract ions from water into immiscible organic media rapidly on the human time scale, and complexation–decomplexation rates are very fast. For example in $CDCl_3$ saturated with D_2O, the rate constant for complexation of $(CH_3)_3CNH_3^+$ picrate with **16** is $\sim 10^{12}$ $mol^{-1} s^{-1}$ (diffusion controlled), and that for decomplexation is $\sim 10^3 s^{-1}$, while the association constant, K_a, is about $10^9 M^{-1}$.[15]

4.2.5 Highly Preorganized Hosts

Two other very highly preorganized hosts containing two cyclic urea units incorporated into an 18-membered macrocyclic ring were prepared and were found to be very powerful binders of the eight standard picrate salts at 25 °C in $CDCl_3$ saturated with D_2O. Thus **28** bound the picrate salts with $-\Delta G^0$ values in kcal mol^{-1} as follows: Li^+, 19.3; Na^+, 19.5; K^+, 16.6; Rb^+, 15.2, Cs^+, 14.7; NH_4^+, 14.9; $CH_3NH_3^+$, >14; $(CH_3)_3CNH_3^+$, >14. Its sulfur containing analogue **29** gave somewhat lower values: Li^+, 15.8; Na^+, 19.1; K^+, 13.7; Rb^+, 12.5; Cs^+, 12.1; NH_4^+, 13.3; $CH_3NH_3^+$, 12.9; $(CH_3)_3CNH_3^+$, 14.2. The crystal structure of **28** shows it to be a spherand, completely preorganized for binding. Stereoviews of the crystal structures of **28**, **28**·Na^+, and **29**·Na^+ are drawn in **30**, **31**, and **32**, respectively (see Chart 4.4).[16]

---

[14] R. Taylor and O. Kennard, *Acc. Chem. Res.*, 1984, **17**, 320; P. Murray-Rust and J.P. Glusker, *J. Am. Chem. Soc.*, 1984, **106**, 1018.
[15] T. Anthonsen and D.J. Cram, *J. Chem. Soc., Chem. Commun.*, 1983, 1414.
[16] J.A. Bryant, S.P. Ho, C.B. Knobler, and D.J. Cram, *J. Am. Chem. Soc.*, 1990, **112**, 5837.

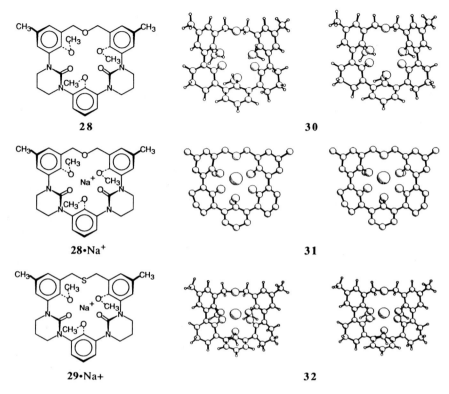

Chart 4.4

4.2.6 An Unusual Color Indicating System

An unusual color indicator for the presence of alkali metal, ammonium, and alkylammonium ions was developed based on the strongly binding properties of host **33**[17], which, like **25**, contains three cyclic urea units, a $(CH_2)_3$ bridge and an additional aryl substituent. The $-\Delta G^0$ values in kcal mol^{-1} for **33** binding the standard eight picrate salts in CDCl$_3$ saturated with D$_2$O at 25 °C were as follows: Li$^+$, 13.6; Na$^+$, 16.7; K$^+$, 15.7; Rb$^+$, 14.8; Cs$^+$, 14.5; NH$_4^+$, 15.2; CH$_3$NH$_3^+$, 14.0; (CH$_3$)$_3$CNH$_3^+$, 12.6. The crystal structure of **33**·3H$_2$O is shown in **34** to be beautifully preorganized for binding these ions.[13] Two of the water molecules bridge two carbonyl oxygens intramolecularly in **33**·3H$_2$O, and one water bridges intermolecularly.

Molecular model examination of **33**·$^+$N$_2$Ar indicated the N$_2$ group to be complementary to the enforced cavity of **33**. This host was found to bind 4-(CH$_3$)$_3$CC$_6$H$_4$N$_2$BF$_4$ with a $-\Delta G^0$ value of 5.9 kcal mol^{-1} in ClCH$_2$CH$_2$Cl at 25 °C, which is 2.3 kcal mol^{-1} higher than the corresponding value of 3.6 kcal mol^{-1} for 18-crown-6.[17]

Addition of five equivalents of host **33** to one equivalent of 4-(CH$_3$)$_3$CC$_6$H$_4$N$_2$BF$_4$ in CH$_2$Cl$_2$ inhibited completely the coupling reaction of the diazo compound with ten equivalents of (CH$_3$)$_2$NC$_6$H$_5$ to give the orange azo dye **35**. However when an aqueous solution of Na$_2$CO$_3$ was shaken with this solution, the orange color of the dye developed immediately.[17] The complexation of diazo compounds with **33** is visualized as arising from favorable dipole–dipole interactions between spherand **33** and ArN$_2^+$, and is depicted in **36**.

4.3 An Incremental Approach to Serine Protease Mimics

The active site of chymotrypsin combines a binding site, a serine hydroxyl, a histidine imidazole, and an aspartate carboxyl in an array preorganized largely by hydrogen bonds as sketched in **37**. With CPK models, we designed **38** as an ultimate target molecule which has roughly the same organization of groups as that of **37**. Compound **38** possesses the desirable features of relatively low molecular weight, of potential synthetic viability, and of providing an incremental approach to its preparation through model compounds **16**, **40**, and **41** of successively increasing complexity. Compound **39** serves as a noncomplexing model compound.

Section 4.2 describes the binding of **16** with (CH$_3$)$_3$CNH$_3^+$ picrate at 25 °C in CDCl$_3$ saturated with water giving $-\Delta G^0$ = 14.4 kcal mol^{-1}. In the crystal structure of **16**·(CH$_3$)$_3$CNH$_3^+$ (**21**), the OCH$_3$ at 12 o'clock is proximal to one CH$_3$ group of the guest, and in CPK models, the hydroxyl group of **40** is proximal to the carbonyl group of **40**·RCH(NH$_3^+$)CO$_2$Ar. Host **40** was synthesized by the reaction of **42** with **43**, to give the methyl ether of **40** which was

[17] D.J. Cram and K.M. Doxsee, *J. Org. Chem.*, 1986, **51**, 5068.

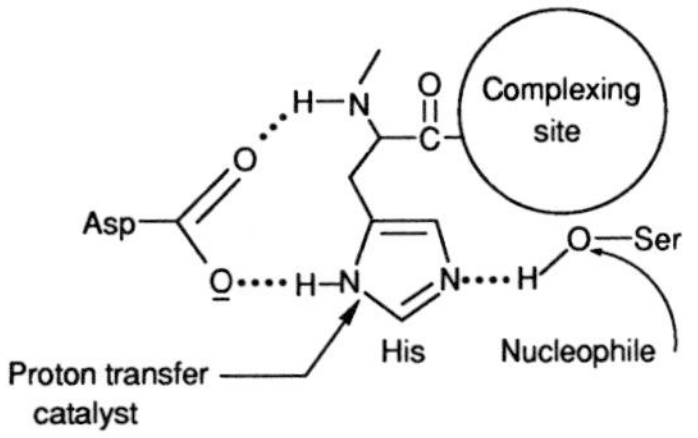

demethylated to give **40** in 60% for the two steps. In CDCl$_3$ saturated with D$_2$O at 25 °C, the K_a values for **40** binding CH$_3$NH$_3{}^+$, (CH$_3$)$_3$CNH$_3{}^+$, and Na$^+$ picrates were 3.5×10^{11}, 4.5×10^9, and 1.2×10^{11} M^{-1}, respectively.[18]

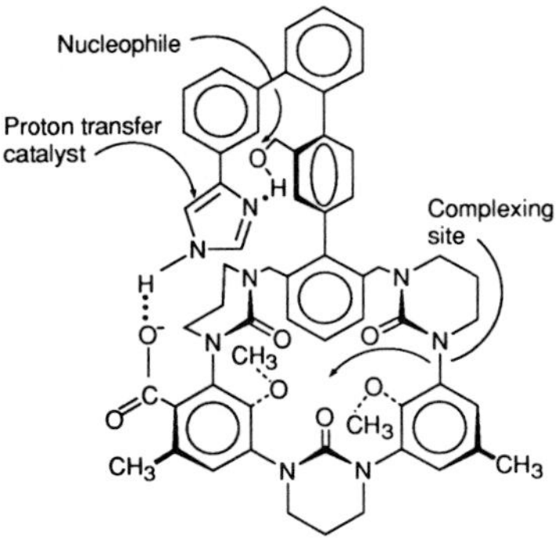

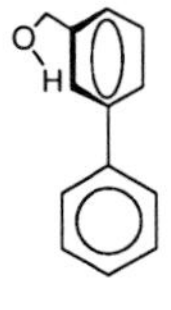

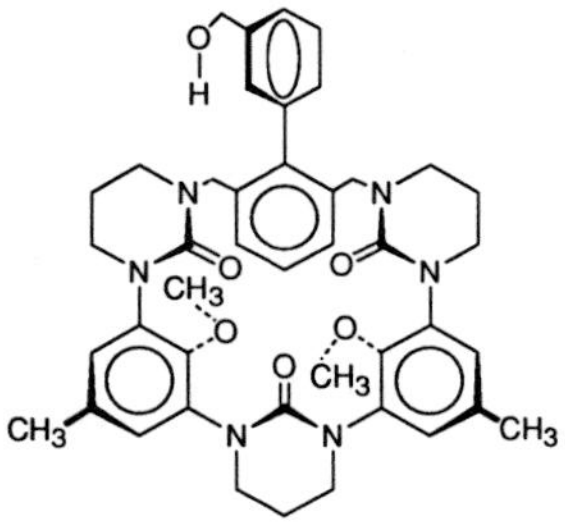

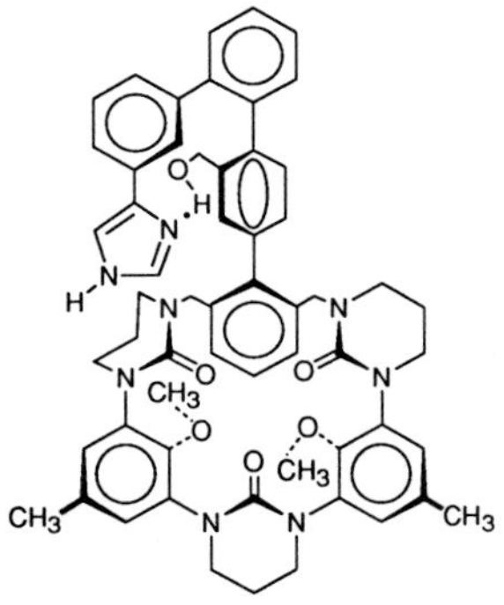

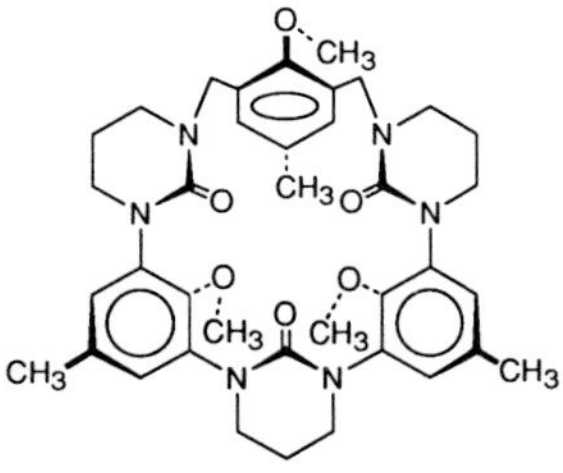

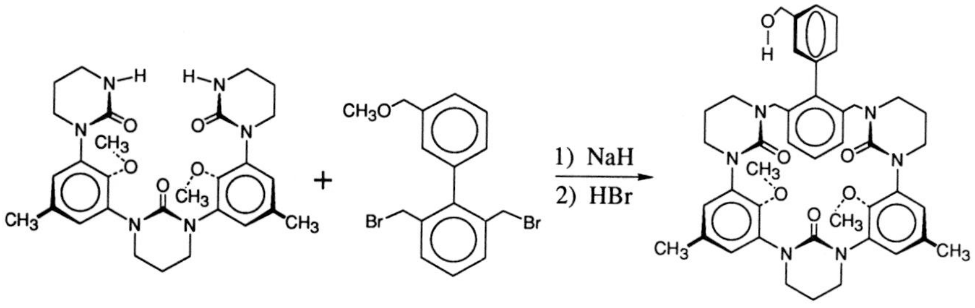

4.3.1 Kinetics of Transacylations

Equal molar quantities of (L)-alanine *p*-nitrophenyl ester perchlorate, **44**, and host **40** were mixed in 10:1 CH$_2$Cl$_2$–pyridine (v:v) at 25 °C to give *p*-nitrophenol

[18] D.J. Cram, H.E. Katz, and I.B. Dicker, *J. Am. Chem. Soc.*, 1984, **106**, 4987.

and ester **45**, which was fully characterized. (For convenience the (D)-alanine ester salt **44** is formulated here and elsewhere.) In the drawing of ester **45**, the dihedral angle for the groups attached to the C—N atoms are a comfortable 60°. Molecular models of the presumed *ortho* ester intermediates leading to **45** are sterically compatible with leaving groups as large as $p\text{-NO}_2\text{C}_6\text{H}_4\text{O}^-$, and amino ester substituent groups as large as $\text{C}_6\text{H}_5\text{CH}_2$.[18]

The kinetics of transacylation of **40** by amino ester salt **44**[18] were followed by the appearance of $p\text{-NO}_2\text{C}_6\text{H}_4\text{O}^-$ at 25 °C in $CDCl_3$ diluted with 0.15% to 10% $(CH_3)_2NCHO$ or CH_3CN by volume buffered by diisopropylethylamine (R_3N) and its salt R_3NHClO_4. The ratio of host to guest was ~15, so that good pseudo-first-order kinetics were observed. The reaction was found to be first-order in specific-base concentration (R_3N of the buffer) in 0.15% $(CH_3)_2NCHO$ in $CDCl_3$, pointing to CH_2O^- as the active nucleophile in the rate-limiting transition state for the transacylation. In water, the pK_a of $CH_3CH(CO_2Ar)NH_3^+$ is ~7 and that of $ArCH_2OH$ ~15, whereas the pK_a of buffer R_3NH^+ is ~10.8. Clearly, complexing $CH_3CH(CO_2Ar)NH_3^+$ with **40** in this medium dramatically raises the pK_a of the NH_3^+ group relative to the CH_2OH group, probably by many units. This hypothesis is compatible with the 13–14 kcal mol^{-1} range of $-\Delta G^0$ values for which $CH_3NH_3^+$ picrate is bound in $CDCl_3$ by tris-urea systems, **40** and **16**. The presence of $NaClO_4$ in a concentration identical to that of **40** decreases the rate constant by a factor of 57. When the concentration of $(CH_3)_2NCHO$ or CH_3CN in the medium was increased to 10% (v:v), the rate constant decreased by factors of about 4 and 2, respectively. Addition of $NaB(C_6H_5)_4$ to reactions in these media decreased these rate constants by factors ranging from 20 to >100. Thus the reactions are subject to competitive inhibition.

4.3.2 Rate Enhancements Due to Complexation

The rate enhancements due to complexation were measured by the ratios of rate constants for host **40** being acylated (k_a) vs. that for the non-complexed *m*-phenylphenol, **39**, being acylated (k_b). Since the two acylations go by different mechanisms, the comparisons are based on the solutions of uncomplexed **39**, **40**, and **44** as the standard starting states and the rate-controlling transition states as the standard final states. The resulting rate-constant enhancement factors measure all of the effects associated with complexation,[19] and are expressed by the k_a:k_b ratio of equation 4.1. In this equation $k_{a_{obsd}}$ is the pseudo-first-order rate constant for **40** reacting with **44**, $k_{b_{obsd}}$ is that for **39**, K_a is the association constant for **44** complexing **40**, and [ROH] is the concentration of **39**. The virtue of k_a:k_b is that it is dimensionless.

$$k_a/k_b = (k_{a_{obsd}}/k_{b_{obsd}})K_a[\text{ROH}] \qquad (4.1)$$

The acylation of **39** in CDCl$_3$–0.1% (CH$_3$)$_2$NCHO by **44** was too slow to measure, but lower limits of k_{obsd} of 0.005×10^{-3} min^{-1} could have been detected. A conservative estimate of the value for **44** complexing **40** is 10^9 M^{-1}, and is based on the measured values of model compounds. The estimated value for k_a:k_b is 10^{11} for the second-order-rate constant ratio for the acylations of **40** vs. **39** by **44**. Although approximate, there is no doubt that the rate-constant-enhancement factor due to complexation is very large indeed.[18] We visualize the transition state of **44** reacting with **40** as a complex held together and stabilized by N$^+$(H----O=C)$_3$ and (CH$_2$O----C=O)$^-$ as the binding sites, whereas the transition state for the reaction of ester **44** with **39** is held together only by (CH$_2$O----C=O)$^-$ as the binding site. The extra stabilization associated with the extra binding sites in the former transition state lowers its free energy enough to provide the factor of 10^{11} in the rate constant. We believe that K_a of Equation 4.1 provides no measure of the contribution that collection alone makes to the rate constant. Collection, orientation, and stabilization of the rate-limiting transition state are all involved in the 10^{11} rate-constant enhancement.[18]

4.4 Introduction of an Imidazole into a Transacylase Mimic

4.4.1 Synthesis

Host **41** combines a binding site, an imidazole, and a hydroxymethylene group to provide all the elements of a chymotrypsin mimic except the carboxylate. The 30 step synthesis of **41** was accomplished by a route whose key steps are summarized in the retrosynthetic scheme outlined in Chart 4.5. The overall yield of **41** was 0.9%.[20,21]

Preliminary experiments carried out in 20% pyridine-d$_5$–80% CDCl$_3$ (v:v) established that **41** was rapidly acylated by (L)-alanine *p*-nitrophenyl ester perchlorate, **44**, to give **46**, which ~ 10^2 times more slowly gave **47**. Only the N-3 of the imidazole can accept and donate the acyl group without strain in the **41·44** complex, judged by CPK model examination.

---

[19] R.L. Schowen, 'Transition States of Biochemical Processes', Plenum, New York, 1978, Part I, Chapter 2, pp. 77–114.
[20] D.J. Cram, P.Y-S. Lam, and S.P. Ho, *Ann. N. Y. Acad. Sci.*, 1986, **471**, 22.
[21] D.J. Cram and P.Y-S. Lam, *Tetrahedron*, 1986, **42**, 1607.

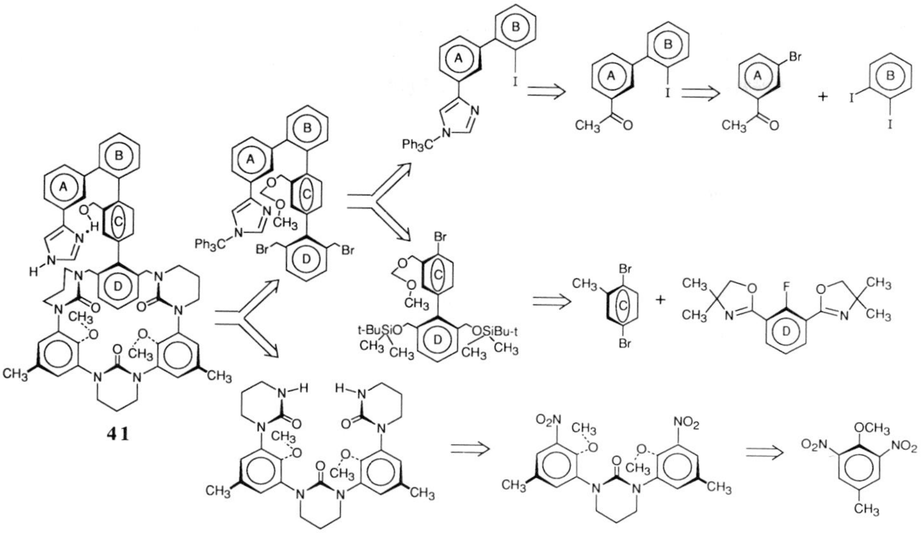

Chart 4.5

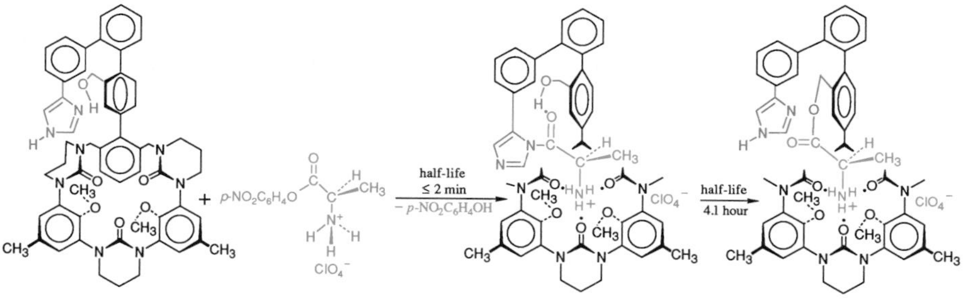

4.4.2 Transacylation Kinetics

The pseudo-first-order rate constants were determined for the reactions of one equivalent of (L)-**44** reacting with fifteen equivalents of various nucleophiles in $CDCl_3$–0.3% $(CD_3)_2NCDO$ (v) at 25 °C. Table 4.2 reports the $k_{a_{obsd}}:k_{b_{obsd}}$ values, where $k_{b_{obsd}}$ is the pseudo-first-order rate constant for standard nucleophile **49** reacting with (L)-**44**, which is set equal to unity, and where $k_{a_{obsd}}$'s are those for nucleophiles, most of which are complexing. Also included in the table are $k_a:k_b$ values defined by Equation 4.2, which are the second-order rate factors by which complexing reactants that combine and orient nucleophiles exceed the rate of non-complexing reactants. In Equation 4.2, K_a is the estimated association constant for the complexing nucleophiles binding to **44**, and [**49**] is constant at 0.0015 M. The use of Equation 4.2 avoids the disadvantages of comparing rate constants for reactions with different molecularities. Thus uncomplexed **44** and the nucleophiles are the *standard starting states, and the* *rate-limiting transition states for transacylation are the standard final states for both the complexing and non-complexing systems in this treatment.* The K_a values for **40**, **41**, and **52** binding **44** are estimated to be, respectively, 10^9, 10^8, and 10^9 M^{-1}, based on measured values for model compounds (see Section 4.2).

48

39

49

50

44

40, R = H
51, R = CH_3

41, R = R' = H
52, R = CH_2OCH_3, R' = H

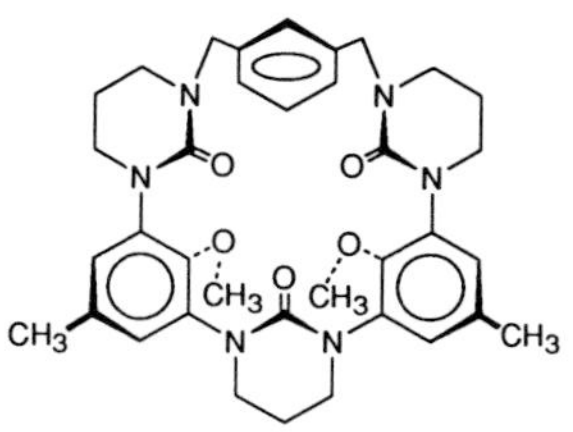

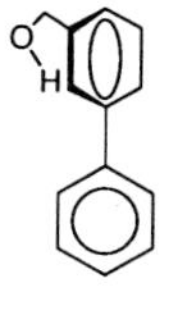

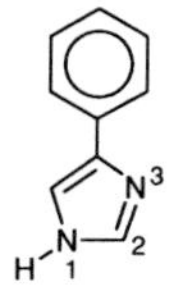

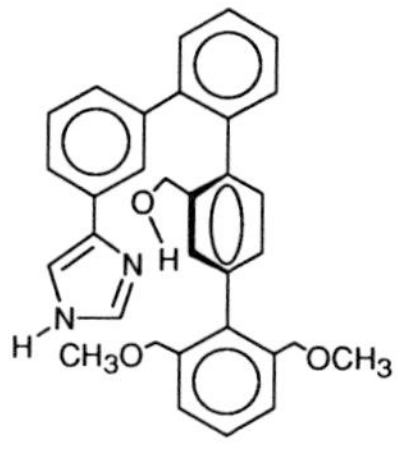

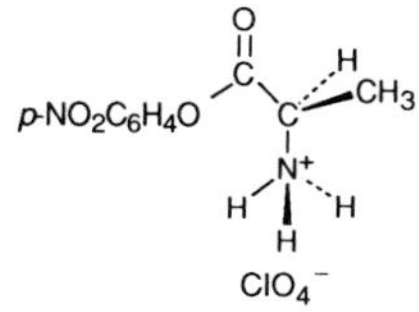

$$k_a/k_b = (k_{a_{obsd}}/k_{b_{obsd}})K_a[\mathbf{49}] \qquad (4.2)$$

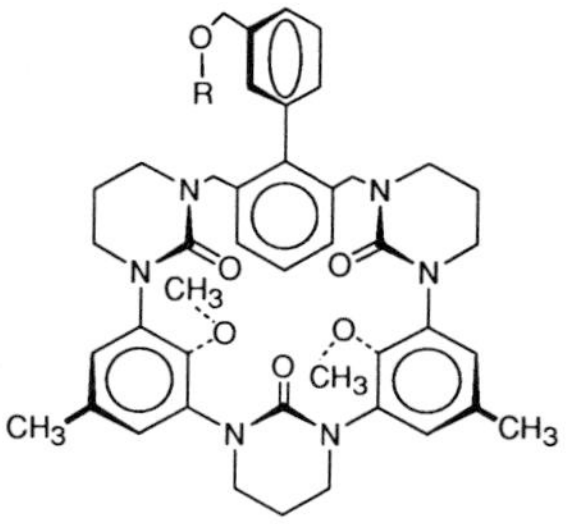

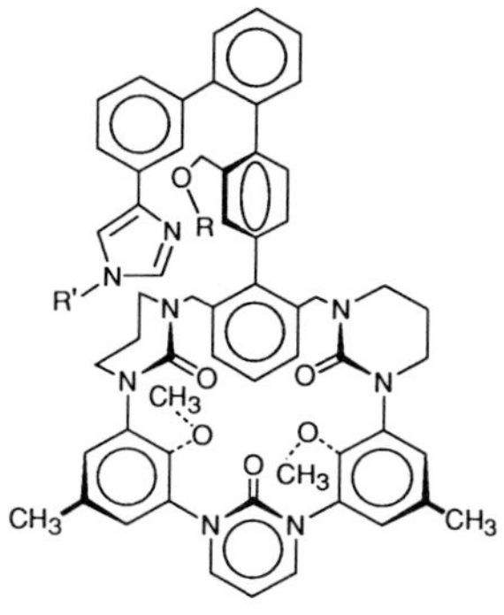

4.4.3 Dependence of Rate Enhancement Factors on Structure

The factors by which the various nucleophiles exceed imidazole **49** in k_{obsd} values ($k_{a_{obsd}}:k_{b_{obsd}}$ of Table 4.2) provide these conclusions. (1) The hydroxyl of **41** participates little in the rate-determining step within the complex **41·44** since both **41·44** and **52·44** (the hydroxyl of **52** is blocked) provide similar factors of $>10^5$ (runs 8 and 10, respectively). Competitive complexation by an added mole of Na^+ per mole of **41** or **52** (runs 9 and 11) depresses these factors by $\geq 10^3$. Actually Na^+ picrate binds **41** and **52** a little better than $CH_3NH_3^+$ picrate does in $CDCl_3$–D_2O at 25 °C.[21] (2) Complexation not only gathers and orients the reactants, but it also activates the acyl donor toward external nucleophiles such as **49**. Thus **51·44** and **40·44** give $k_{a_{obsd}}:k_{b_{obsd}}$ factors of 10^3–10^4 in runs 6 and 7, respectively. (3) Gathering the imidazole and hydroxyl in the same molecule, as in **50**, increases the rate-constant factor by 7 (run 4), not far from the rate-

Partial Enzyme Mimics

Table 4.2 *Rate constant factors in $CDCl_3$–$(CD_3)_2NCHO$ (0.3%, v) for acyl transfer from (L)-alanine p-nitrophenyl ester salt **44** (0.0015M) to nucleophile (0.0001M), additive (0.0015M) at 25 °C*

Run	Nucleophile Kind	Additive	$k_{a_{obsd}}:k_{b_{obsd}}$ (M^{-1})	$k_a:k_b$
1	49	none	1	
2[a]	49	39	1	
3	D_2O	48	5	
4[a]	50	none	7	
5[a]	40	none	390	6×10^8
6	49	51 + 39	1700	
7[a]	49	40	6700	
8	41	none	210 000	3×10^{10}
9	41	$NaClO_4$	170	
10	52	none	290 000	4×10^{11}
11	52	$NaClO_4$	77	

[a] Rate constants were unaffected by the presence of 2,4,6-trimethylpyridine at 0.0003 M.

constant factor of 5 observed when $D_2O \cdot 49$, probably hydrogen bonded, is the nucleophile (run 3).

The greatest increase in rate constant is provided by host–nucleophile **52**, whose $k_a:k_b$ is $\sim 4 \times 10^{11}$. In **52**, the hydroxyl is protected and the imidazole and complexing site co-operate in stabilizing the transition state for transacylation by complexation. This 4×10^{11} rate-constant factor for **44** acylating **52** is comparable to the $\sim 10^{11}$ factor observed for **44** acylating **40** as compared to **39** in the same medium, but with $R_3N:R_3NHClO_4$ buffer present to deprotonate the hydroxyls of **40** and **39**. The R_3N present is $> 10^4$ stronger as a base than the imidazole group of **41** or **52**. Thus, covalently bonding a complexing site to an imidazole, as in **41** or **52**, provides a large kinetic transacylation rate-constant factor without addition of bases stronger than those present in the transacylase enzymes.[22]

Semi-quantitative experiments, carried out at 25 °C in $CDCl_3$ saturated with D_2O, provided catalytic turnover for hydrolysis of **44** catalyzed by **49** or **52**. Without catalyst, the hydrolysis of **8** had a 50-hour half-life. With **44** present, 1.5 equivalents of **44** hydrolyzed in 2 hours. Host **52** provided a catalytic rate initially three times that of **49**, but the alanine produced acted as an inhibitor which slowed the rate until its crystallization maintained a steady state of about one equivalent every 3–4 hours. In the same medium, **41** reacted with **44** initially faster than did **49**, but slower than **52**. Spectral experiments suggested that conformationally isomeric esters of the acylated host **47** were produced at about 5–10 times the rate at which alanine was generated.[22] Others have found that the imidazole group of chymotrypsin was acylated first by esters of non-specific substrates.[23] Acylimidazoles are intermediates in many reactions

[22] D.J. Cram, P.Y-S. Lam, and S.P. Ho, *J. Am. Chem. Soc.*, 1986, **108**, 839.
[23] C.D. Hubbard and J.F. Kirsch, *Biochemistry*, 1972, **11**, 2483.
[24] Y.I. Ihara, M. Nango, Y. Kimurai, and N. Kuraki, *J. Am. Chem. Soc.*, 1983, **105**, 1252.

involving protease model systems.[24] However, to our knowledge, no evidence exists that acylimidazoles intervene on the catalytic pathway when alcohols or amines are leaving groups in reactions catalyzed by the transacylases.

This research has demonstrated that large rate accelerations for acyl-transfer reactions can be realized through complexation using the catalytic functional groups of the serine transacylases supported on a designed and synthesized binding site. Our synthetic systems fall far short of mimicking the full range of transacylase properties, but our high complexing abilities and large rate enhancements offer strong encouragement to investigators interested in this research field.

CHAPTER 5

Cavitands

'Research is an adventure much of whose appeal is uncertainty and risk. Failure and ambiguity are the constantly challenging companions of investigators of the unknown. After the unknown is reduced to the known, the subject loses much of its charm. How dull experiments would be should they always give a fully predictable result. Paradoxically, this kind of reproducibility is one of the objectives of research.'

5.1 Origins of the Cavitand Concept

Cavitands are synthetic organic compounds with enforced cavities large enough to complex complementary organic compounds or ions. In the evolution of the cavitand concept, we assembled with CPK models a sphere from 12 benzenes and 24 methylenes, different views of which are formulated in **1**, **2**, and **3** (see Chart 5.1). Notice that each benzene in **1**–**3** contains only two hydrogens that are located 1,4 to one another. Other spheres are easy to assemble in which simple aryl bridging groups, such as O, S, C=O, NH, *etc.*, are substituted for the CH_2 groups. In order to show bonding sequences, we resorted to Mercator projection formula **1**. Gerhardus Mercator was a Flemish geographer in the sixteenth century who invented the Mercator two-dimensional projection maps of the world for navigation. The spirit of drawing **1** is that it is useful in molecular navigation by showing the connectivity sequence of the system. Drawings **2** and **3** are three-dimensional drawings of **1** that have been deformed to allow the viewer to 'look inside'.[1]

Notice that in **2** and **3**, the surface is composed of two kinds of saucer-shaped modules, one containing three and the other four benzenes arranged in cycles, as in **4** and **5**, respectively. Various subtractions of benzenes from sphere **3** give interesting structures: one benzene from **3** gives the pot-shaped model **6**, two benzenes centered 180° from one another gives the collar-shaped model **8**, while subtraction of one three-benzene module gives bowl-shaped model **7**. Subtraction of one four-benzene module from sphere **2** leaves vase-shaped model **9**. Saucer **4** is rigid, whereas saucer **5** is conformationally mobile. Models of **6**–**8** are rigid, while that of **9** has other conformations that appear much less stable than **9**.[1]

A literature search revealed that derivatives of both **4** and **5** had been known for a long time. The hexamethoxy derivative of **4**, cyclotriveratrylene **10**, was

[1] D.J. Cram, *Science*, 1983, **219**, 1177.

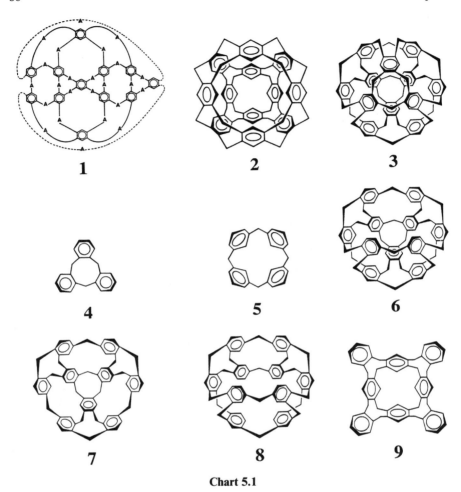

Chart 5.1

identified by Erdtman[2] and Lindsey,[3] and several studies revealed that the ring system of **10** exists only in the conformation drawn.[2] The compound is made by

[2] H. Erdtman, F. Haglid, and R. Ryhage, *Acta Chem. Scand.*, 1964, **18**, 1249.
[3] A.S. Lindsey, *J. Chem. Soc.*, 1965, 1685; A. Lüttringhaus and K.C. Peters, *Angew. Chem., Int. Ed. Engl.*, 1966, **5**, 593.

treating veratrole (3-methoxy-4-hydroxybenzyl alcohol) with formaldehyde and acid. Following the lead of Erdtman,[2] Högberg[4] reported that resorcinol and acetaldehyde in an acidic medium produced diastereomerically pure **11** in 57% yield. These early studies provided a route into the saucer-shaped building blocks **4** and **5**, out of which bowl- and vase-shaped cavitands have been made.

5.2 Systems Based on Cyclotriveratrylene Ring Assemblies

With the help of CPK molecular models, cavitands **13**, **14**, and **98** were designed and prepared[5] using Kellogg's macroring closure, $(CH_3)_2NCHO$, or $(CH_3)_2SO$-Cs_2CO_3,[6] as applied to the reactions of **12** with the respective dichlorides **15**, **16**, and **17**. The yields of **13**, **14·**CH_2Cl_2, and **18·**CH_2Cl_2 were 25, 17, and 42%, respectively. A face stereoview of the crystal structure of **18·**CH_2Cl_2 is shown in **19**, in which one Cl of the CH_2Cl_2 guest is deep in the cavity and the other Cl is high in the cavity. Two of the *m*-xylyl bridges in **19** extend upward in an axial (*a*) conformation, and one extends outward in an equatorial (*e*) conformation to provide an (*aae*)-structure. Molecular models of **13** and **18** show the [1.1.1]*ortho*cyclophane and six attached oxygens to be rigid, but the 1,3-xylyl-type bridges to be conformationally mobile, while the model of **14** appears to be rigid with all three phenanthroline bridges axial to give the molecule with enforced C_{3v} symmetry. The ^{1}H NMR spectrum of **13** is temperature dependent in CD_2Cl_2 or $(CD_3)_2SO$ from 180–400K. Equilibration between conformers is fast on the spectral time scale and provides an averaged C_{3v} spectrum.

Application of molecular mechanical calculations to the enthalpic stability of sample conformations of **18·**CH_2Cl_2 showed **20·**CH_2Cl_2 to be one of the two most stable (*aaa* and *aae*) structures. For comparison, the calculated structure of **18·**CH_2Cl_2 and the observed crystal structure of **18·**CH_2Cl_2, shown in the respective views of the host only, are found in **20** and **21**. The near identity of the crystal and calculated structures provided an example of how the techniques of molecular design, synthesis, crystal structure determination, and molecular mechanics merge to provide coherence to the caviplex **18·**CH_2Cl_2.[5]

5.3 Octols From Resorcinols

In an extension of the work of Högberg,[4] we explored the scope and limitations of the acid-catalyzed reaction of resorcinol and 2-substituted resorcinols with a series of aldehydes to give the *cis*, *cis*, *cis* isomer of the cyclic tetrameric condensation product of general structure **22**. The octols are key starting materials for the synthesis of many hosts described in the next sections and chapters. The condensation itself is interesting because the *cis*, *cis*, *cis* diastereomer is only one of four that are possible, and because the C_{4v} conformation

---

[4] A.G.S. Högberg, *J. Org. Chem.*, 1980, **45**, 4498.
[5] D.J. Cram, J. Weiss, R.C. Helgeson, C.B. Knobler, A.E. Dorigo, and K.N. Houk, *J. Chem. Soc., Chem. Commun.*, 1988, 407.
[6] D. Piepers and R.M. Kellogg, *J. Chem. Soc., Perkin Trans.*, 1978, 383.

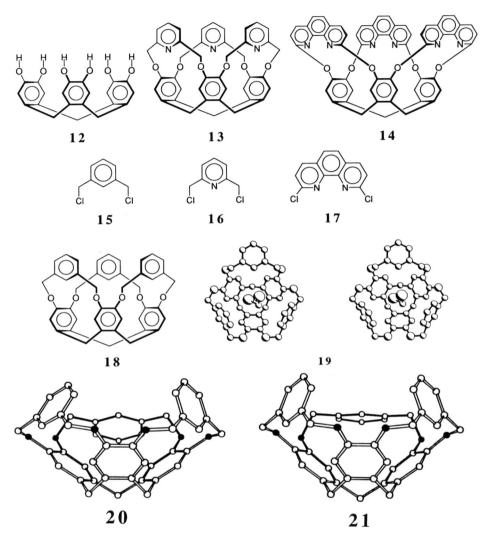

drawn for **22** appears to predict the reaction products of **22**, although several others are possible. For example, with CPK models one can show an easy interconversion of **22** through C_{2v} conformers to **23**, which possesses the same C_{4v} symmetry and concatonated hydrogen bonding pattern as **22**. However, in **23** the R groups are all equatorial rather than axial as in **22**. Of the many crystal structures reported in the following pages, the other configurations of **22** have never been encountered. Another important feature of **22** is the solubility properties that the nature of the R group confers on the octols and their derivatives.[7]

---

[7] L.M. Tunstad, J.A. Tucker, E. Dalcanale, J. Weiser, J.A. Bryant, J.C. Sherman, R.C. Helgeson, C.B. Knobler, and D.J. Cram, *J. Org.Chem.*, 1989, **54**, 1305.

Cavitands

Table 5.1 Syntheses of octols of general structure 1

Compound number	R of 1	A of 1	% yield
11	CH_3	H	73
24	CH_3CH_2	H	88
25	$CH_3(CH_2)_2$	H	92
26	$CH_3(CH_2)_3$	H	89
27	$CH_3(CH_2)_4$	H	77
28	$CH_3(CH_2)_{10}$	H	68
29	$(CH_3)_2CHCH_2$	H	95
30	$HO(CH_2)_4$	H	80
31	$Cl(CH_2)_5$	H	67
32	$C_6H_5CH_2$	H	70
33	$C_6H_5CH_2CH_2$	H	69
34	$4\text{-}O_2NC_6H_4CH_2CH_2$	H	64
35	$4\text{-}BrC_6H_4CH_2CH_2$	H	56
36	CH_3	CH_3	79
37	$CH_3(CH_2)_4$	CH_3	80
38[a]	$CH_3(CH_2)_4$	CH_3CH_2	87
39[b]	$CH_3(CH_2)_4$	C_6H_5	65
40	CH_3	CO_2H	0
41	CH_3	Br	0
42	CH_3	NO_2	0
43	C_6H_5	H	80
44	$4\text{-}CH_3C_6H_4$	H	93
45	$4\text{-}CH_3CH_2C_6H_4$	H	70
46	$4\text{-}BrCH_2C_6H_4$	H	40

[a] Reference 8. [b] Reference 9.

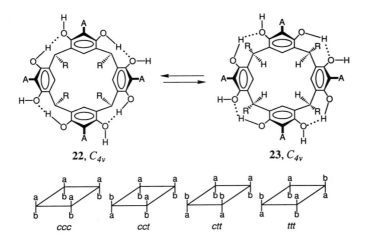

Table 5.1 lists compounds **11** and **24–46** synthesized by the condensation of various aldehydes with resorcinol, and the yields of products of general structure **22** produced. Usually the product crystallized from the 2:2:1 respective

volumes of EtOH, H$_2$O, and concentrated H$_3$OCl used as a reaction mixture, and was recrystallized and characterized. Other diastereomers and intermediates were in the filtrates. With electron-withdrawing A groups CO$_2$H, Br, and NO$_2$ (compound numbers **40**, **41**, and **42**), products were not isolated, but with electron-releasing A groups CH$_3$, CH$_3$CH$_2$, and C$_6$H$_5$ (**36–39**) good yields were observed.[7-9]

Högberg[4] has shown that in the reactions of resorcinol with benzaldehyde and 4-bromobenzaldehyde that the C_{2v} diastereoisomer formulated generally in **48** is formed faster than the C_{4v} diastereomer **47**, but that longer times favored formation of the C_{4v} isomer. The crystal structure of **47** derivatives demonstrated it to have the C_{4v} configuration.[10]

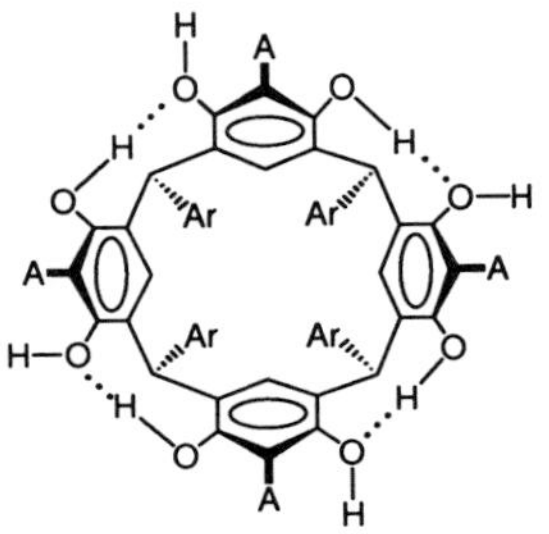

47, C_{4v} 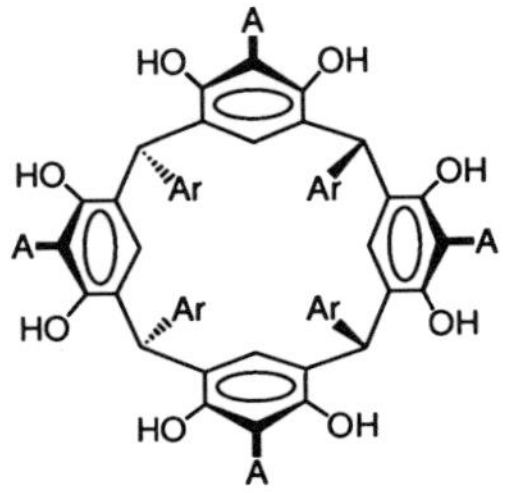   **48**, C_{2v}

Chart 5.2 portrays both side and face stereoviews of the crystal structures of **11**·2.5CH$_3$CN·3H$_2$O and **26**·EtOH·(CH$_3$)$_2$CO as **49** and **50**, and **51** and **52**, respectively. As expected, they are highly hydrogen bonded, both intra- and inter-molecularly. Correlations of patterns of ^{1}H NMR chemical shifts of **11**, **26**, and **46** with those for the other octols provided the basis for the initial configurational assignments for the other octols. Crystal structures of numerous compounds made from **11**, **27**, **29**, **31**, **33**, and **36–38** found in future sections of this book confirmed the assignments. A crystal structure of **30** established its configuration as well.[11]

We attribute the stereoselectivity shown in the syntheses of these octols to be due to several factors; (1) The *cis, cis, cis* isomers appear to be the most insoluble products of a series of reversible reactions, and in most cases the product crystallizes from the medium, driving the reactions toward that product; (2) The condensation medium is hydrophilic. In runs involving long aliphatic R groups (see **28–30** and **34**), the *cis, cis, cis* configuration with the four R groups in polar conformations are the only ones that allow these lipophilic moieties to contact one another, driven together by a hydrophobic effect.

[8] J.A. Bryant, C.B. Knobler, and D.J. Cram, *J. Am. Chem. Soc.*, 1990, **112**, 1254.
[9] R.C. Helgeson and D.J. Cram, unpublished results.
[10] B. Nilsson, *Acta Chem. Scand.*, 1968, **22**, 732.
[11] R.C. Helgeson, C.B. Knobler, and D.J. Cram, unpublished results.

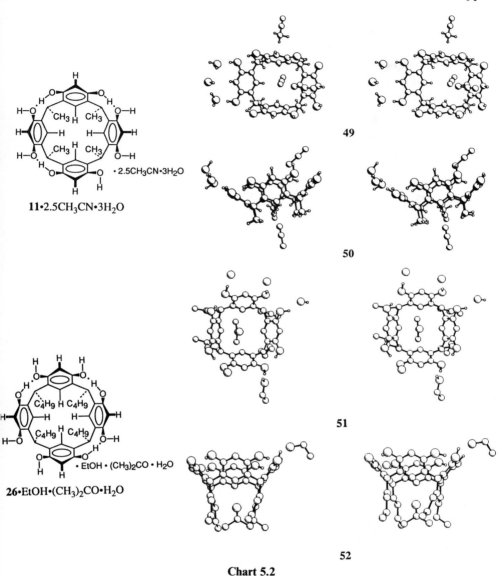

Chart 5.2

5.4 Bowls from Octols

5.4.1 Syntheses

The octols of the last section are conformationally mobile, although their hydrogen-bonding network predisposes them in nonhydrogen-bonding solvents to occupy bowl-like structures. In this section are described cavitands prepared from octols in which four sets of OH hydrogens are replaced by CH_2,

CH_2CH_2, or $CH_2CH_2CH_2$ groups as in general formulas **53** and **54**. The first of these provides a side and the second a face view of the structure. The R group in these formulae varies from H to CH_3 to Br to I, the last three groups increasing the depth of the bowl. In the initially prepared compounds, octol **11** served as the starting material. The four CH_3CH groups in **53** provide four methyl groups which together constitute a pedestal for CPK models of **53**. We refer to these groups as feet, *e.g.* **53** has methyl feet.[12]

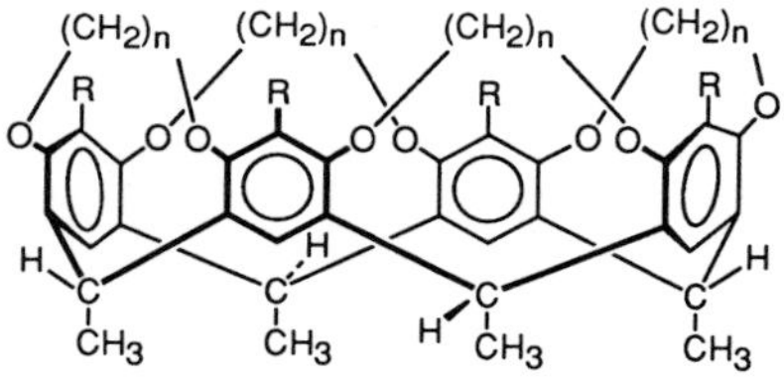

53, R is H, CH_3, Br, or I; n is 1, 2, or 3

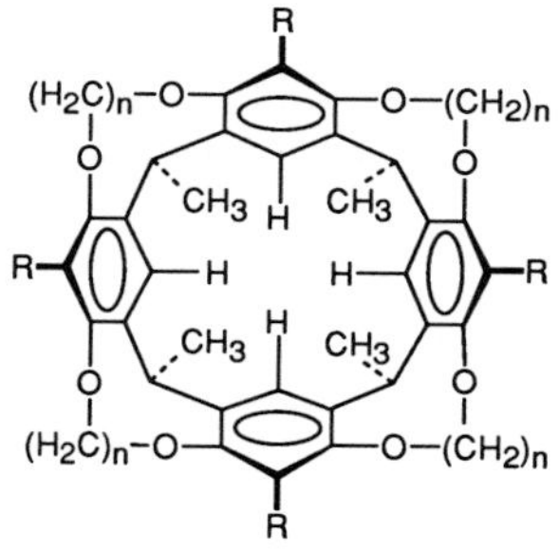

54

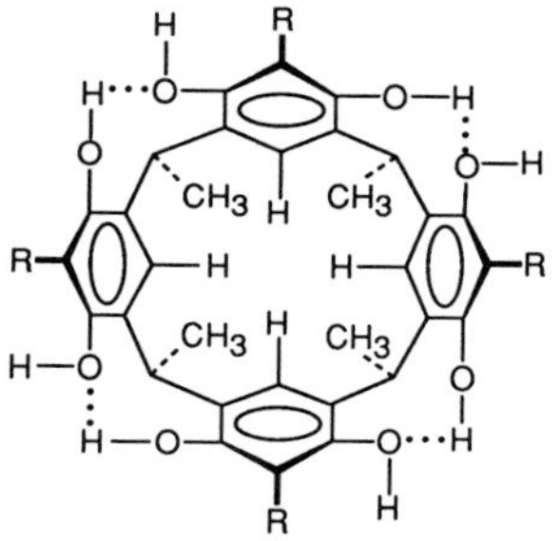

11, R = H; **36**, R = CH_3; **55**, R = Br

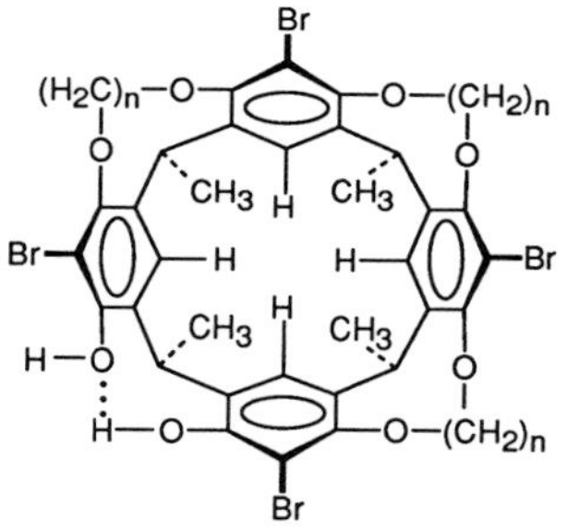

56, n = 1; **57**, n = 2; **58**, n = 3

The three conformationally mobile octols, **11**, **36**, and **55**, were converted to bowl-shaped cavitands **54** by four-fold ring closures to build methylene or polymethylene bridges anchored by the four sets of proximate oxygens. The reagents involved were $BrCH_2Cl$, $TsOCH_2CH_2OTs$, or $TsO(CH_2)_3OTs$ with Cs_2CO_3–$(CH_3)_2SO$, and the yields ranged from 23 to 63%. Higher yields were always observed when R = CH_3 or Br than R = H, probably because with the two substituents present steric inhibition of intermolecular reactions leading to non-cyclic oligomers was greater than that of their intramolecular reactions leading to ring closure. Octol **55**, in which the R of **53** is Br, underwent the first three ring closures faster than the fourth, which allowed good yields of tris-bridged **56–58** to be obtained. Practical amounts of mono- and bis-bridged compounds were also synthesized by appropriate manipulation of the reaction conditions.[12]

---

[12] D.J. Cram, S. Karbach, H-E. Kim, C.B. Knobler, E.F. Maverick, J.L. Ericson, and R.C. Helgeson, *J. Am. Chem. Soc.*, 1988, **110**, 2229.

5.4.2 Crystal Structures

Nine crystal structures of caviplexes derived from these cavitands are provided in Chart 5.3 in both face and side stereoviews. Their compositions are **59**·CH$_2$Cl$_2$; **60**·CH$_3$CN; **60**·(CH$_2$)$_6$·C$_6$H$_6$; **61**·CHCl$_3$; **62**·CH$_3$C$_6$H$_5$; **63**·(CH$_2$)$_6$; **63**·C$_6$H$_6$·C$_6$H$_6$; **64**·CH$_2$Cl$_2$; and **65**·CH$_2$Cl$_2$. These cavitands are generally conical, with the cone supported on a square framework of four methyl feet, as suggested in diagram **66**. The nearly closed bottoms of their cavities are defined by a 16-membered [1.1.1.1]metacyclophane macroring. The open tops are defined by [*n.n.n.n*]metacyclophane macrorings containing eight oxygens and from 24 to 32 ring members (that of diol **65** is incomplete).[12]

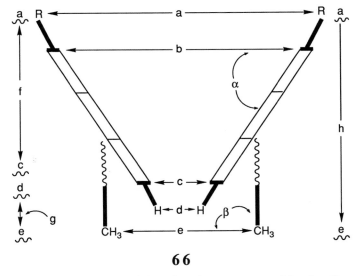

66

The depths of the cavities of these bowls are measured by the distance **f** from the plane **a**, defined by the centers of the R groups, to plane **c**, defined by the lowest carbons of the four aryls (see **66**). The **f**$_{av}$ values vary with the values of *n* and the nature of the R groups of general formula **54**. Cavitands **59–62**, with *n* = 1, provide **f**$_{av}$ values as follows: R = H, 3.34 Å; CH$_3$, 3.86 Å; Br, 4.15 Å; I, 4.21 Å. Cavitands **63** and **64**, with *n* = 2, give **f**$_{av}$ values of 3.14 Å for R = CH$_3$ and 3.33 Å for R = Br. Host **65**, with *n* = 3 and R = Br, has an **f**$_{av}$ value of 3.15 Å. Thus the cavity depths increase with increasing Ar—R bond lengths, but decrease with increasing values of *n*, as expected.

Although the bases of the cones are essentially square, the tops vary from being almost square in **59–62** (*n* = 1) to more rectangular in **63–65** (*n* = 2 or 3). A measure of the extent to which the cavitand tops depart from a square geometry is indicated by how much the two diagonal distances for **a** (in diagram **66**) differ from one another. The two **a** values for **59**·CH$_3$C$_6$H$_5$ are 9.11 and 8.96 Å, which differ by 1.7% of the average of the two **a** values. The percentage differences for the other caviplexes are as follows: **60**·CH$_3$CN, 0%;

94   Chapter 5

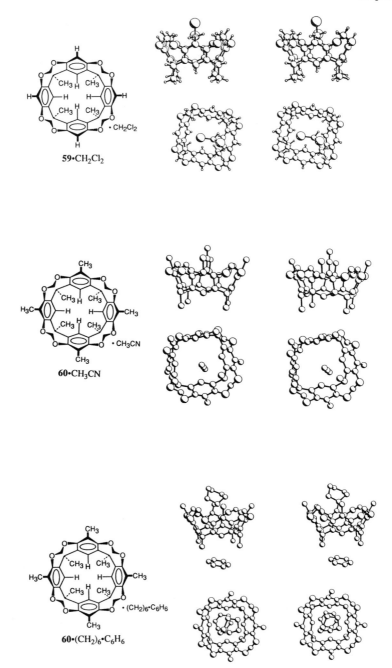

59·CH₂Cl₂

60·CH₃CN

60·(CH₂)₆·C₆H₆

Chart 5.3

Cavitands

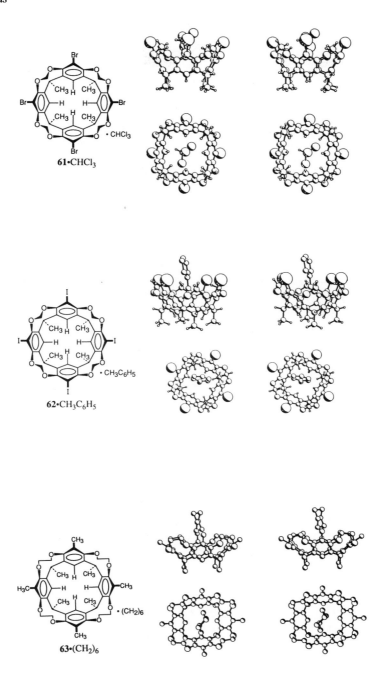

Chart 5.3 (*continued*)

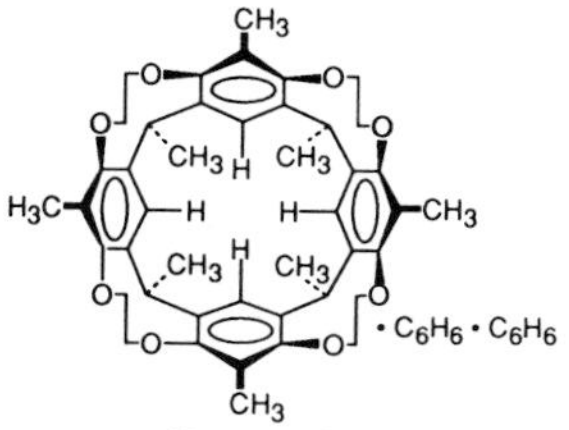

63•C$_6$H$_6$•C$_6$H$_6$

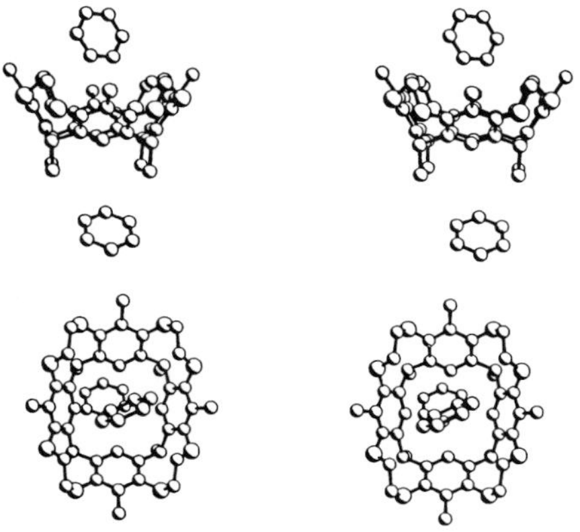

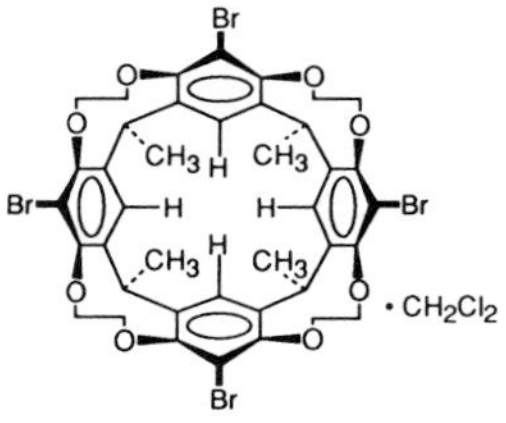

64•CH$_2$Cl$_2$

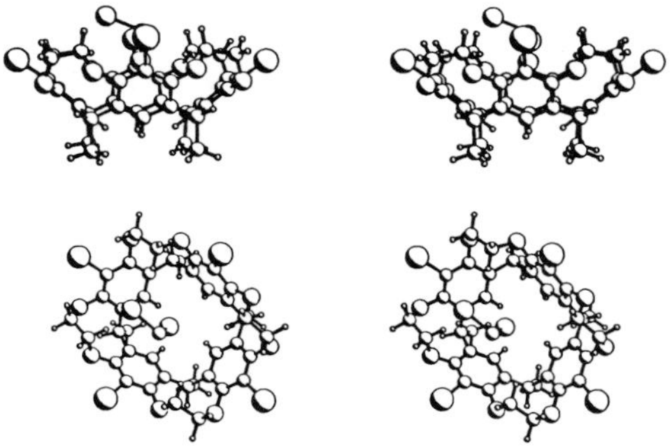

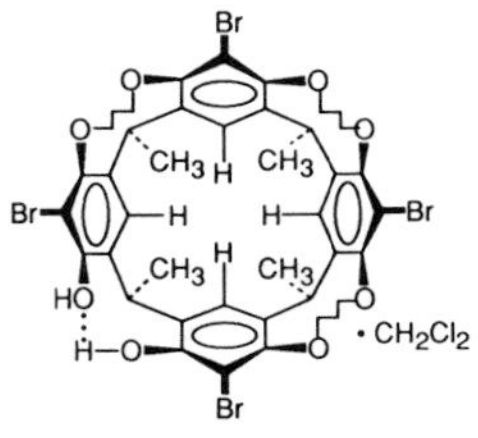

65•CH$_2$Cl$_2$

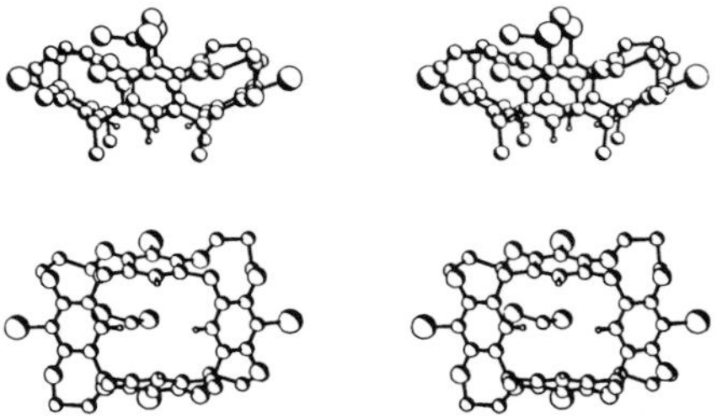

Chart 5.3 (*continued*)

60·$(CH_2)_6$·C_6H_6, 1.4%; **61**·$CHCl_3$, 4.9%; **62**·$CH_3C_6H_5$, 1.6%; **63**·$(CH_2)_6$, 21.4%; **63**·C_6H_6·C_6H_6, 22.6%; **64**·CH_2Cl_2, 24.7%; and **65**·CH_2Cl_2, 49.8%. Thus the cavitands containing the longer O-to-O bridges use the increased conformational mobility to lengthen the cavity size in one dimension at the expense of shortening it in another dimension.

A measure of the relative rigidity of the support structure for the bowls is found in the averaged low value of 0.03 ± 0.05 Å by which the methyl carbons of the feet deviate from coplanarity, the largest deviation being for diol **65**·CH_2Cl_2

Cavitands

(± 0.08). The values of $(\alpha)_{av}$ (see **66**) provide a measure of how much the cones tilt. These values are 60.9°, 62.8°, 60.3°, 61.2°, and 61.2° for the cavitands with OCH_2O bridges; 47.9°, 49.0°, and 49.3° for those with OCH_2CH_2O bridges; and 49.0 for the one with $OCH_2CH_2CH_2O$ bridges. The parameter $(\beta)_{av}$ provides a measure of angles of the $Ar-CH_3$ bond to the best plane of the methyl feet (see **66**). This angle varies from 86 to 90° for the bowls containing OCH_2O bridges, from 86.4 to 88.0° for those with OCH_2CH_2O bridges, and is 83.8° for the one with $OCH_2CH_2CH_2O$ bridges. The (H-to-H)$_{av}$ distances at the bottom of the cone (see **66**) ranges from 3.64 to 4.19 Å, which after subtracting the diameter of a hydrogen leaves a hole too small for molecules other than O_2 or possibly H_2O to pass through.

The crystals examined were the result of many experiments with each host and a variety of solvent guests. Generally the hosts with the larger cavities complexed the larger solvent molecules, or the larger parts of the guests. For example, in **59**·CH_2Cl_2, the CH_2 inserts into the cavity, whereas in **61**·$CHCl_3$, one Cl and part of a second occupy the cavity. In the two caviplexes with methyl-containing guests (**60**·CH_3CN and **62**·$CH_3C_6H_5$), the methyl groups are inserted into the bottom of the cavity. In **60**·$(CH_2)_6$ the two hydrogens and carbon (CCH_2C) of a *boat* cyclohexane penetrate the top of the cavity. Boat cyclohexanes are rarely encountered without additional bridges or substituents. In **63**·$(CH_2)_6$ the *chair* cyclohexane lies across the face of the mouth in the smaller dimension of the flattened cone so that only two equatorial hydrogens penetrate the cavity.

No complexing studies in solution were carried out with these hosts because of their low solubilities. Their importance lies in their use as intermediates in the syntheses of hosts with multiple cavities or much larger cavities discussed in later sections and chapters of this book.[12]

5.5 Cavitands with Cylindrical Cavities

Ordinarily, when host and guest complex in solution, each must desolvate. Most cavitands contain holes large enough to admit solvent. New cavitands **67–69** contain cylindrical wells of varying depths whose limited diameters deny occupancy to all but slim linear guests. These hosts were prepared by treating the conformationally mobile octol **11** with the appropriate dialkyldichlorosilanes in $(CH_2)_4O-Et_3N$ to close four 8-membered rings to give **67** (37%), **68** (9%), and **69** (7%). Side and face views of the crystal structure of **67**·S=C=S are drawn in **70** and **71**. The S=C=S guest fully occupies a cavity whose sides are lined with four inward-turned methyl groups attached to the four $(CH_3)_2Si$ bridges between the aryl oxygens and by the bases of the four aryl groups.[13]

These three hosts, **67–69**, were soluble enough in $CDCl_3$ to allow their association constants, K_a, to be determined by ^1H NMR titrations. Upon complexation of CS_2, the 'a' protons moved as much as 0.18 p.p.m. upfield, the 'b' protons lining the upper sides of the cavity moved downfield maximally by 0.44 p.p.m., whereas the 'c' protons remained little changed. At 250 K, K_a

[13] D.J. Cram, K.D. Stewart, I. Goldberg, and K.N. Trueblood, *J. Am. Chem. Soc.*, 1985, **107**, 2574.

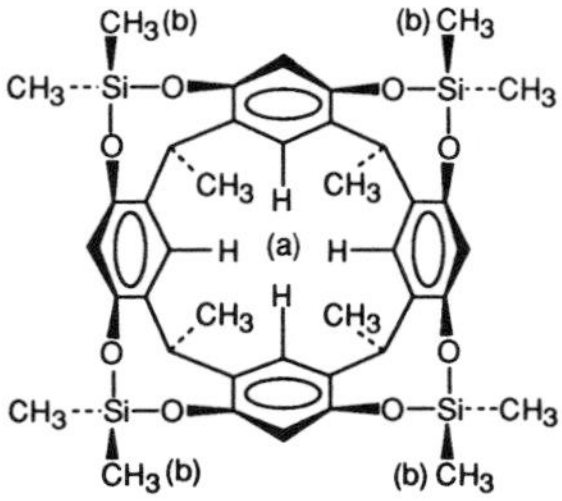

values, M^{-1}, for **67**, **68**, and **69** binding CS_2 were 0.82, 8.1, and 13.2, respectively. From the changes in K_a values at 212, 250, 275, and 300 K for **67** binding CS_2, the following thermodynamic parameter values were estimated (212 K): ΔG^0, -0.4; ΔH, -3.5; $T\Delta S$, -3.1 kcal mol^{-1}. Thus enthalpy favors and entropy disfavors complexation. Similar experiments with **67** and $CH_3C{\equiv}CH$ in $CDCl_3$ produced similar chemical shifts, which indicates that the $C{\equiv}CH$ end was fully inserted into the well of the host. Similar experiments with **67** and C_6H_6, I_2, CH_3I, $K_3Fe(CN)_6$, t-$BuNH_3CSN$, H_2O, CO_2, or CH_2Cl_2 produce little change in the chemical shifts of protons 'a' and 'b'. However, these signals *only* were significantly broadened when **67** in $CDCl_3$ was saturated with O_2 at 250 K. Displacement of O_2 with N_2 reproduced sharp signals. Thus **67–69** show high structural recognition for linear molecules of appropriate dimensions.[13]

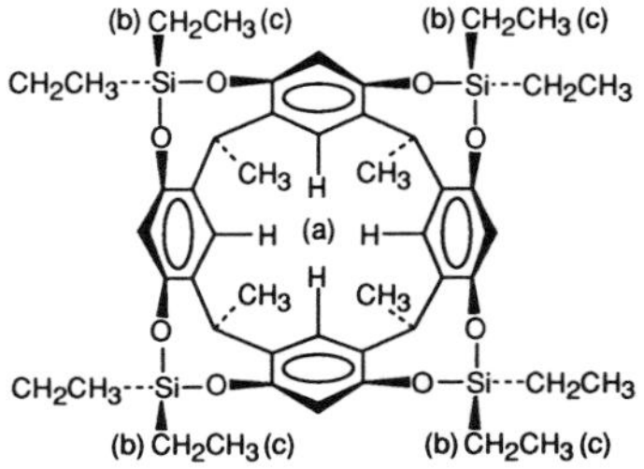

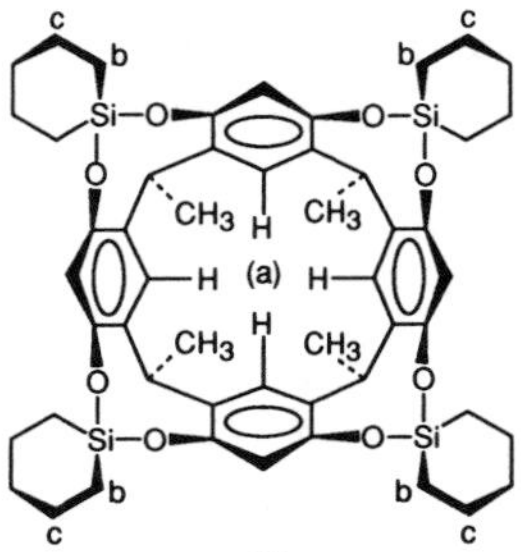

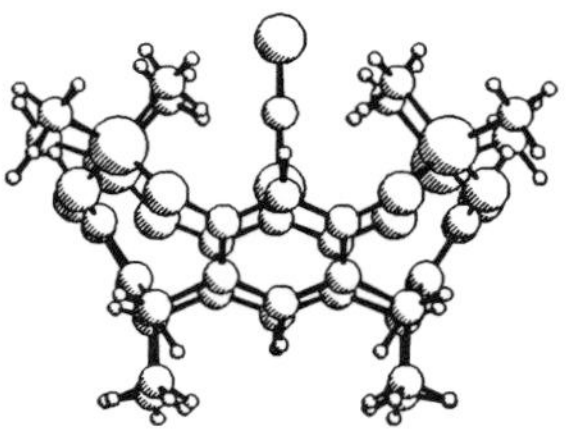

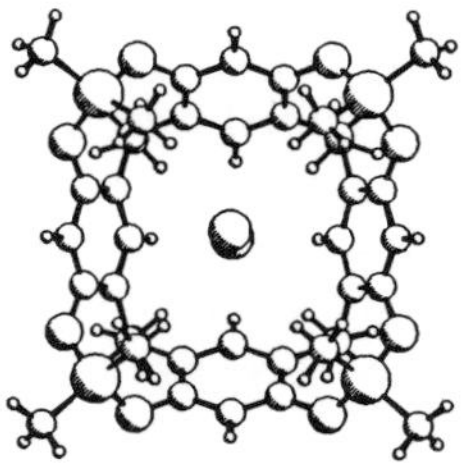

In the cavitation model for dissolution, the free energy for cavity formation in the solvent, ΔG_c, is of opposite sign and similar magnitude to the free energy for solvent–solute interactions, ΔG_i.[14] Hosts **67–69** are the equivalent of benzene–alkane solvent molecules organized to form a cavity whose ΔG_c is supplied during synthesis, leaving ΔG_i unopposed. Solvolytic driving forces for expelling CS_2 from $CDCl_3$ are likely negligible. Possibly ΔG_i is mainly composed of overall attractive dipole–dipole interactions. The fact that K_a for the three hosts binding CS_2 follows the order **69 > 68 > 67** correlates, as expected, with the increase in shared surfaces between host and guest. Thus the principles of preorganization and complementarity are applicable to the design of complexes between lipophilic entities.[15]

---

[14] O. Sinanoglu, 'Molecular Associations in Biology', ed. E. Pullman, Academic Press, New York, 1968, pp, 427–445.
[15] D.J. Cram, *Science*, 1988, **240**, 760.

5.6 Cavitands Containing Two Binding Cavities

5.6.1 Syntheses

Cavitands of general structures **70** (side view) or **71** (face view) are interesting particularly because they contain two fused cavities, one shaped like a box and the other like a bowl. Their structures lend themselves to design for desired properties in the following ways: (1) the rims of the bowls can be varied by different choices of substituents A and bridging groups B for shaping the bowl cavity, for manipulating the solubilities of the cavities, or introducing potentially co-operating functional groups to act as catalysts; (2) the rims of the box cavity can be independently varied by changes in the R groups for similar purposes as applied to the box cavity. By bromination of octols **43–46** (Section 5.1, general structure **72**), octols **73–76** were produced, respectively. These eight octols, **43–46** and **73–76**, served as starting materials for the sequential syntheses of cavitands **77–87** by standard reactions.[16]

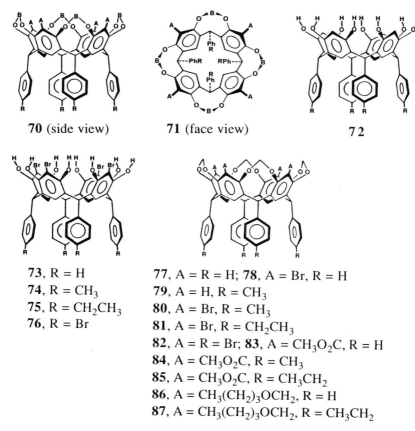

70 (side view)　　**71** (face view)　　**72**

73, R = H
74, R = CH$_3$
75, R = CH$_2$CH$_3$
76, R = Br

77, A = R = H; **78**, A = Br, R = H
79, A = H, R = CH$_3$
80, A = Br, R = CH$_3$
81, A = Br, R = CH$_2$CH$_3$
82, A = R = Br; **83**, A = CH$_3$O$_2$C, R = H
84, A = CH$_3$O$_2$C, R = CH$_3$
85, A = CH$_3$O$_2$C, R = CH$_3$CH$_2$
86, A = CH$_3$(CH$_2$)$_3$OCH$_2$, R = H
87, A = CH$_3$(CH$_2$)$_3$OCH$_2$, R = CH$_3$CH$_2$

---

[16] J.A. Tucker, C.B. Knobler, K.N. Trueblood, and D.J. Cram, *J. Am. Chem. Soc.*, 1989, **111**, 3688.

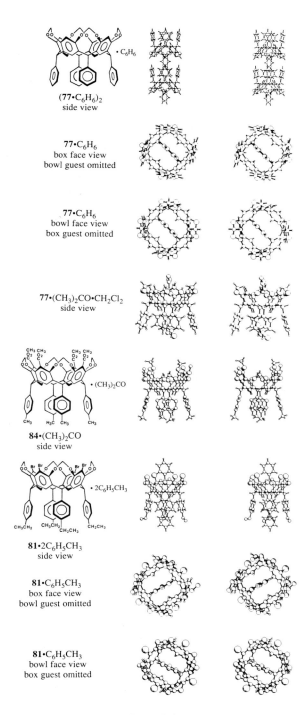

Chart 5.4

5.6.2 Crystal Structures

Chart 5.4 contains stereoviews of $77 \cdot (C_6H_6)_2$, $77 \cdot (CH_3)_2CO \cdot CH_2Cl_2$, and $84 \cdot (CH_3)_2CO$. Host **77** crystallizes as polymeric stacks of alternating cavitands and C_6H_6 guest molecules, half of the latter occupying the bowl of one molecule and the other half the box of a second. The benzene is completely encapsulated within the large void formed by the rim-to-rim placement of the two kinds of cavities. The benzene molecule is arranged in a manner that maximizes the effective width of each of the two constituent cavities. There is positional disorder of this guest with respect to rotation 90° about an axis perpendicular to the plane of the page in the bowl and box face views in $77 \cdot C_6H_6$. Crystals of $77 \cdot (CH_3)_2CO \cdot CH_2Cl_2$ were grown from a mixture of CH_2Cl_2 and $CH_3CO \cdot CH_3$ as solvent. Only CH_3COCH_3 deposited in the bowl, but either CH_2Cl_2 or CH_3COCH_3 occupied the box cavity, only the former being drawn. One methyl of the CH_3COCH_3 guest points inward to contact 25 carbons of the host at distances less than 4.20 Å leaving the other CH_3 and CO groups packed against the inward-oriented H's of the OCH_2O groups. The geometry of host **77** is very similar in $77 \cdot C_6H_6$ and $77 \cdot (CH_3)_2CO \cdot CH_2Cl_2$ complexes. The CH_2Cl_2 guest of the latter is almost completely enclosed in the box cavity. The inward pointing Cl is only 2.98 Å from one of the aryl hydrogens of the host.

In $84 \cdot (CH_3)_2CO$, the bowl cavity contains either disordered $CHCl_3$ or CH_3COCH_3, which is not shown. The CH_3COCH_3 of the box cavity lies in a plane perpendicular to the best plane of the methine carbons of the host. The four phenyl rings defining the box cavity are related pairwise by a C_2 axis which is colinear with the C=O bond of the guest. The average value of the angles defined by each phenyl and the plane of the methine carbons is 100.4°.

Chart 5.4 displays side and two face stereoviews of $81 \cdot 2C_6H_5CH_3$. A methyl of each of the two $CH_3C_6H_5$ guests points inward into each of the two cavities of the host in such a way as to maximize host–guest contacts. The plane of the aryl carbons of the toluene in the box cavity is oriented 90.5° to the best plane of the methine carbons of the host, and this guest lies along a diagonal of the box. The four ethyl groups of the host wrap around the guest to maximize area of host–guest contacts. The angles formed between the phenyl rings of the box and the best plane of the methine carbons is 98.8°. When **81** was recrystallized from C_6H_5Br (instead of $C_6H_5CH_3$), $81 \cdot 2C_6H_5Br$ was formed whose crystal structure was isomorphous to $81 \cdot 2C_6H_5CH_3$. Each of the two Br atoms faced inward in this structure, which is not shown.[16]

5.6.3 Complexations

The K_a values for both the bowl and the box of polyether host **87** (the most soluble host) binding CD_3CN in CCl_4 at 328, 300, and 273 K were determined by ^1H NMR titrations. The calculations involved the assumption that each of the two kinds of cavities behaved independently of one another with respect to its binding. At 273 K the K_a for the bowl was 313 M^{-1}, and the K_a for the box was 159 M^{-1}. The following thermodynamic parameters were deduced from the temperature dependences of the K_a's: for the bowl, $\Delta H = -6.2$ kcal mol^{-1} and $\Delta S = -11.4$ cal mol^{-1}T^{-1}; for the box, $\Delta H = -6.4$ kcal mol^{-1} and $\Delta S = -13.6$ cal mol^{-1}T^{-1}. Thus the enthalpy of binding is opposed by the entropy of binding to give, at 273K, ΔG^0 values of -3.1 kcal mol^{-1} for the bowl and -2.7 kcal mol^{-1} for the box cavity of **87** complexing CD$_3$CN. A little more than half of the enthalpy of binding goes into the cost in entropy of collecting and orienting the guest molecules in the complexes. Each cavity binds CD$_3$CN strongly enough to result in double occupation for over half the host molecules at 0.0022 M host and 0.052 M guest concentrations.[16]

This study provides an entrée into the design and synthesis of potential organic catalysts in which either bowls or boxes are used as binding and orienting sites, and catalytic functional groups occupy the rims of the cavities.

5.7 Hosts Based on Fused Dibenzofuran Units

5.7.1 Cavitands Containing Two Clefts

Highly preorganized cavitands containing cleft- and collar-shaped cavities were designed and synthesized by fusing dibenzofuran units at the positions α to their oxygens to form macrocyclic ring hosts **88** and **89**. The two ethyl groups substituted in each dibenzofuran unit were required to impart solubility to the systems. The critical ring closures involved dilithiation of **90** in their α-positions followed by Fe(acac)$_3$ oxidative coupling of the diorganometallic compound to give **88** in 11% and **89** in 1.6% yields.[17]

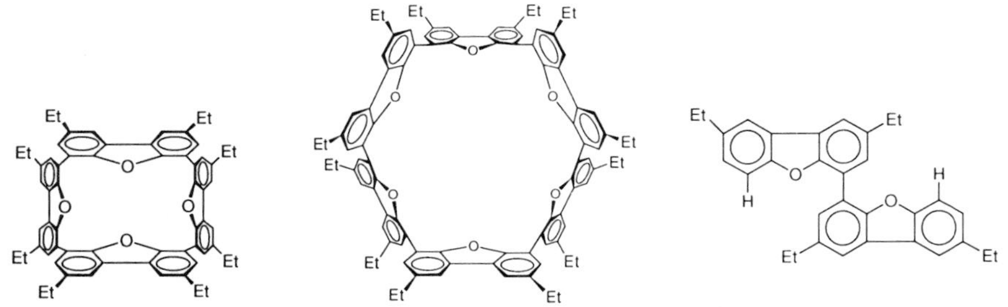

88   **89**   **90**

In CPK molecular models, **88** is saddle-shaped, contains two cleft-shaped cavities approximately 12 Å long, 3.4 Å deep, and 4.3 Å wide, and possesses D_{2d} symmetry. The four oxygens are nearly coplanar and line a hole just large enough to embrace a model of CH$_4$, NH$_4^+$, or Cs$^+$. In models, **89** possesses D_{3d} symmetry, is collar-shaped, and forms a large cavity about 11 × 7 × 7 Å. Cross sections of the model that are perpendicular to its C_3 axis are hexagonal. The six oxygens lie in one plane neatly embracing a model of a benzene ring in the same plane which divides the cavity into halves. Each half can accommodate three additional benzene rings stacked in planes perpendicular to the central benzene.

[17] R.C. Helgeson, M. Lauer, and D.J. Cram, *J. Chem. Soc, Chem. Commun.*,1983, 101.

Cavitands

The twelve ethyl groups nearly close the gaps on the surfaces of the collar-shaped cavity. Without the ethyl groups, the macroring hosts are too insoluble for isolation and purification.[17]

In another study, **91** was prepared, in 7% yield, by a method similar to that for **88**. A stereoview of the complex of **91** with two molecules of benzene, one in each cavity, is drawn in **92**. The benzene guests are not centered in the cleft in two respects: the benzene is 0.2 Å closer to one end of the long axis of the cleft than the other; the benzene is closer to one of the dibenzofuran units than the other. Accordingly the guests are disordered. The plane of the benzene guest is roughly parallel to the plane of the dibenzofuran unit to which it is closest, the distances between the planes ranging from 3.43–3.82 Å.[18]

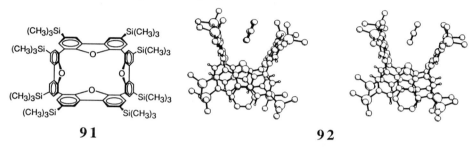

91   **92**

The association constants of **91** with π-acids 1,3-$(NC)_2C_6H_4$ and 1,3-$(O_2N)_2C_6H_4$ in $CDCl_3$ at 25 °C were determined by ^1H NMR titrations. The dicyanobenzene guest provided K_a = 2.0 M^{-1}, which gives ΔG^0 = -0.40 kcal mol^{-1} binding free energy. The dinitrobenzene guest gave K_a = 3.7 M^{-1} and ΔG^0 = -0.80 kcal mol^{-1} of complexation. These low values correlate with the low surface areas common to host and guest in the complexes involved.[18]

5.7.2 Cavitands Containing a Single Cleft

Molecular models of macrocycles **93–99** possess single clefts, the floors of which contain potential complexing functional groups. In principle, X and Y might consist of catalytic functional groups, and Z remote groups that might impart desired solubility properties to the compounds. The syntheses of **93–97** were undertaken to establish feasibility for obtaining these types of hosts. Dichlorides **98** and **99** were key compounds in these syntheses. Chart 5.5 summarizes the synthesis of **99**, and illustrates the ease of manipulation of the dibenzofuran nucleus, which is subject to electrophilic substitution in its 2- and 8-positions and metalation in its 4- and 6-positions.

Dibenzofuran was iodinated to give **100**, which was metalated, and the product trimethylsilylated to give **101**. Dibenzofuran was also dimetalated with s-BuLi-tetramethylethylenediamine (TMEDA), and the bis-organolithium compound produced was brominated to give dibromide **102**. This material was monolithiated with phenyllithium, the product carbonated, and the carboxyl produced was reduced with $BH_3 \cdot O(CH_2)_4$ to provide bromoalcohol **103**. Di-

---

[18] E.B. Schwartz, C.B. Knobler, and D.J. Cram, *J. Am. Chem. Soc.* 1992, **114**, 10775.

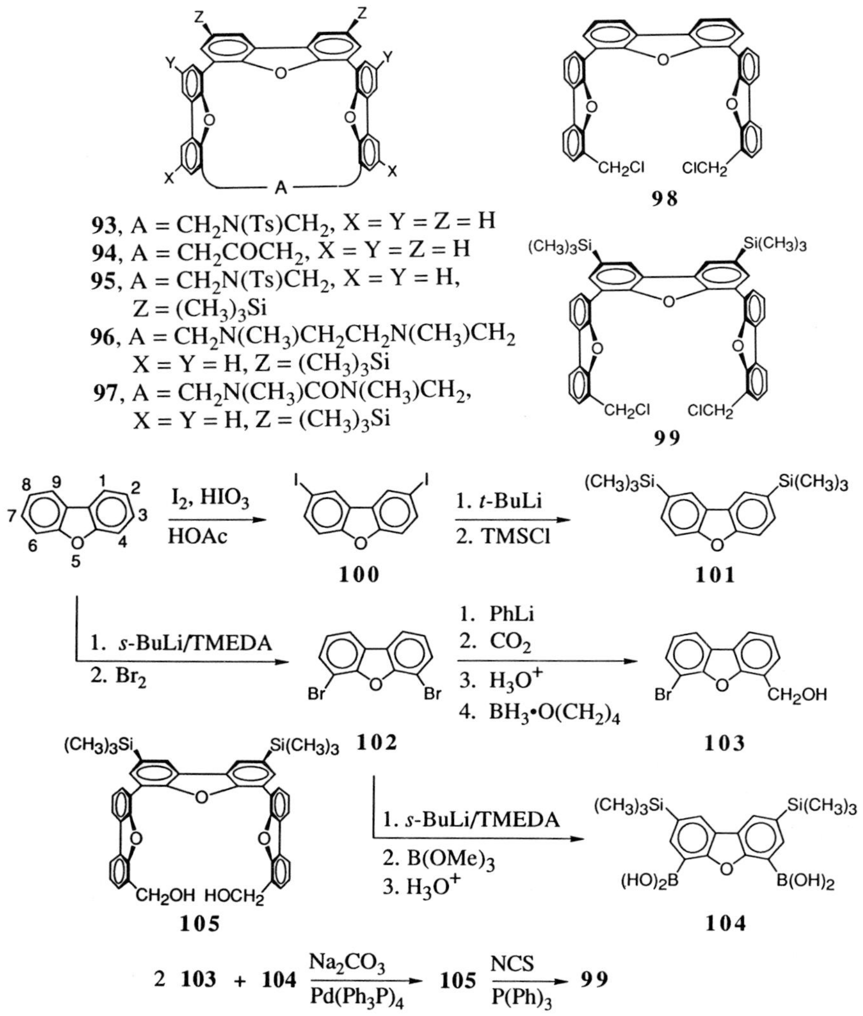
Z
Z
Y
Y
O
O
O
X
X
A
93, A = CH2N(Ts)CH2, X = Y = Z = H
94, A = CH2COCH2, X = Y = Z = H
95, A = CH2N(Ts)CH2, X = Y = H, Z = (CH3)3Si
96, A = CH2N(CH3)CH2CH2N(CH3)CH2 X = Y = H, Z = (CH3)3Si
97, A = CH2N(CH3)CON(CH3)CH2, X = Y = H, Z = (CH3)3Si
CH2Cl
ClCH2
98
(CH3)3Si
Si(CH3)3
CH2Cl
ClCH2
99
I2, HIO3
HOAc
100
1. t-BuLi
2. TMSCl
101
1. s-BuLi/TMEDA
2. Br2
Br
Br
102
1. PhLi
2. CO2
3. H3O+
4. BH3•O(CH2)4
CH2OH
103
CH2OH HOCH2
105
1. s-BuLi/TMEDA
2. B(OMe)3
3. H3O+
(HO)2B
B(OH)2
104
2 103 + 104
Na2CO3
Pd(Ph3P)4
105
NCS
P(Ph)3
99

Chart 5.5

bromide **102** was bis-lithiated with *s*-BuLi-TMEDA; the dilithio product was boronated to give boronic ester which was hydrolyzed to **104**. Two moles of bromoalcohol **103** and one of the diboronic acid **104** were submitted to the Suzuki coupling reaction in EtOH–C_6H_6–H_2O–Na_2CO_3–$Pd(Ph_3P)_4$ to give diol **105** (60% yield), which when treated with *N*-chlorosuccinimide (NCS) in $(CH_2)_4O$-$P(Ph)_3$ produced dichloride **99**. A similar synthesis provided dichloride **98**.[18]

The critical macroring closures proceeded in yields (not maximized) that ranged from very good to disappointing. Dichlorides **98** and **99** reacted with *p*-$CH_3C_6H_4SO_2NH_2$ in $(CH_3)_2NAc$–Cs_2CO_3 to give sulfonamides **93** and **95** in 76 and 74% yields, respectively. Macrocyclic ketone **94** was prepared by

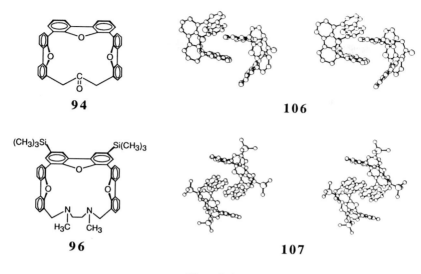

Chart 5.6

treating dichloride **98** with isocyanide p-CH$_3$C$_6$H$_4$SO$_2$CH$_2$NC in (CH$_2$)$_4$O, and the initially formed cyclic isocyanide intermediate was hydrolyzed with C$_6$H$_6$–H$_3$OCl to **94** (7%). Recrystallization of **94** from CH$_2$Cl$_2$ gave crystals suitable for crystal structure determination. The reaction between dichloride **99** and CH$_3$NHCH$_2$CH$_2$NHCH$_3$ in (CH$_3$)$_2$NAc–Cs$_2$CO$_3$ gave macrocyclic diamine **96** (12%). Crystals of X-ray quality were grown from CH$_2$Cl$_2$–CH$_3$OH. Treatment of dichloride **99** with CH$_3$NHCONHCH$_3$–NaH–(CH$_2$)$_4$O gave macrocyclic urea compound **97**. As expected, the three hosts containing the two trimethylsilyl units were much more soluble in organic solvents than were **93** and **94** without them.

Chart 5.6 provides line structures of cyclic ketone **94** and diamine **96**, and stereoviews of their respective structures, **106** and **107**. In each crystal structure, two molecules dimerize, one dibenzofuran of each molecule reciprocally occupying a cleft of the other molecule. In **106** the inserted dibenzofuran moiety resides closer to one of the two facing dibenzofurans defining the cleft, the distance between the planes of the aryl units varying from 3.34–3.75 Å. The inserted dibenzofuran unit does not fill the cleft entirely, but extends as far as the short aliphatic carbonyl bridge at the base of the cleft will allow. The keto oxygens of the inserted dibenzofuran units are oriented *anti* to the ether oxygens of the two clefts, but generally *syn* to one another.

In **107**, the crystal structure of cyclic diamine **96**, each molecule behaves both as a host and a guest by inserting and accepting a dibenzofuran unit into clefts. Each inserted unit resides closer to one facing dibenzofuran unit of the cleft than to the other, in a nearly parallel fashion, with the two diamine bridges remote from one another. The distances between the near planes of the two dibenzofuran units vary from 3.40–3.80 Å. The distance between the deepest atom of the inserted dibenzofuran unit and the plane defined by the three oxygens at the floor of the cleft is 2.29 Å. The methyl groups of the diamine bridge point away from the cleft and each other, providing room for insertion. Two interstitial water molecules reside between dimers in **107** (not shown). The dimensions of the clefts of **107** are: length, 5.10–7.15 Å, width, 5.76–8.57 Å, and depth, 2.88–4.41 Å.

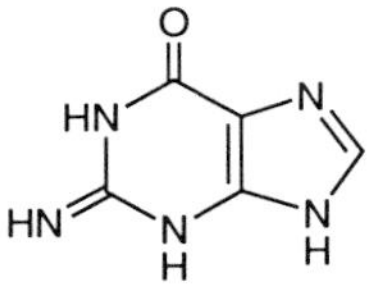

108 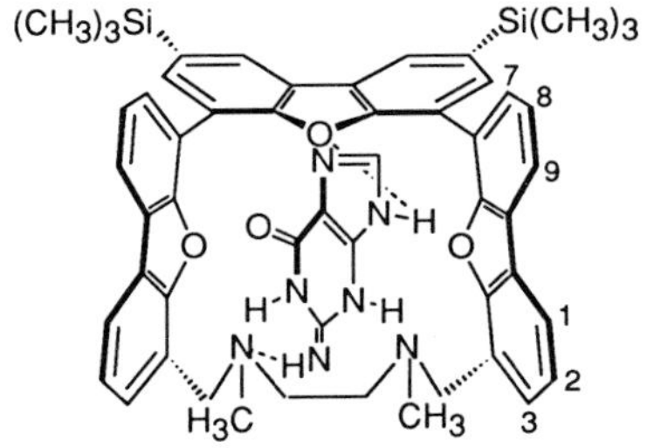   **109**

In CPK models, guanine, **108**, is complementary to host **96**, there being one N-H----O and two N----H-N hydrogen bonds between host and guest as visualized in **109**. Experimentally, a 0.003 M solution of **96** in CD_3OD was prepared and sonicated with solid guanine to give a solution of complex whose ^1H NMR signals were much sharper than those of the host in the absence of guest. In a control experiment, guanine did not detectably, by ^1H NMR, dissolve in CD_3OD in the absence of host upon heating or sonication. In the ^1H NMR spectrum of the complex **109**, the broad benzyl doublet of doublets of **96** taken alone was resolved into a sharp doublet of doublets, and one of the doublets was shifted upfield from δ 3.85 to δ 3.65. Upon complexation, the NCH_3 proton signal of the host also moved upfield from δ 2.38 to δ 2.23 and sharpened. Most of the aromatic proton signals of the host moved upfield slightly upon complexation, but the protons *ortho* to the benzyl groups shifted from δ 7.33 to δ 7.18. The remote trimethylsilyl protons did not shift upon complexation of the host.

This study demonstrates that hosts with rigidly preorganized clefts with binding groups located at the bottom of the clefts can be prepared, which bind in a predictable manner. An attractive feature of compounds such as **96** is that positions 1–3 and 7–9 of each flanking dibenzofuran unit is potentially available to carry functional groups that catalyze the reactions of potential guests.

CHAPTER 6

Vases, Kites, Velcrands, and Velcraplexes

'The virtue of the principle of minimum explanation for research results is manifold: the alternative of maximum explanation leads to chaos; when explanations no longer fit the data, the least change is the least troublesome. The problem with the principle of minimum explanation is that nature is unaware of the principle. As experimental results unfold, new layers of complexity appear, and more elaborate explanations usually are demanded.'

6.1 Vases, Kites, and Temperature

6.1.1 Synthesis

As part of the search for easily-made cavitands possessing large internal volumes, the conformationally mobile octol **1**, described in Chapter 5, was treated with aryl compounds containing two *ortho* leaving groups subject to nucleophilic aromatic substitution, such as the dichloroquinoxaline **2**. Examination of CPK molecular models indicated that **3** should be the expected product, and that its *vase* conformation should be enforced. No other conformation of **3** could be easily assembled. The reaction went in KOH–(CH₃)₂NCHO to give **3**·(CH₃)₂NCHO (34%) which when crystallized from CHCl₃ gave **3**·1.3 CHCl₃. The use of Cs₂CO₃–(CH₃)₂SO as the reaction medium raised the yield to 83%. When three equivalents of quinoxaline **2** and mild conditions were used, an easily-separated mixture of 30% **3** and 40% **4**, with only three aryl

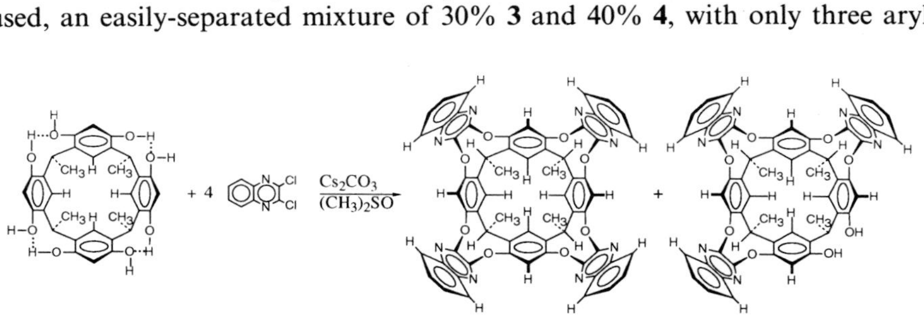

[1] J.R. Moran, J.L. Ericson, E. Dalcanale, J.A. Bryant, C.B. Knobler, and D.J. Cram, *J. Am. Chem. Soc.*, 1991, **113**, 5707.

bridges, was obtained. Therefore the fourth aryl bridge appears to form more slowly than the first three, and **4** is available for introducing a second kind of bridge into the same molecule.[1]

Similarly, **1** was bridged with tetrachloropyrazine **5**, whose first two *ortho* chlorines are more easily substituted than the third and fourth. The product **6** (75%) again possessed the expected vase structure. Molecular models of a product **8** in which four methyl groups are substituted for the four hydrogens between the four sets of oxygens of **6** cannot be assembled in the vase (*aaaa*) conformation, because of steric repulsions between the four methyls and the eight unshared electron pairs of quinoxaline or pyrazine groups. Experimentally when tetramethyl octol **7** was treated with pyrazine **5**, a rather strained kite (*eeee*) conformation of the product **8** was formed (83%) in which the cavity collapsed by elongating in one dimension and narrowing in the other. Molecular models of kite-shaped **8** can be assembled only with extreme difficulty because of repulsions between the four methine hydrogens and the four pyrazines.

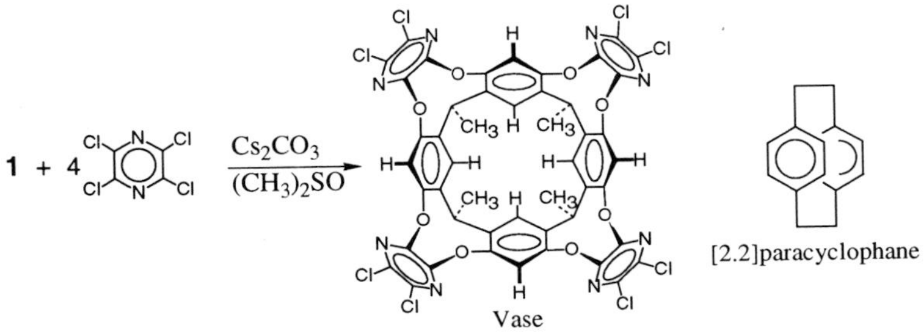

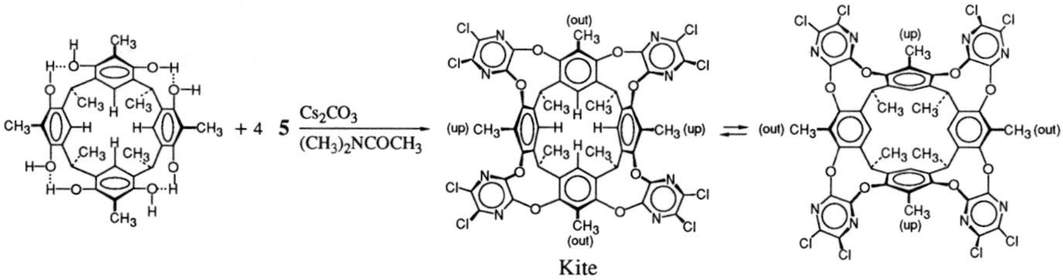

6.1.2 Conformational Analysis

Pyrazines or quinoxalines in **3**, **6**, or **8** act as flaps which when 'up' in the fully axial (*aaaa*) conformation resemble a vase with a large interior cavity, that in the case of **3** is complementary to a molecular model of [2.2]paracyclophane **9** or ten molecules of water. When the four flaps are down as in the fully equatorial (*eeee*) conformation of **8**, the model resembles a Boeing 747 with its flaps down coming in for a landing, which in turn looks like a kite. In models of possible kite forms of **3** or **6** and the enforced kite form of **8**, there are two small cavities, complementary to methyl groups, each of which is lined with two oxygens and the sloping face of a benzene. In **8a** the two methyl groups located at 12 and 6 o'clock are oriented outward, and those at 3 and 9 o'clock are oriented upward.

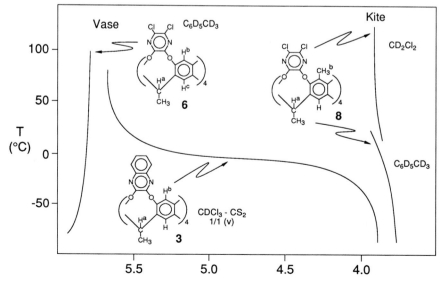

3-vase
aaaa conformation

3-kite
eeee conformation

The isolation of tetramethyl product **8**, whose four ArCH$_3$ groups imposed a kite conformation on the molecule, suggested that under some conditions compounds such as **3** or **6**, without the aryl methyls, might be composed of equilibrating vase and kite forms, as drawn for 3-vase and 3-kite. Accordingly, the 500 MHz ^{1}H NMR spectra of **3**, **6**, and **8** were examined as a function of temperature in solvents dictated somewhat by the solubility of the compounds. Figure 6.1 shows the effect of temperature changes on the chemical shift of H^a, the four methine protons common to all three compounds. The H^a signal of cavitand **6** occurred at δ 5.8–5.9, and changed little from 90 to −65 °C. Its whole spectrum was consistent with a molecule of C_{4v} symmetry.

Figure 6.1 *Effect of temperature changes in the chemical shifts of H^a protons (500 MHz ^{1}H NMR spectra) of **3**, **6**, and **8***

The spectrum of **3** proved to be more complicated and interesting. The H^a signal was a sharp quartet at δ 5.7 at 45 °C and above, but broadened and moved upfield continuously as the temperature was lowered reaching δ 3.92 at −62 °C, but changed little from −62 to −72 °C. A coalescence occurred at about −5 °C, with a $\Delta G\ddagger$ of about 11.6 kcal mol^{-1}. The two benzene proton singlets for H^b at δ 8.04 and for H^c at δ 7.27 at 45 °C moved *much less* when the temperature was lowered to −62 °C, and each divided into two singlets found at δ 7.38 and 7.28, and δ 7.41 and 7.21, respectively. The entire spectrum of **3** conformed to C_{4v} symmetry above 45 °C and to C_{2v} symmetry below −62 °C. The H^a proton signal of **8**, as shown in Figure 6.1, was remarkably independent of solvent and the temperature from 115 to −90 °C. It remained between δ 3.7 and 3.9 over the entire solvent–temperature range, with no splitting of the signal. However in $C_6D_5CD_3$, the Ar—CH_3^b signal splits into two signals at about 70 °C. The relatively invariant chemical shift of H^a and the splitting of H^b indicate that **8** exists in the C_{2v} kite structure, whose two identical forms, **8a** and **8b**, are pseudorotating rapidly on the ^1H NMR time scale above 70 °C. The insensitivity of H^a to this pseudorotation is compatible with the fact that a CPK model of the C_{2v} conformer indicates a similar proximal environment for all four of these methine protons. Thus the effect of the Ar—CH_3 groups in **8** is to force the molecule into the degenerate C_{2v} kite form and to avoid the vase form altogether.

6.1.3 Rationalization of Temperature Effects on Conformation

These temperature effects on conformation are explained as follows. There is no doubt that the vase forms of **3** and **6** are less strained than their respective kite conformations. In CPK models of the kite conformers, the four H_a protons are forced so strongly into the faces of the four aryl flaps that the models are difficult to assemble and preserve, whereas models of the vase form appear strain free. The ^1H NMR spectrum of **6** shows it to be only in the vase form at all temperatures, but the spectrum of **3** indicates that only the vase form of **3** exists above 45 °C, and only the kite form below −62 °C. Although the vase and kite forms of **3** possess equal molecular surfaces, the more *extended surfaces* of the kite form must contact and orient more solvent molecules than the more *confined surfaces* of the vase conformer. These contacts are expected to be enthalpy-stabilizing but entropy-destabilizing for both systems. In the conversion of vase to kite, more solvent molecules are collected and oriented. At low enough temperatures, the kite form is the more stable conformer in solution because its more favorable enthalpy of solvation overrides the sum of the unfavorable $T\Delta S$ of solvation and the greater strain energy of this conformer. As the temperature rises, the unfavorable $T\Delta S$ cancels more of the favorable enthalpy until the free energy of solvation no longer overrides the greater internal strain of the kite form. Above 45 °C, the vase form dominates, since its stability in solution does not depend as much as that of the kite form on its free energy of solvation. In contrast to **3**, **6** remains in the vase form over the whole temperature range of 125 °C because it lacks the four outer benzo groups of **3**.

The surface difference between **3** and **6** is enough to lower the importance of the solvation contribution to the $T\Delta S$ part of the free energy of **6**, so that the intrinsically greater enthalpy of stabilization of the vase compared to the kite form determines the geometry of **6** over the whole temperature range.

6.1.4 Crystal Structure

The crystal structure of **10·2CH$_2$Cl$_2$** was determined, and **11** is a stereoview of the caviplex. Host **10** is identical to **3** except that four Cl(CH$_2$)$_5$ feet in **10** replace the four CH$_3$ feet of **3**. In crystal structure **11**, the host is in the vase form (side view), one CH$_2$Cl$_2$ guest molecule lies in the lower part of the vase cavity, and the second CH$_2$Cl$_2$ lies in a cavity formed by the four Cl(CH$_2$)$_5$ feet. The host possesses approximate C_{4v} symmetry, and the two guest CH$_2$ groups lie close to the C_4 axis of the host. The upper part of the host cavity is lined with the four quinoxaline units arranged as if their faces were the sides of a box. The planes of the quinoxalines are tilted an average of 6° inward as their 6- and 9-hydrogen atoms essentially touch one another. The eight ether oxygens are arranged with their unshared electron pairs facing outward. The colored drawing of **10** shown in Figure 6.2 provides an impression of the rigid box-shaped upper cavity and the conformational flexibility of the shape of the lower cavity formed by the feet.

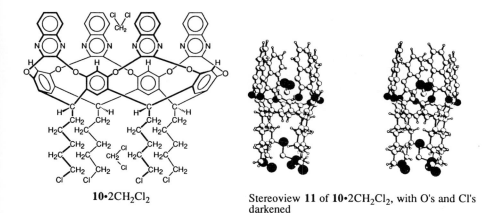

10·2CH$_2$Cl$_2$ Stereoview **11** of **10·2CH$_2$Cl$_2$**, with O's and Cl's darkened

The two cavities of crystal structure **11** closely resemble those of an analogue by Dalcanale[2] in which four CH$_3$(CH$_2$)$_5$ feet replace the four Cl(CH$_2$)$_5$ feet of **11**, and where three CH$_3$COCH$_3$ molecules replace the two CH$_2$Cl$_2$ guest molecules of **11**. In the crystal structure of the caviplex, the vase cavity contains two acetone molecules, and the cavity formed by the feet contains the third acetone guest.[2] These crystal structures closely conform to those predicted based on CPK molecular model examinations.

---

[2] E. Dalcanale, P. Soncini, G. Bacchilega, F. Ugozzoli, *J. Chem. Soc., Chem. Commun.*, 1989, 500.

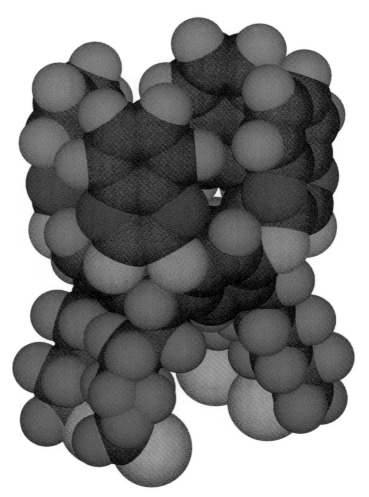

Figure 6.2

6.2 Design and Synthesis of Kite-shaped Systems

6.2.1 Structural Basis for Dimer Formation

In the ^{1}H NMR spectrum of kite-shaped pyrazine compound **8** of the previous section, there were hints that monomer molecules were in equilibrium with their dimers. An examination of CPK models of **8** revealed that it possesses a roughly planar rectangular face, 15×20 Å, containing two methyl groups protruding upward and oriented essentially perpendicular to this face. The face also contains two methyl-sized indentations lined by a sloping aryl face, a methyl pointing out horizontally, and the unshared electron pairs of two oxygens. These protrusions and indentations are arranged as in A of Chart 6.1.

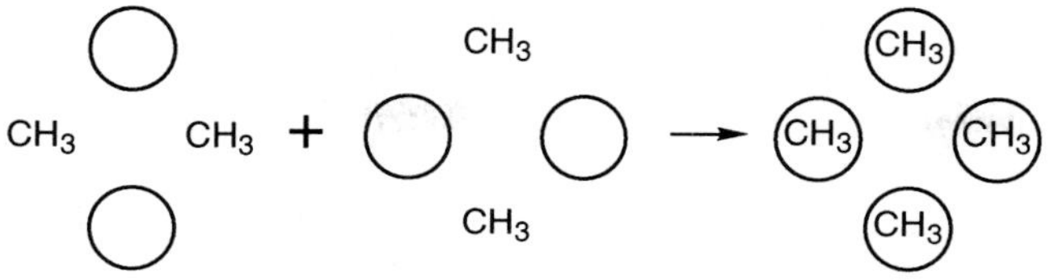

Vases, Kites, Velcrands, and Velcraplexes 113

Monomer A + Monomer B → Dimer C

Chart 6.1

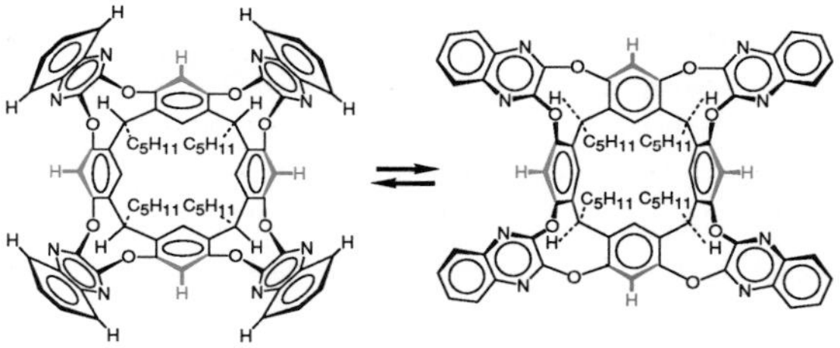

12a, vase (C_{4v})   12b, kite (C_{2v})

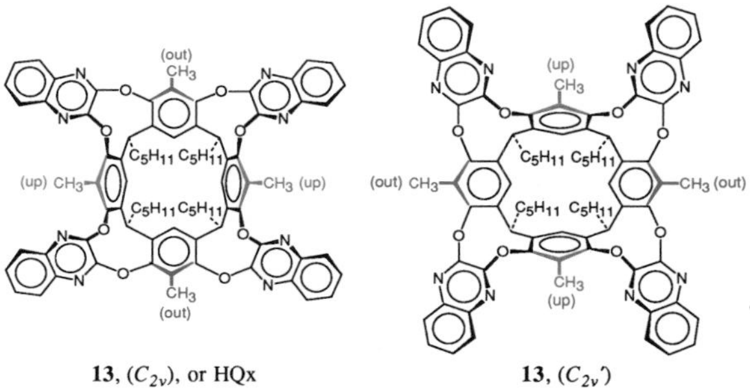

13, (C_{2v}), or HQx   13, (C_{2v}')

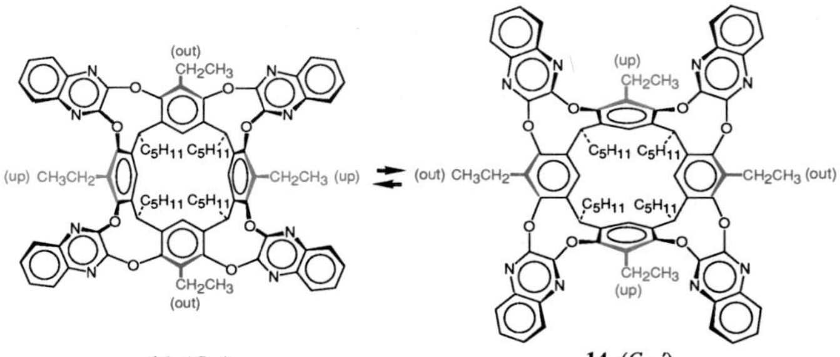

14, (C_{2v})   14, (C_{2v}')

Chart 6.2

Rotation of A upside down and 90° in the plane of the page produces B, which then fits on top of A to give C, in which four methyl groups as guests occupy four host cavities. When two molecular models of **8** are fitted to one another face-to-face as in C, they share a large common surface composed of the tetramethyl lock and their large rectangular areas which occupy roughly parallel planes. Because **8** and its analogues were too insoluble for study of the dimerization phenomenon, analogues **12–14**, of Chart 6.2, and **27–42** were synthesized with four pentyl feet substituted for the four methyl feet of **8**, which increased their solubilities.[3]

6.2.2 Synthesis

In most of these syntheses, octol **15** was condensed with pyrazines and quinoxalines **17–26** or **2**, shown in Chart 6.3, to produce four new nine-membered rings, each involving the making of eight bonds between oxygen and carbon. Remarkably high yields were obtained which usually ranged between 50 and 86% for these reactions. Compound **13** was synthesized by the condensation of octol **15** with quinoxaline **2**; compound **16** with only three quinoxaline bridges was isolated as a by-product of the reaction. The syntheses of analogues **12** and **14** of Chart 6.2 parallel the synthesis of **13** with the appropriate octols substituted by octol **15**. The two *ortho* aryl-halogen bonds of pyrazines and quinoxalines **17–26** of Chart 6.3 are all activated by electron-withdrawing groups in these nucleophilic aromatic substitution reactions.[3]

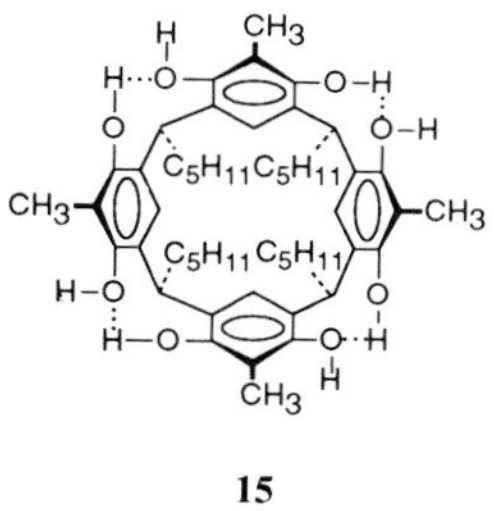

15

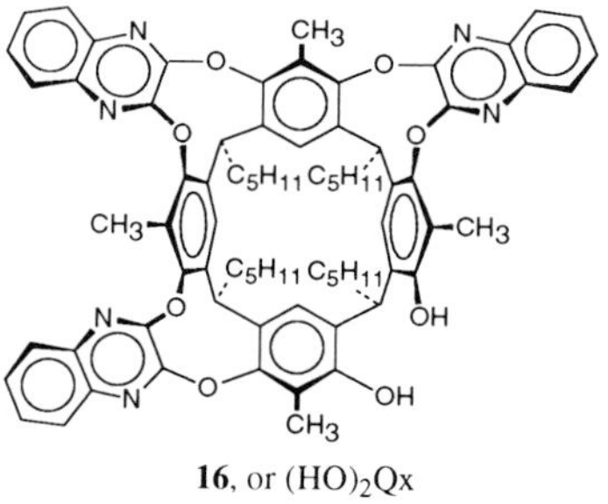

16, or $(HO)_2Qx$

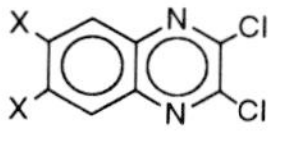

2, X = H
17, X = CH_3
18, X = Cl
19, X = Br

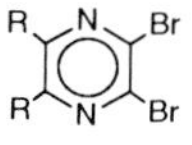

20, R = H
21, R = CH_3
22, R = CH_3CH_2

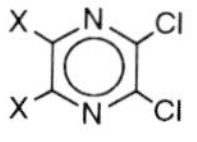

5, X = Cl
23, X = CN

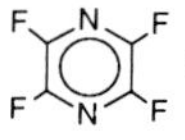

24

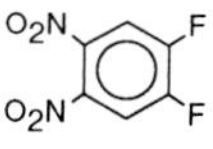

25

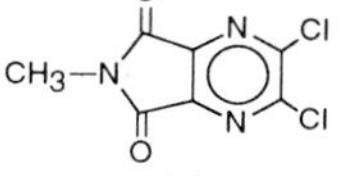

26

Chart 6.3

[3] D.J. Cram, Heung-Jin Choi, J.A. Bryant, and C.B. Knobler, *J. Am. Chem. Soc.*, 1992, **114**, 7748.

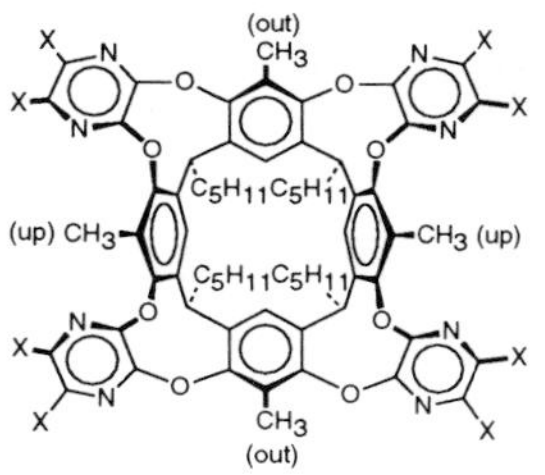

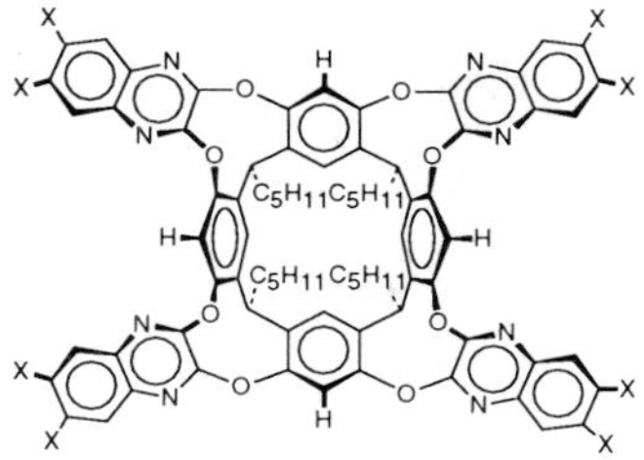

Vases, Kites, Velcrands, and Velcraplexes

Chart 6.4

27, X = Cl, or ClPz
28, X = H, or HPz
29, X = CH₃, or MePz
30, X = CH₃CH₂, or EtPz
31, X = F, or FPz
32, X = OCH₃, or MeOPz
33, X = SCH₃, or MeSPz
34, X = N(CH₃)₂, or Me₂NPz
35, X = CN, or CNPz
36, X = CH₃, or MeQx
37, X = Cl, or ClQx
38, X = Br, or BrQx

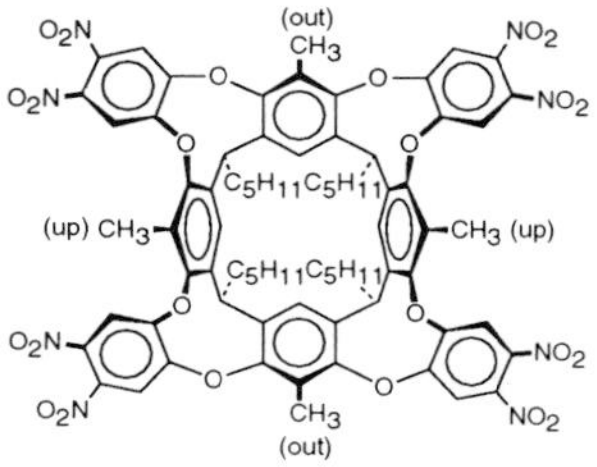

39, or NO₂Bz
40, or ImidePz
41, or UreaBz 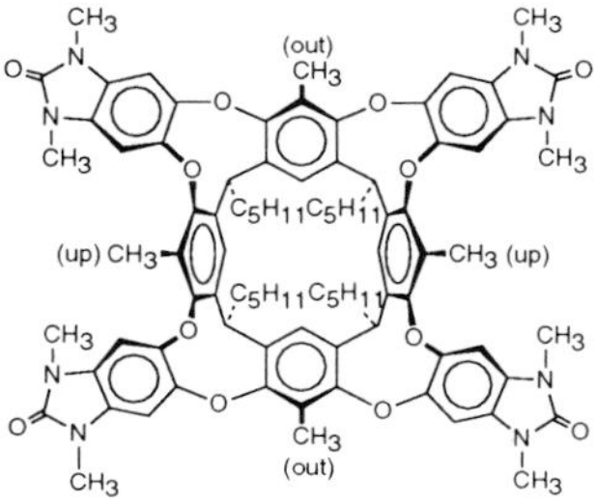
42, or FMeNPz 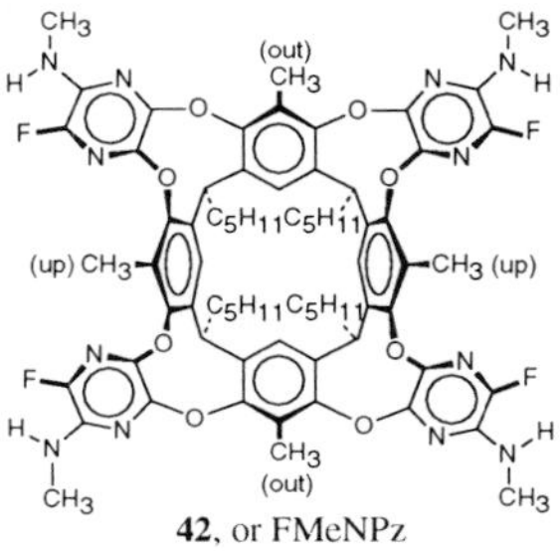

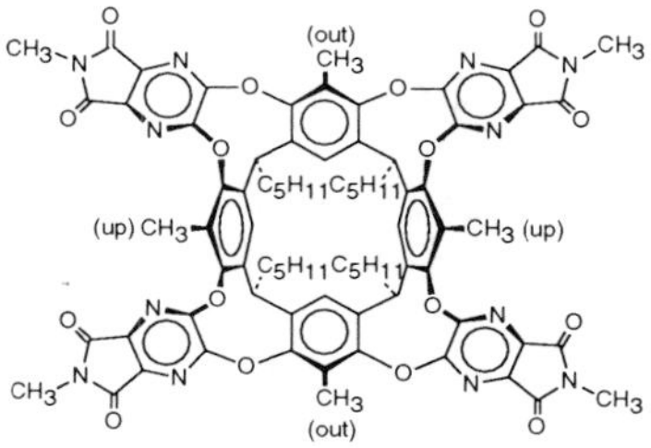

The remaining new monomers of Chart 6.4 were prepared from those already synthesized. The eight fluorine atoms of **31** were readily displaced by strong nucleophiles. Thus treatment of **31** with CH_3O^-, CH_3S^-, $(CH_3)_2N^-$, and CH_3NH^- gave compounds **32**, **33**, **34**, and **42**, respectively. A crystal structure determination of **42** established its substitution pattern (Section 6.3). The tetraurea compound **41** was synthesized from octanitro compound **39** utilizing five conventional reactions in 11% overall yield.

Most of the compounds described above dimerize. We proposed the terms velcrands and velcraplexes for these molecular velcro-like systems. For convenience, we refer to the pyrazine velcrands by the Pz abbreviation, to the quinoxaline by the Qx abbreviation, and to the benzene by the Bz abbreviation, with their substituent types as prefixes. These short names are placed beside the compound numbers below the formulas in Chart 6.4.

6.3 Evidence That Velcrands Form Velcraplexes

6.3.1 Crystal Structures

The crystal structures of dimers **13·13**, **27·27**, **28·28**, **29·29**, **42·42**, and monomer **14** were determined. Drawing **43** is a ball and stick face view of **13·13** based on

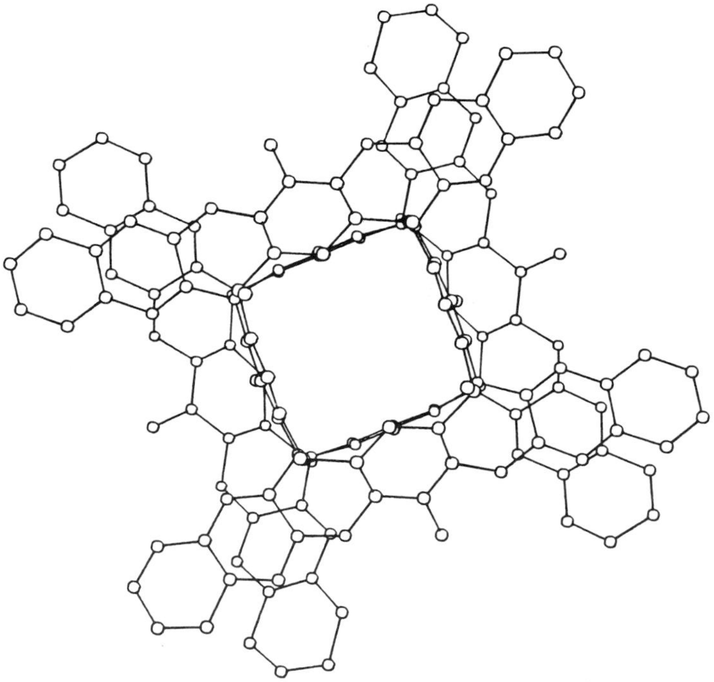

43

its crystal structure. Chart 6.5 portrays face and side stereoviews of the six crystal structures as well as identifying line structures of their monomers. In the crystal structure of **29·29**, the atoms of the pentyl feet most distant from the central surfaces were disordered and not located. Disordered or partially disordered interstitial solvent molecules were present in these crystal structures which made them difficult to refine. These structures were determined to obtain the numbers of short interatomic, non-bonded distances across the surface common to each monomer unit in the dimers. Table 6.1 identifies the kinds of close atoms, sets limiting distances for their being counted, the numbers of short distances of each kind, and the sums of these short contacts for each dimer.[3]

Vases, Kites, Velcrands, and Velcraplexes

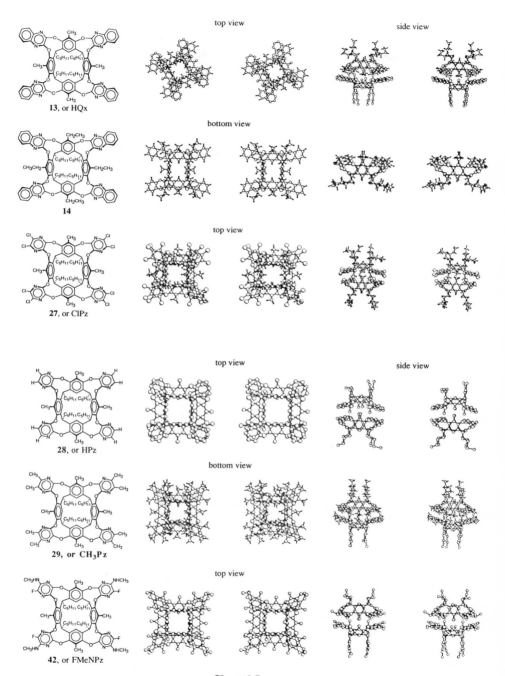

Chart 6.5

Table 6.1 *Short intermonomer atomic distances in crystalline dimers*

Kinds of atoms	Distance (Å) less than	Number of short distances in dimers of				
		13	27	28	29	42
H----H	2.7	32	7	8	18	8
N----H	3.0	0	0	9	7	4
C----H	3.2	58	17	27	46	16
O----H	3.0	10	11	15	9	8
C----N	3.5	6	8	10	6	14
C----O	3.5	8	8	8	8	8
N----O	3.3	0	0	0	0	0
C----C	3.7	18	24	7	13	22
O----O	3.3	0	0	0	0	0
Cl----N	3.6	0	4	0	0	0
Cl----H	3.3	0	3	0	0	0
Cl----C	3.7	0	3	0	0	0
Cl----O	3.6	0	1	0	0	0
F----N	3.4	0	0	0	0	2
C----F	3.5	0	0	0	0	0
F----H	3.0	0	0	0	0	0
total close distances		132	86	84	107	82

Examination of the six crystal structures provides the following conclusions. (1) Four aryl-methyl groups are inserted into four cavities that are stereoelectronically complementary to the methyls to provide a lock–key arrangement which prevents rotation or slipping of one complexing partner relative to the second (see Chart 6.1). (2) Substitution of the four aryl-methyls of **13** (HQx) with four aryl-ethyl groups of **14** gives a crystal of monomer whose core structure, except for Et *vs.* Me substituents, is conformationally identical to that of any one of the complexing partners of the five dimers. Clearly the five monomers containing methyl groups are preorganized for complexation, whereas in **14**, the four Et groups are too large to enter the four holes to form dimer. Furthermore, **12b**, which contains four aryl-H in place of four aryl-CH_3 groups of **13**, shows no tendency to dimerize. Taken together these results provide a striking example of structural recognition in complexation, one in which potential complexing partners can distinguish between H, CH_3, and C_2H_5 substituents. (3) In all five dimers, four aryls of one partner largely share common interfaces with the four aryls of the second partner (see **43**). (4) In all dimers, four pairs of peripheral aryl substituents are close (pseudo-*gem*) to one another. For example, in **27·27**, four pairs of chlorine atoms are close, four other pairs are distant from one another. Thus four sets of Ar—X dipoles of one complexing partner are roughly aligned with four sets of Ar—X dipoles of the second. (5) If the peripheral substituents are different, as in **42·42**, several isomeric complexes might have formed, although only the one with the MeNH substituents close and the F substituents distant was observed (see top view of **42·42**). This structural recognition in complexation probably reflects the more stable structure for monomer in which each $NHCH_3$ group can sterically

Vases, Kites, Velcrands, and Velcraplexes

Figure 6.3

approach coplanarity with its attached aryl to provide stabilizing charge delocalization symbolized by structure D. Moreover, four sets of Ar—F aligned dipoles are avoided in the structure adopted by **42·42**. If the differences in the conformations and disorder in the pentyl feet are disregarded, all of the velcraplexes possess approximate S_4 symmetry.

The colored side and top views of **13·13** found in Figure 6.3 beautifully illustrate the complementarity of the two binding partners relative to one another. Notice in the side view how the pairs of quinoxalines lie in nearly parallel planes in contact with one another, while the four yellow methyl groups (three are visible) fit into the four holes lined with oxygens and aryl faces. In the top view, notice the very small unoccupied cavity that runs through the center of the dimer, the outer aryl groups of the quinoxalines lying only partially over one another, and how the four yellow methyl groups (three are visible) fit into cavities lined with oxygens and the faces of tilted aryl groups.

6.3.2 Solution Dimerization

Molecular weight determinations by vapor pressure osmometry confirmed that dimers are present in $CHCl_3$ solution as well as in the crystalline state. However the best criterion of solution structure came from ^1H NMR spectra. As with its tetramethyl footed homologue **3** of Section 6.1, **12** is conformationally mobile, existing in the vase C_{4v} conformation **12a** at 25 °C and above, and in the kite C_{2v} conformation **12b** at temperatures below -50 °C. The aryl-methyls of **13** (HQx) and aryl-ethyls of **14** sterically inhibit formation of vase conformations, but undergo degenerate equilibrations between two identical kite structures as is illustrated by the $C_{2v} \leftrightarrows C_{2v'}$ interconversions of **14** in Chart 6.2. The activation free energy for this equilibration in **14** is $\Delta G\ddagger \sim 17.8$ (T_c, 100 °C) in $CDCl_2CDCl_2$ and ~ 15.3 (T_c, 42 °C) kcal mol^{-1} in $C_6D_5CD_3$; for **27**, $T_c = 70$ °C. These T_c temperatures, and others, are well above those used to distinguish between dimer and monomer, and therefore this conformational mobility of the monomers can be disregarded. Most of the chemical shifts of the proton signals were different in monomer and in dimer, but generally those of the aryl-methyl groups provided the largest $\Delta\delta$ values for monomer and dimer. In some systems, homo-dimerization was so weak (EtPz, MeOPz, MeSPz, Me_2NPz, CNPz, NO_2Bz, **30**, **32–35**, **39**, respectively) dimer could not be detected, whereas with UreaBz *only dimer* could be detected in solution.[3]

In systems such as ClPz, MePz, ImidePz, UreaPz, HQx, MeQx, and ClQx (**27, 29, 40, 41, 13, 36, 37**, respectively) velcraplexes were detected even in the FAB-MS gas phase. Their abundance relative to monomer correlated roughly with the free energies of formation of their homo-dimers in solution. Thus many of these velcrands form velcraplexes in solid, solution, and gas phase.

Vases, Kites, Velcrands, and Velcraplexes

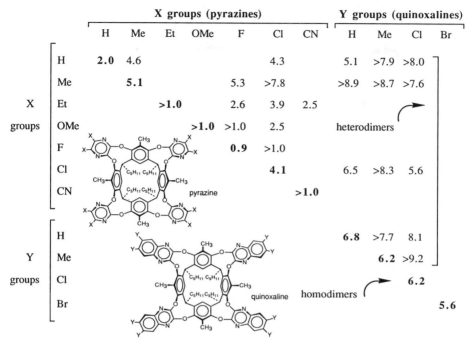

		X groups (pyrazines)							Y groups (quinoxalines)			
		H	Me	Et	OMe	F	Cl	CN	H	Me	Cl	Br
X groups	H	2.0	4.6				4.3		5.1	>7.9	>8.0	
	Me		5.1			5.3	>7.8		>8.9	>8.7	>7.6	
	Et			>1.0		2.6	3.9	2.5				
	OMe				>1.0	>1.0	2.5		heterodimers			
	F					0.9	>1.0					
	Cl						4.1		6.5	>8.3	5.6	
	CN							>1.0				
Y groups	H								6.8	>7.7	8.1	
	Me									6.2	>9.2	
	Cl										6.2	
	Br						homodimers					5.6

For X = MeS or Me$_2$N, <1 in homodimerization, and <1 in heterodimerization with X = F, Cl, or MeO.

Figure 6.4 *Free energies ($-\Delta G^0$, kcal mol^{-1}) in CDCl$_3$ of dimerization, $\Delta S = (\pm)$ 5 entropy units, of pyrazine and quinoxaline velcrands*

6.4 Free Energies of Formation of Velcraplex Dimers in Solution

6.4.1 Determination

The association–dissociation rates for monomers and dimers at temperatures between +30 to −50 °C were usually slow on the ^{1}H NMR time scale, so the populations of the two species were determined directly by integration of their respective peaks, which provided a convenient means of determining association constants and free energies of association. Determination of the variation of K_a values with temperature provided enthalpies and entropies of association of velcrands to form velcraplexes. Values of $-\Delta G^0$ as low as 1 kcal mol^{-1} and in favorable cases as high as 9.2 kcal mol^{-1} were the limits of measurement. Figure 6.4 relates the $-\Delta G^0$ values of formation in CDCl$_3$ of the homo- and hetero-dimers containing pyrazine and quinoxaline moieties in which X and Y groups, respectively, are attached to their outer aromatic nuclei. These particular systems gave $\Delta S = \pm 5$ entropy units, and therefore the $-\Delta G^0$ values were relatively insensitive to the usually small differences in temperatures at which the measurements were made. The $-\Delta G^0$ values for the homo-dimers are placed in boldface along a diagonal line and those for the hetero-dimers are found above this line in the diagram.

6.4.2 Correlation of Binding with Structure

Correlations of binding free energies with structure of these systems in $CDCl_3$ reflect an interesting variety of different effects. The most remarkable feature of the $-\Delta G^0$ values is the range from <1 to >9 kcal mol^{-1}, depending on the peripheral substituents. The high values for binding in the absence of hydrogen bonding, metal ligation, ion-pairing, or hydrophobic effects are without parallel. This leaves dipole–dipole, van der Waals attractions, and solvophobic effects (other than hydrophobic) as the driving forces for complexation. These unusually high binding values are attributed to the unusually large surfaces shared by the binding partners, and to the high degree of preorganization for binding of each monomer. These features provide the potential for a large number of small but simultaneously acting attractive forces to be generated by dimerization. If a large number of small effects are present, their sum can produce large effects. The numbers of the sums of short intermolecular atomic distances listed in Table 6.1 taken from the crystal structures correlate roughly with the magnitudes of the $-\Delta G^0$ values. Table 6.2 lists these numbers, as well as the average contribution each contact makes to the $-\Delta G^0$ of binding. Although the contributions of each kind of short contact must vary widely, these averages are remarkably close together: for HQx·HQx, 0.052 (0.042); for MePz·MePz, 0.047 (0.042); for ClPz·ClPz, 0.047 (0.044); and HPz·HPz, 0.024 (0.023) kcal mol^{-1}. The corresponding averaged $-\Delta H$ values are listed parenthetically to the $-\Delta G^0$ values, and provide a little wider spread, from 0.055 to 0.023 kcal mol^{-1} for each average contact. These averaged values contain much cancellation of various effects, particularly since they involve both the liquid and crystalline phases. All five of these dimers are very similar in their interplanar relationships involving pairs of pyrazine or quinoxaline groups. The four group pairs in each dimer deviate from being parallel to one another no more than 6°, the interplanar distance being approximately 3.5 Å in each dimer.

Table 6.2 *Close contacts in crystalline dimers*

Number	Dimer Structure	Number of close contacts	$-\Delta G^0$ (kcal mol^{-1}) Total	Average per contact
13·13	HQx·HQx	132	6.81	0.052
29·29	MePz·MePz	107	5.06	0.047
27·27	ClPz·ClPz	86	4.06	0.047
28·28	HPz·HPz	84	2.02	0.024

6.4.3 Importance of Preorganization to Binding

The importance of preorganization to binding is indicated by the fact that none of the pyrazine or benzene monomeric systems containing potentially non-coplanar substitutents gave detectable homo-dimers (see Figure 6.4 and Table 6.3). Homo-dimers were not observed for EtPz, MeOPz, MeSPz, Me$_2$NPz, or NO$_2$Bz. In all of these systems, the four pairs of *ortho* substituents sterically inhibit each other from occupying conformations coplanar with their attached aryl rings. This effect is large enough in the Me$_2$NPz and MeSPz systems to inhibit hetero-dimerization as well, but not great enough to disallow hetero-dimerization of EtPz, MeOPz, or NO$_2$Bz with partners containing coplanar substituents. However, coplanarity does not guarantee that homo-dimerization does occur. Thus FPz and CNPz do not homo-dimerize, probably because of the four sets of aligned Ar$^{\delta+}$—X$^{\delta-}$ dipoles that would be created in their homo-dimers.

6.4.4 Binding in Homo- *vs.* Hetero-dimers

Table 6.3 provides the following decreasing order of $-\Delta G^0$ values in kcal mol^{-1} for formation of the various homo-dimers that were studied in CDCl$_3$ at 273 K: UreaBz·UreaBz, >9.0; (HO)$_2$Qx·(HO)$_2$Qx, >8.1; ClQx·ClQx, 6.8; HQx·HQx, 6.8; CH$_3$Qx·CH$_3$Qx, 6.2; BrQx·BrQx, 5.6; ImidePz·ImidePz, 5.6; MePz·MePz, 5.1; ClPz·ClPz, 4.1; HPz·HPz, 2.0; EtPz·EtPz, CNPz·CNPz, FPz·FPz, MeOPz-·MeOPz, NO$_2$Bz·NO$_2$Bz, <1.0. The order appears to be governed by a mixture of effects. Examination of a CPK molecular model of UreaBz·UreaBz suggests that the intermolecular proximal dipoles associated with the cyclic urea groups are largely compensating, which should enhance binding. This effect combined with solvophobic forces that favor binding might provide a system in which ΔH and $T\Delta S$ contributions to ΔG^0 are both negative and additive (see Section 6.5). Similar CPK model examination of the *syn* isomeric dimer of (HO)$_2$Qx-·(HO)$_2$Qx indicates the dimer possesses a geometry very favorable for the existence of intermolecular hydrogen bonding of hydroxyl to hydroxyl, and of hydroxyl to ether oxygens. Such attractive forces combined with the other contact attractions account for the >8.1 kcal mol^{-1} value for $-\Delta G^0$. All of the $-\Delta G^0$ values for the homo-dimers of the quinoxaline type are higher than any of the pyrazine or benzene types (an exception is that both BrQx·BrQx and Imide Pz·ImidePz have $-\Delta G^0 = 5.6$ kcal mol^{-1}). We attribute this to the larger number of attractive contacts in the surfaces common to the complexing partners in the quinoxaline dimers as compared to the others (see Section 6.3).

Examination of Figure 6.4 and Table 6.3 provides the generalization that with the one important exception, hetero-dimers are more strongly bound than the homo-dimers of the more weakly-binding of the two partners, and frequently more than either of the homo-dimers formed from the partners. Seven examples of the former and fifteen of the latter are listed. We interpret this effect as reflecting the presence in all homo-dimers of four aligned pairs of proximal and identical dipoles. In the hetero-dimers, these aligned dipoles

Table 6.3 *Thermodynamic parameters for complexation at 273 K in CDCl$_3$*

Velcraplex	$-\Delta G^0$ (kcal mol^{-1})	ΔH (kcal mol^{-1})	ΔS (cal mol^{-1} K^{-1})	Velcraplex	$-\Delta G^0$ (kcal mol^{-1})	ΔH (kcal mol^{-1})	ΔS (cal mol^{-1} K^{-1})
HPz·HPz	2.22	−1.95	1.0	FPz·FPz	<1.0		
HPz·MePz	4.60			ClP·ClPz	4.06	−3.76	1.1
HPz·ClPz	4.41			ClPz·HQx	6.69	−7.21	−1.9
HPz·HQx	5.02			ClPz·MeQx	>8.3		
HPz·MeQx	>7.9			ClPz·ClQx	5.66	−6.84	−4.2
HPz·ClQx	>8.0			ClPz·FPz	<1.0		
MePz·MePz	5.06	−4.51	2.0	HQx·HQx	6.81	−7.28	−1.9
MePz·FPz	5.36	−4.21	4.2	HQx·MeQx	7.65	−6.78	3.0
MePz·ClPz	>7.8			HQx·ClQx	8.03	−7.20	3.0
MePz·HQx	>8.9			MeQx·MeQx	6.20	−5.76	1.6
MePx·MeQx	8.88			MeQx·ClQx	>9.2		
MePz·ClQx	7.39	−7.63	−0.4	ClQx·ClQx	6.28	−6.08	0.5
EtPz·EtPz	<1.0			BrPz·BrQx	5.60	−4.98	2.3
EtPz·FPx	2.32	−3.82	−5.5	NO$_2$Bz·NO$_2$Bz	<1.0		
EtPz·ClPz	3.76	−4.69	−3.4	NO$_2$Bz·MePz	4.70	1.47	22.6
EtPz·CNPz	2.50	−1.67	3.0	ImidePz·ImidePz	5.63	5.35	40.4
CNPz·CNPz	<1.0			ImidePz·HQx	>9.1		
MeOPz·MeOPz	<1.0			ImidePz·UreaBz	<1.0		
MeOPz·ClPz	2.51	−2.48	0.1	UreaBz·UreaBz	>9.0		
MeOPz·FPz	<1.0			(HO)$_2$Qx·(HO)$_2$Qx	>8.1		

differ enough from one another to favor binding in some cases, or to be less unfavorable to binding in others. The exception involves ImidePz and UreaBz, *which only homo-dimerize but do not hetero-dimerize in the presence of one another*. That case will be discussed in Section 6.5. A few examples are found in which a system incapable of homo-dimerizing can be induced to hetero-dimerize when mixed with one that does homo-dimerize. Thus MeOPz and EtPz, which do not detectably homo-dimerize, form MeOPz·ClPz ($-\Delta G = 2.5$ kcal mol^{-1}), and EtPz·ClPz (3.8 kcal mol^{-1}) when mixed with ClPz, which does homo-dimerize. Similarly NO$_2$Bz, which does not homo-dimerize, forms NO$_2$Bz·MePz (4.7 kcal mol^{-1}) when mixed with MePz, which does homo-dimerize. More striking are the hetero-dimers that are formed from two monomers neither of which homo-dimerize detectably. Examples are EtPz·FPz (2.3 kcal mol^{-1}) and EtPz·CNPz (2.5 kcal mol^{-1}).

An arrangement of the hetero-dimers of Table 6.3 in decreasing order of their binding strengths shows MeQx·ClQx with the highest measured $-\Delta G^0 = 9.2$ kcal mol^{-1}. Generally the hetero-dimers fall into three groups with respect to their stabilities. Those composed of two unlike quinoxalines, or one quinoxaline and a pyrazine have $-\Delta G^0$ values that range from 6.7 to >9.2 kcal mol^{-1} (MeQx·ClQx). Those hetero-dimers containing two pyrazines with electronically unlike but coplanar substituents provide $-\Delta G^0$ values from 4.4 to >7.8 kcal mol^{-1} (MePz·ClPz). The hetero-dimers containing two electronically similar pyrazines or a pyrazine and a benzene with one set of coplanar substituents provide $-\Delta G^0$ values that range from <1.0 up to 4.7 kcal mol^{-1} (NO$_2$Pz·MePz). The three effects reflected in these three general classes appear to be the numbers of close atom contacts between the two complexing partners, the electronic complementarity between the substituents, and the extent of preorganization for binding of the substituents.

6.4.5 Effect of Solvent Changes on Binding

Table 6.4 lists the $-\Delta G^0$ values for MePz·MePz formation in a variety of solvents, along with the solvents' polarizability parameters [E_T (30)].[4] For the seven solvents that allowed $-\Delta G^0$ to be measured, the values ranged from a low of 4.25 for CCl$_4$ to a high of 5.19 kcal mol^{-1} for (CD$_2$)$_4$O, covering a change from 32.4 to 40.7 in E_T (30) units. No correlation between $-\Delta G^0$ and E_T (30) is evident, in fact the relationship is essentially random. Furthermore, for both (CD$_2$)$_6$ and (CD$_3$)$_2$CO whose respective E_T (30) values are 30.9 and 42.2, $-\Delta G^0$ values are both >6.3 kcal mol^{-1}. In more polar solvents such as (CD$_3$)$_2$NCDO, CD$_3$CN, CD$_3$NO$_2$, CD$_3$CD$_2$OD, and CD$_3$OD, only dimer was detected, indicating very high inaccessible values for $-\Delta G^0$ in these solvents. We conclude that the linear free energy relationship of Diederich and Smithrud[5] can apply only in systems where the entropies of solvation of both host and guest remain relatively constant, or vary in such a way as to remain proportional to ΔH.

---

[4] C. Reichardt, 'Solvents and Solvent Effects in Organic Chemistry', 2nd Edn., VCH, Weinheim, 1988, Chapter 7, pp. 339–405.
[5] D.B. Smithrud and F.J. Diederich, *J. Am. Chem. Soc.*, 1990, **112**, 339.

Table 6.4 *Variation of free energies of formation of MePz·MePz with solvent and solvent polarizability parameter*

Solvent	T (°C)	$-\Delta G^0$ (kcal mol^{-1})	$E_T(30)$
$(CD_2)_6$	25	>6.34	30.9
CCl_4	0	4.25	32.4
$C_6D_5CD_3$	0	5.18	33.9
C_6D_6	10	4.82	34.3
C_6D_5Cl	0	4.82	36.8
$(CD_2)_4O$	0	5.19	37.4
$CDCl_3$	0	4.92	39.1
CD_2Cl_2	0	5.14	40.7
$(CD_3)_2CO$	0	>6.3	42.2

The results of a second study of the effect of solvent on $-\Delta G^0$ values are reported in Table 6.5, in this case involving the dimerization of ClPz. The $-\Delta G^0$ values with $CDCl_3$, CD_2Cl_2, and $C_6D_5CD_3$ were essentially the same at ~4.1 kcal mol^{-1}, and changed little with temperature. With a mixture of $CDCl_3$ and more polar solvents, the $-\Delta G^0$ was determined at -18 °C; by the time 25% of $(CD_3)_2CO$, CD_3NO_2, or CD_3OD had been added, they had risen to 5.1, 5.3, and 5.6 respectively. When 50% or 100% of $(CD_3)_2CO$ was used, the value was >5.9. Thus large increases in solvent polarizability induce large increases in the free energy of binding with this system, presumably reflecting increased solvophobic driving forces as the solvent becomes more polar.

Table 6.5 *Free energies of dimerization of ClPz in various solvents compared to $CDCl_3$ as a standard solvent*

Solvent	T (°C)	$-\Delta G^0$ (kcal mol^{-1})	Solvent	T (°C)	$-\Delta G^0$ (kcal mol^{-1})
$CDCl_3$	25	4.1	1% CD_3OD	-18	4.1
$CDCl_3$	-18	4.1	5% CD_3OD	-18	4.2
$CDCl_3$	-46	4.0	10% CD_3OD	-18	4.8
CD_2Cl_2	-46	4.3	25% CD_3OD	-18	>5.6
10% $C_6D_5CD_3$	-46	3.5	50% CD_3OD	-18	>5.6
10% $C_6D_5CD_3$	-18	4.1	1% CD_3NO_2	-18	4.4
10% $(CD_3)_2CO$	-18	4.6	5% CD_3NO_2	-18	4.4
25% $(CD_3)_2CO$	-18	5.1	10% CD_3NO_2	-18	4.6
50% $(CD_3)_2CO$	-18	>5.9	25% CD_3NO_2	-18	5.3
100% $(CD_3)_2CO$	-18	>5.9	10% $(CD_2)_4O$	-18	4.3

6.5 Enthalpies and Entropies of Complexation of Velcrands

6.5.1 Partition of Driving Forces Between Two Parameters

The enthalpies and entropies of complexation listed in Table 6.3 in $CDCl_3$ as solvent show an *unparalleled spread of values with changes in structures of the complexing partners*. At one extreme of the ΔH values is -7.6 kcal mol^{-1} (ΔS -0.4 cal mol^{-1} K^{-1}) for MePz·ClQx, signifying the complexing process is *enthalpy driven* and *entropy neutral*. At the opposite extreme for ΔH is $+5.4$ kcal mol^{-1} (ΔS, $+40.4$ cal mol^{-1} K^{-1}) for ImidePz·ImidePz, indicating a dimerization process which is *entropy driven* ($-T\Delta S$, -11.0 kcal mol^{-1} at 273 K) and *enthalpy opposed*, giving a ΔG^0 value of -5.6 kcal mol^{-1}. Most host–guest complexations in non-hydroxylic media are *enthalpy driven* and *entropy opposed*, with ΔH values ranging from -3 to -8 kcal mol^{-1}, and ΔS from -12 to -18 cal mol^{-1} K^{-1}.

Formation of ImidePz·ImidePz provides one extreme for the entropy of complexation. The only other example in Table 6.3 of an entropy-driven process involves formation of NO_2Bz·MePz, whose entropy is 22.6 cal mol^{-1} K^{-1} to give at 273 K a $-T\Delta S$ value of -6.2 kcal mol^{-1} opposed by ΔH of $+1.5$ kcal mol^{-1}. Thus this dimerization is *entropy driven* but *enthalpy weakly opposed*. The most extreme example of a negative entropy is -5.5 cal mol^{-1} K^{-1} for the formation of EtPz·FPz, with a $-T\Delta S$ at 273 K of $+1.5$ kcal mol^{-1} and ΔH of -3.8 to give ΔG^0 of -2.3, thus illustrating an *enthalpy driven* and *entropy opposed* process. The closest example to a *combined enthalpy* and *entropy driven* dimerization involves formation of MePz·FPz, whose $\Delta H = -4.2$, $\Delta S = +4.2$ ($T\Delta S$ at 273 K is -1.1) and $\Delta G^0 = -5.3$ kcal mol^{-1}. Most of the other fourteen dimers reported in Table 6.3 are *enthalpy driven* and *entropy neutral* (ΔS values vary from extremes of $+3.0$ to -4.2, but nine are ± 2 cal mol^{-1} K^{-1}). UreaBz·UreaBz is bound so strongly that monomer was never detected, but the large changes of [UreaBz·HQx] to [HQx·HQx] ratios compared to changes of [HQx·HQx] itself with temperature in equal molar mixtures of the two systems strongly suggest that the formation of UreaBz·UreaBz is entropy driven. Thus all of the dimers studied possess entropies of formation that are either much less negative than usual or are positive, in some cases strongly positive, enough to make dimerization entropy driven.

6.5.2 Solvolytic Effects on Dimerization

These effects are correlated as follows. The binding surfaces of two monomers are large enough to be solvated by sixteen to eighteen molecules of $CDCl_3$, each molecule of which is weakly bound and somewhat oriented, depending on the polarity of the substituent groups that compose the surfaces. When dimerization occurs, these solvent molecules are detached from the surface and become part of randomly oriented solvent. The desolvation process involves overcoming solvent-to-surface attractions that produce a positive enthalpy, but the liberation of many more or less oriented solvent molecules to bulk solvent increases randomness that produces a positive entropy. With polar substituents such as the sixteen nitro groups of $NO_2Bz \cdot NO_2Bz$, the sixteen carbonyl groups of ImidePz·ImidePz, or the sixteen N═C═O moieties of UreaBz·UreaBz, these effects outweigh the *negative enthalpies* of two monomers sharing a common rigid surface, and the *negative entropies* of collecting and locking two monomers into a dimer. These same effects are present in all the dimerizations, but the enthalpic effects of solvent-to-surface attractions and accompanying entropic effects of solvent orientations are outweighed particularly by the monomer-to-monomer attractions in the dimer to make enthalpically driven, entropy neutral processes the most commonly encountered. In other words, 'solvophobic' binding driving forces are present in all of these dimerizations, and sometimes they are dominant.

6.5.3 Balancing of Entropic and Enthalpic Dimerization Effects

A few specific cases are particularly interesting and instructive. As noted above, EtPz·FPz has the most negative entropy of formation, -5.5 cal mol^{-1} K^{-1}. Examination of a CPK model of this dimer shows that the conformational freedom of eight ethyl groups has to be limited in this dimerization, entailing an additional entropic cost. In the formation of MePz·FPz and $NO_2Bz \cdot MePz$, both the enthalpy and entropy favor complexation by at least -1 kcal mol^{-1}, which is unusual. Even more unusual is the fact that although ImidePz and UreaBz form strongly bound homo- and hetero-dimers, they give no detectable ImidePz·UreaBz when mixed. We can only point to the complexities of the multiple dipoles in each system, which appear to be overall electronically unfavorable for attractions in this mixed dimer, or at least, much less attracting than in their homo-dimers.

The free energies, enthalpies, and entropies of homo-dimerization of ClPz and of ImidePz in a number of different solvents are listed in Table 6.6. Both ΔH and ΔS contribute to increases in $-\Delta G^0$ values for dimerization as $CDCl_3$ was diluted 10% with either $(CD_3)_2CO$ or CD_3OD at 255 K, and to about the same extent with either diluent. Velcrand ClPz is one of the few systems whose

Table 6.6 *Thermodynamic parameters for dimerization of ClPz in three media at 255 K, and imidePz in two media at 273 K*

Monomer	Media	$-\Delta G^0$ (kcal mol^{-1})	ΔH (kcal mol^{-1})	ΔS (cal mol^{-1} K^{-1})
ClPz	$CDCl_3$	4.05	-3.77	1.1
ClPz	10% $(CD_3)_2CO$ in $CDCl_3$	4.57	-4.06	2.0
ClPz	10% CD_3OD in $CDCl_3$	4.80	-4.04	2.8
ImidePz	$CDCl_3$	5.68	5.35	40.4
ImidePz	CD_2Cl_2	4.13	5.07	33.7
ImidePz	20% CD_2Cl_2 in CCl_4	6.98	3.34	37.8

Vases, Kites, Velcrands, and Velcraplexes

dimerization is favored by both enthalpy and entropy parameters in CDCl$_3$. ImidePz dimerization, in CDCl$_3$ at 273 K, is entropy favored and enthalpy opposed. In CD$_2$Cl$_2$, both entropy and enthalpy decrease to provide a lower binding free energy. In 20% CCl$_4$ in CD$_2$Cl$_2$, the enthalpy decreased but the entropy decreased much less to give a higher free energy of dimerization. The declining ΔH in passing from CDCl$_3$ to CD$_2$Cl$_2$ to CCl$_4$ suggests this is the order of binding ability of solvent to the face of the monomer unit, which must be overcome during dimerization. Of course in all three media, the binding is entropy driven and enthalpy opposed.

6.6 Thermodynamic Activation Parameters for Association of Velcrands and Dissociation of Velcraplexes

The activation free energies for association of five velcrands and for dissociation of their velcraplexes were determined at the coalescence temperatures of their ^{1}H NMR spectral signals. The results are given in Table 6.7 for $-\Delta G^0_a$ values and their $-\Delta G^\ddagger$ for association and dissociation at their coalescence temperatures. The five systems, MePz·FPz, ClPz·ClPz, EtPz·ClPz, EtPz·FPz, and HPz·HPz, are arranged in decreasing order of their ΔG^0, $\Delta G^\ddagger_a$, $\Delta G^\ddagger_d$, and T_c values, which all correlate to produce a monotonic trend. The total spread in $-\Delta G^0$ is 3.4 kcal mol^{-1}, whereas that for $\Delta G^\ddagger_a$ is 1.9 and for $\Delta G^\ddagger_d$ is 5.4 kcal mol^{-1}. This suggests the transition state structures are closer to the dimers than to the monomers.[3]

The activation free energies for both the associative and dissociative processes are remarkably high for dimer held together only by dipole–dipole, van der Waals, and solvophobic forces. We attribute these high values to the fact that dimer formation involves exchanging attractions between up to eighteen molecules of solvent and two large monomeric faces for attractions between two faces with approximately one hundred close contacts. The rigid preorganization of these surfaces makes them non-adaptive to an incremental and simultaneous exchange of solvent-to-monomer attractions for monomer-to-

Table 6.7 *Free energies of association in CDCl$_3$ at 273 K and activation free energies for association and dissociation at coalescence temperatures (T_c)*[a]

Velcraplex	$-\Delta G^0$ (kcal mol^{-1})	$\Delta G^\ddagger_a$ (kcal mol^{-1})	$\Delta G^\ddagger_d$ (kcal mol^{-1})	T_c (K)
MePz·FPz	5.4	10.0	15.5	315
ClPz·ClPz	4.1	9.95	14.0	285
EtPz·ClPz	3.8	9.6	13.0	273
EtPz·FPz	2.3	9.2	11.6	243
HPz·HPz[b]	2.0	8.1	10.1	213

[a] The subscripts a and d to $\Delta G^\ddagger$ refer to association and dissociation, respectively;
[b] CH$_2$Cl$_2$ was the solvent.

monomer attractions during association. The four methyl-into-cavity locks when engaged prevent one partner from slipping or rotating with respect to the other until most of the close contacts have disappeared during dissociation. The absence of strong partner-to-partner or monomer-to-solvent attractions in one or the other starting state persists during the transition state, thus accounting for the 8 to 15 kcal mol^{-1} transition state energies. Were it not for the solvophobic effect of drowning most of the original solvating molecules in the bulk solvent by the time the transition is reached from the monomer side, $\Delta G^{\ddagger}_a$ would be much higher.

These results taken in sum show that velcraplex formation from velcrands is unique in many respects. The high binding in the velcraplexes reflects the highly preorganized complementary surfaces shared by the two velcrand partners. Although the highest binding free energy observed is >9.2 kcal mol^{-1}, we believe it probable that in polar media with high affinity of solvent molecules for each other, *e.g.* CH_3OH, values of 15–20 kcal mol^{-1} might be reached, driven largely by solvophobic forces.

CHAPTER 7
Carcerands and Carceplexes

> *'Simple objects with concave surfaces can be arranged in order of their increasing concavity, for example, saucers, bowls, pots, vases, and spheres. If organic compounds are to resemble such objects, provision must be made to enforce their shapes by limiting their conformational mobility. In principle, spheres can be assembled from rigid saucers by simple connecting groups. Once the concept of spherical compounds is in mind, fascinating questions about possible occupancy of their interiors seek answers, and a new field of research is born.'*

7.1 Conception

Cubane,[1] pentaprismane,[2] and dodecahedrane[3] are closed-surface compounds, whose interiors are much too small to contain organic compounds or inorganic ions. In 1983 we formulated what to our knowledge was the first closed-surface hydrocarbon sphere **1** with an enforced interior of sufficient size to embrace simple organic compounds, inorganic ions, or gases.[4]

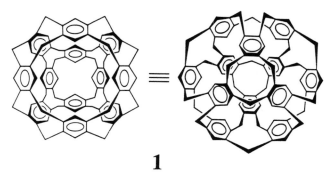

1

Interesting questions regarding the physical and chemical properties of imprisoned guests were posed and discussed.[4] Retrosynthetic thought processes led to the cavitands (see Section 5.1), whose preparations provided an entrée to viable syntheses of what we call carcerands. Carcerands are closed-surface,

---

[1] P.E. Eaton and T.W. Cole, *J. Am. Chem. Soc.*, 1964, **86**, 962, 3157.
[2] P.E. Eaton, Y.S. Or, S.J. Branka, and B.K.R. Shanker, *Tetrahedron*, 1986, **42**, 1621.
[3] L.A. Paquette, R.J. Ternansky, D.W. Balogh, and G. Kentgen, *J. Am. Chem. Soc.*, 1983, **105**, 5446.
[4] D.J. Cram, *Science*, 1983, **219**, 1177.

globe-shaped molecules with enforced hollow interiors large enough to incarcerate simple organic compounds, inorganic ions, or both. Carceplexes are carcerands whose interiors are occupied by prisoner molecules or ions that cannot escape their molecular cells without breaking covalent bonds between atoms that block their escape.

7.2 The First Closed Molecular Container Compound

7.2.1 Synthesis

The synthesis of the first carcerand involved the sequence **2** → **3** → **4** → **5** → **6** → **7**, which is outlined in Chart 7.1, the yields appearing beside the compound numbers. The critical shell closure is formulated in two different ways, **5** + **6** → **7**, and **5a** + **6a** → **7a**. Although top views of **5**, **6**, and **7** show the compact character of the molecule and emphasize the relationship of **7** to the inspirational structure **1**, side views **5a**, **6a**, and **7a** better show the connectedness of the atoms of the carcerand. The carceplex product was a mixture of very insoluble compounds, which was purified by prolonged extraction with H_2O, EtOH, CH_2Cl_2, $CHCl_3$, and EtOAc, during which small amounts of the complexes were undoubtedly lost. The resulting material proved to be virtually insoluble in hot naphthalene, anisole, nitrobenzene, pyridine, or xylene.[5,6]

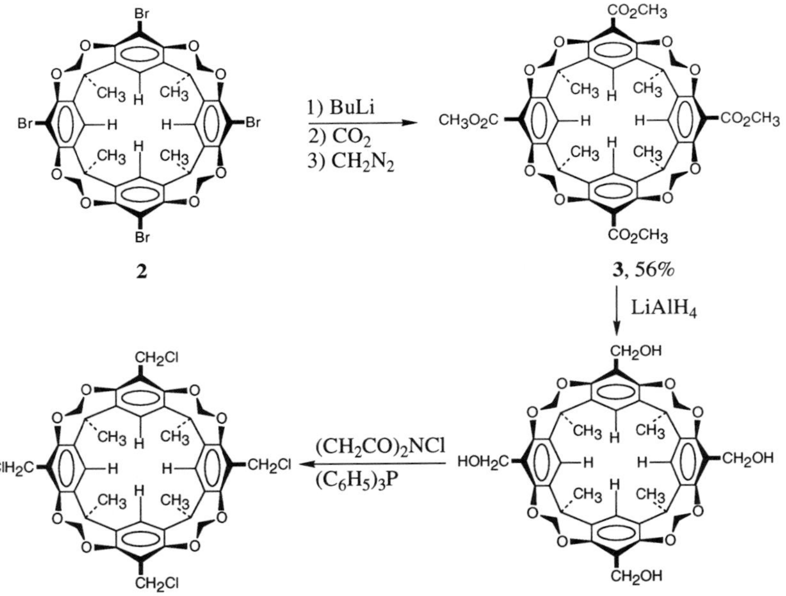

---

[5] D.J. Cram, S. Karbach, Y.H. Kim, L. Baczynskyj, and G.W. Kalleymeyn, *J. Am. Chem. Soc.*, 1985, **107**, 2575.

[6] D.J. Cram, S. Karbach, Y.H. Kim, L. Baczynskyj, K. Marti, R.M. Sampson, and G.W. Kalleymeyn, *J. Am. Chem. Soc.*, 1988, **110**, 2554.

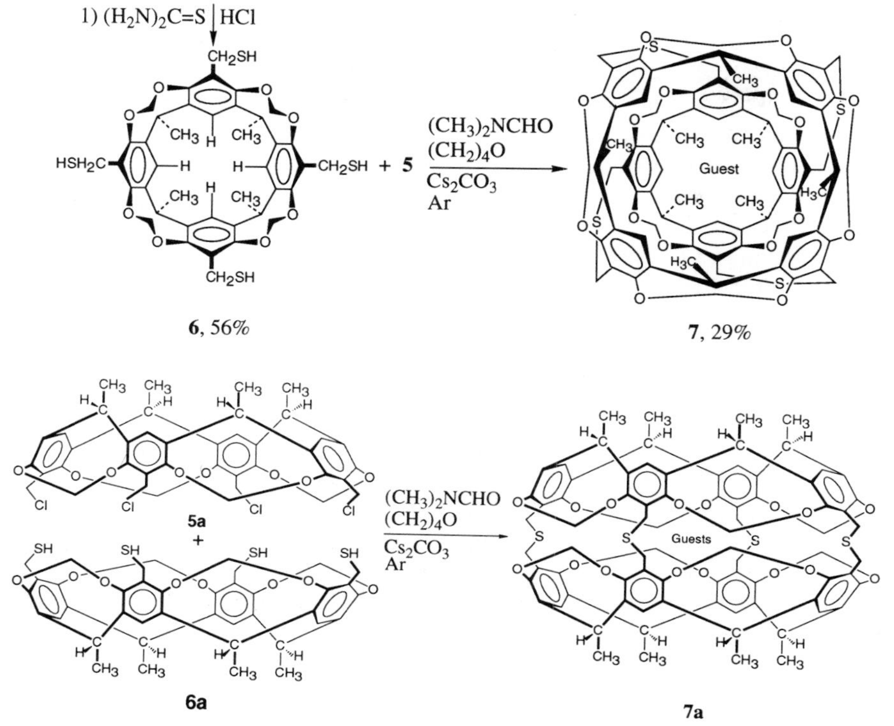

Carcerands and Carceplexes 133

Chart 7.1

7.2.2 Characterization

The insolubility and infusibility (m.p. > 360 °C) of this mixture prohibited further purification and separation into its component parts. Therefore the mixture was characterized as such. Elemental analyses demonstrated the presence of C, H, N, S, Cl, Cs, and Ar, whose sum was 79.05%. The remaining 20.95% was treated as oxygen. The Cs content was determined by neutron activation analysis, and the Ar by combustion coupled with static noble gas spectrometry. The Cs and Cl analyses indicated the elements to be present in stoichiometric amounts, thus as CsCl. The sulfur analysis showed how much host was present, the nitrogen how much $(CH_3)_2NCHO$, and the Ar how much argon. Taken together, the analyses fitted the molecular formula $C_{80}H_{72}O_{16}S_4 \cdot 0.39(CH_3)_2NCHO \cdot 0.69(CH_2)_4O \cdot 0.35CsCl \cdot 0.0065Ar \cdot O_{4.15}$, but not uniquely. An FTIR spectrum indicated the presence of the $(CH_3)_2NCH=O$ carbonyl group, and a solid state ^{13}C NMR spectrum was consistent with structural expectations.[5,6]

The best structural evidence for the carceplexes involved high resolution fast atom bombardment mass spectra, FAS-MS (Xe gun). The ten most prominent m/z peaks, in decreasing order of intensity, gave ions within ± 10 p.p.m. of theory (M is $C_{80}H_{72}O_{16}S_4$): $(M + Cs)^+$, 100; $[M \cdot (CH_3)_2NCHO + H]^+$, 98;

$(M + H)^+$, 86; $(M + Cs + H_2O)^+$, 61; $[M + (CH_2)_4O + H_2O]^+$, 39; $(M + O + H)^+$, 33; $[M + (CH_3)_2NCHO + Cs]^+$, 32; $[M + (CH_3)_2NCHO + Ar]^+$, 28; $(M + Cs + 2 O)^+$, 20; and $(M + 2 Cs + H_2O + H)^+$, 14. These results indicate that every species present in the medium during shell closure except carbonate was incarcerated in the host. Molecular models of the above carceplexes demonstrate that the interior of the host **7** amply accommodates the guests observed in the mass spectrum, and no peaks were observed corresponding to masses for host–guest combinations that could not be prepared with CPK models, except those involving CO_3^{2-}, HCO_3^-, and Cl^-. No masses above 1717 were observed. The peak intensities correlate roughly with expectations based on the elemental analyses of the mixture of carceplexes.[6]

The presence of argon in the mixture indicates the inability of the noble gas to escape its prison. The diameter of argon is about 3.10 Å, considerably greater than the diameter of 2.59 Å estimated for the portals at the constricted ends of **7**. The amount of argon indicates that about 1 in every 150 shell closures encapsulates a molecule of this noble gas.[6]

Molecular model examination of **7** indicates its cavity is complementary to $ClCF_2CF_2Cl$ (Freon 114). When this gas was substituted for argon in a shell closure, a FAB-MS of the product gave a faint peak for $(M + ClCF_2CF_2Cl)^+$ at 1584. The whole sample of the mixture after extensive washing was decomposed at 120 °C in CF_3CO_2H, and the gases liberated were analyzed by GC–MS, which revealed the presence of $ClCF_2CF_2Cl$ and its fragmentation ions. Thus a small amount of Freon 114 was trapped inside the carcerand during shell closure.[6]

Whereas the elemental analysis of **7·Guests** indicated that equivalent amounts of Cs and Cl were present, the FAB-MS indicated that Cs was inside the carcerand and Cl was largely outside. Furthermore, the amount of Cs incarcerated (~0.35 equivalent per equivalent of host) is much greater than expectations based on the statistical presence of dissolved Cs^+ present during the shell closure. The following explanation is plausible. The S—C bonds made during the shell closure probably involve an S_N2 reaction mechanism between $\sim CH_2$-S^- Cs^+ contact ion pairs and $\sim CH_2Cl$ groups. After two or three of the $\sim CH_2$-S-$CH_2 \sim$ bonds are in place, no room exists for escape of a Cs^+ located in the cavity of the incipient host **7·G**. Examination of models of the S_N2 linear transition state for making the third or fourth bond shows that a linear arrangement of S----C----Cl is possible only in a conformation in which the nucleophile S^- is oriented toward the cavity and the leaving group Cl^- diverges from the cavity. If Cs^+ is ion-paired to the S^-, it must reside in the cavity, from which it can not escape, as is suggested by **8**. Thus the CsCl is produced as a shell-separated ion pair.[6]

To determine whether the H_2O of the carceplex $(M + Cs + H_2O)^+$ had entered the interior cavity during the extraction process after the shell closure, **7·G** was suspended in refluxing D_2O for three days, recovered, and submitted to FAB-MS. The peak intensity ratio $(M + Cs + D_2O)^+ : (M + Cs + H_2O)^+$ changed from 0.34 before treatment to 0.85 after treatment, an increase in D_2O over H_2O by a factor of 2.4. We conclude that water molecules at least can enter

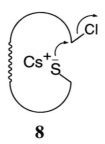

8

the interior of **7**, probably with high activation energy. The opening between the four CH₃ groups at each end of the long axis of **7** in CPK models is just large enough to allow a model of H_2O to be pushed inside without breaking bonds if considerable pressure is applied.[6]

7.3 Soluble Carceplexes with CH_2SCH_2 Connecting Groups

To overcome the insolubility of the carceplexes **7·G** discussed in the last section, new analogues **9–15** (Chart 7.2) were designed, synthesized, and characterized, in which the eight CH_3 groups of **7·G** were replaced by either eight $(CH_2)_4CH_3$ or eight $CH_2CH_2C_6H_5$ groups.[7,8] We hoped that the conformational flexibility of these 'feet' would provide the carceplexes with enough additional solvation sites to render them soluble. The choice of the alkyl group $CH_3(CH_2)_4$ was

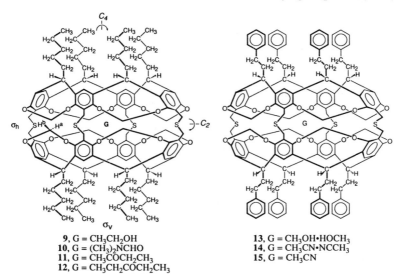

9, G = CH₃CH₂OH
10, G = (CH₃)₂NCHO
11, G = CH₃COCH₂CH₃
12, G = CH₃CH₂COCH₂CH₃

13, G = CH₃OH·HOCH₃
14, G = CH₃CN·NCCH₃
15, G = CH₃CN

Chart 7.2

---

[7] J.A. Bryant, M.T. Blanda, M. Vincenti, and D.J. Cram, *J. Chem. Soc., Chem. Commun.*, 1990, 1403.
[8] J.A. Bryant, M.T. Blanda, M. Vincenti, and D.J. Cram, *J. Am. Chem. Soc.*, 1991, **113**, 2167.

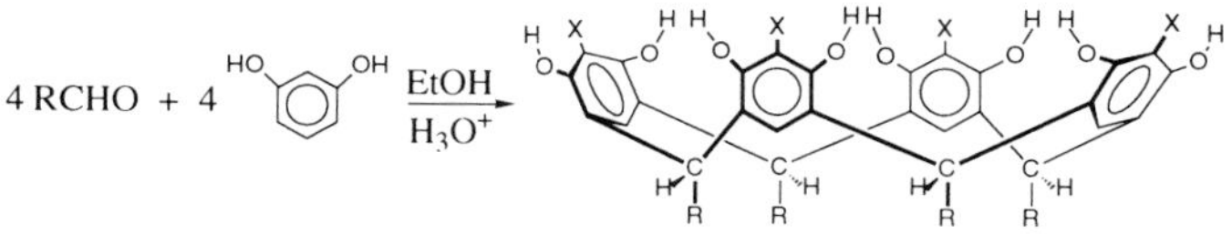

$$4 \text{ RCHO} + 4 \text{ resorcinol} \xrightarrow[\text{H}_3\text{O}^+]{\text{EtOH}}$$

16, R = CH$_3$(CH$_2$)$_4$
17, R = C$_6$H$_5$CH$_2$CH$_2$

18, X = H, R = (CH$_2$)$_4$CH$_3$
19, X = Br, R = (CH$_2$)$_4$CH$_3$
20, X = H, R = CH$_2$CH$_2$C$_6$H$_5$
21, X = Br, R = CH$_2$CH$_2$C$_6$H$_5$

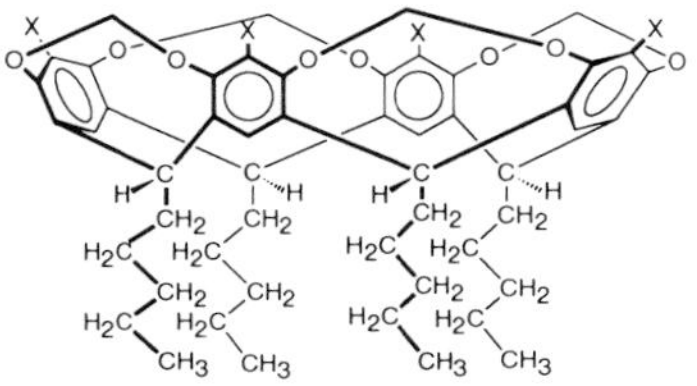

22, X = Br
23, X = CO$_2$CH$_3$
24, X = CH$_2$OH
25, X = CH$_2$Cl
26, X = CH$_2$SH

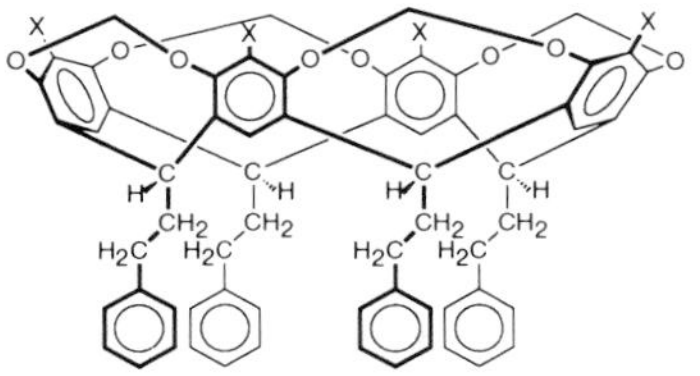

27, X = Br
28, X = CO$_2$CH$_3$
29, X = CH$_2$OH
30, X = CH$_2$Cl
31, X = CH$_2$SH

Chart 7.2 (*continued*)

based on the ready availability of hexanal, **16**, as a starting material, the ^{1}H NMR signals of the CH$_3$(CH$_2$)$_4$ protons concentrate in a narrow section of the spectra of compounds carrying these groups, and because the solubility of octol **18** in organic solvents tended to approach a maximum as the chain lengths of its homologues reached five carbon atoms.[9] The choice of the C$_6$H$_5$CH$_2$CH$_2$–group was made because dihydrocinnamaldehyde, **17**, is readily available as a starting material, its ^{1}H NMR spectrum tends to be simple, and because we thought it likely that crystalline carceplexes suitable for X-ray crystal structure determinations were more likely to form than those with purely aliphatic groups.[7,8]

7.3.1 Syntheses

The syntheses of **9–12** involved the simple reaction sequence **16** → **18** → **19** → **22** → **23** → **24** → **25** → **26** followed by the shell closures **25** + **26** → **9–12** in which the reaction solvent became the incarcerated guest molecule. The four-fold cyclizations of tetrabromooctol **19** with CH$_2$BrCl–(CH$_3$)$_2$NCHO–K$_2$CO$_3$ to give cavitand **22** went in 56% yield. The four shell closures to produce carceplexes were carried out with either Cs$_2$CO$_3$ or Rb$_2$CO$_3$ as base at moderately high dilution. The product was isolated by chromatography on silica gel, which removed any metal-ion containing carceplex. The yield varied with the solvent for shell closure, **9** (20%) in CH$_3$CH$_2$OH, **10** (20%) in (CH$_3$)$_2$NCHO, **11** (32%)

---

[9] L.M. Tunstad, J.A. Tucker, E. Dalcanale, J. Weiser, J.A. Bryant, J.C. Sherman, R.C. Helgeson, C.B. Knobler, and D.J. Cram, *J. Org. Chem.*, 1989, **54**, 1305.

in $CH_3COCH_2CH_3$, and **12** (23%) in $(CH_3CH_2)_2CO$. In all cases the carceplex contained one molecule of solvent.[8]

The syntheses of **13–15** followed the reaction path **17** → **20** → **21** → **27** → **28** → **29** → **30** → **31**, followed by the shell closures **30** + **31** → **13–15**. The four-fold cyclizations of tetrabromooctol **21** with $CH_2BrI–(CH_3)_2NCHO–K_2CO_3$ went in 53% yield. The shell closures to produce carceplexes and their isolations were accomplished as before. When $CH_3OH–C_6H_6$ was the solvent, **13** with two molecules of CH_3OH was the product (22%). With $CH_3CN–C_6H_6$, a mixture of **14** in which 2 molecules of CH_3CN were incarcerated and **15** containing 1 molecule of CH_3CN as a prisoner was isolated (14 and 11% respectively).[8]

Since **25**, **26**, **30**, and **31** were relatively insoluble in CH_3OH, CH_3CH_2OH, and CH_3CN, the reactants were dissolved in C_6H_6 and added to carbonate–alcohol or carbonate–nitrile, but only carceplexes containing the more polar solvents as guests were isolated. When only C_6H_6 was used as solvent, no carcerand or carceplex was produced. Thus shell closures showed high structural recognition in incarcerating the more polar molecules in the medium.[8]

The shell-closing reaction involves a base-catalyzed S_N2 mechanism with either solvated $ArCH_2S^-$ or $ArCH_2S^-M^+$ as nucleophile. The former would lead to incarceration of solvating solvent and the latter to either M^+ or M^+----solvating solvent. In the carceplex syntheses reported in Section 7.1, an inseparable mixture of incarcerated materials (**7·**Guests) was isolated, the major components of which were **7·**Cs^+ and **7·**$Cs^+(CH_3)_2NCHO$.[6] Probably similar guests were incarcerated in the runs leading to **9**, **10**, and **13–15**, but the metal-ion-containing products remained at the top of the chromatograph columns used to purify these compounds. Thus we isolated products formed when solvated $ArCH_2S^-$ acted as nucleophile, in which only the more polar component of the medium solvated $ArCH_2S^-$. Molecular model examination of the product-determining transition state making the second or third bond indicates a linear arrangement of the three atoms involved (as in **32**) is possible only when the solvated S^- is inside the cavity and the Cl^- is outside as shown in **33**.[8]

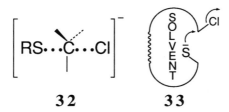

32   **33**

This explanation indicates why empty carcerand was never formed. Its presence among the products would require that $ArCH_2S^-$ be surrounded by a vacuum whose volume was at least equal to that of the carcerand interior. Organic solvents have roughly 30% of their volume as 'empty space' found at places where solvent-to-solvent contacts are non-complementary. Many such small spaces would have to gather in one place to create non-solvated high energy ArCH$_2$S$^-$ species. Thus the production of empty carcerand is unlikely based on both enthalpic and entropic grounds.[8]

7.3.2 Characterization

Many attempts to obtain crystals of these carceplexes of crystallographic quality failed. However, a crystal structure of tetrabromide **27·2H$_2$O** was determined; a stereoview is portrayed in **34**. Notice that one molecule of water occupies the rigid upper cavity, and that a second lies between the four CH$_2$CH$_2$C$_6$H$_5$ feet which conformationally organize to form a cavity.[10]

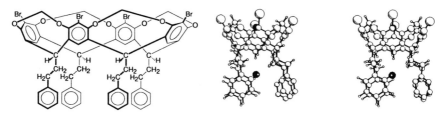

27   **34·2H$_2$O** crystal structure (O's of H$_2$O darkened)

Elemental analyses for all kinds of atoms present were performed on complexes **9–13** and **15**. They each were within 0.25% of theory and the sum of the analyses for each compound was 100 ± 0.25%. These data attest to the purity and establish that the guest-to-host ratios are unity, except for **13** which contains two molecules of CH$_3$OH.[8]

Initially **14** and **15** were obtained as a 1.2:1 mixture, which was converted to pure **15** by heating the mixture at 110 °C in toluene for 72 hours. This treatment resulted in the expulsion of the second molecule of CH$_3$CN from the cavity. However, when **15**, which contains one molecule of CH$_3$CN, was heated in 1,2,4-trichlorotoluene at 215 °C for 5 days, no further loss of guest occurred. The rates of loss of CH$_3$CN from **14** were followed at 110, 100, 90, and 80 °C by ^{1}H NMR spectral changes in Cl$_2$CDCDCl$_2$ as solvent, which provided an activation energy of 20 kcal mol^{-1} for the process. The rates indicated that **14** was initially produced in the shell closure, but partially lost its second guest as the reaction proceeded. Attempts to introduce CH$_3$CN into **15** to form **14** failed (CH$_3$CN-C$_6$H$_6$, 80 °C, 72 hours). In an attempt to remove a molecule of CH$_3$OH from **13**, the carceplex was heated at 110 °C for 5 days in C$_6$D$_5$CD$_3$. No CH$_3$OH was liberated.[8]

We attribute the loss of CH$_3$CN from **14**, when heated, to give **15** to what we call the *billiard ball effect*. At high temperatures, two molecules of CH$_3$CN can collide in the inner phase of the carcerand to provide high kinetic energy to one molecule and low to the other. If these collisions occur with the proper geometry, the high energy guest is expelled. When only one guest is present, its collisions with the sides of the container provide many fewer possible geometries, and lower opportunities for concentrating the energy of the system in

---

[10] J.C. Sherman, C.B. Knobler, and D.J. Cram, *J. Am. Chem. Soc.*, 1991, **113**, 2194.

Carcerands and Carceplexes

a kinetic form for the single guest. In a sense, one guest drives out the other, but the sides of the host are not mobile enough to play the same role for the remaining guest. The hydrogen bonding between the two guest CH_3OH molecules inhibits their exhibiting a billiard ball effect.[8]

Carceplexes **9–13** and **15** all gave strong molecular ions in their desorption chemical ionization mass spectra (DCI-MS) with $(CH_3)_3CH$ as the reagent gas in both their positive and negative modes (80–100%). In the spectrum of **13**, masses of substantial intensity were observed for $(H \cdot CH_3OH)^+$ but not for $(H \cdot CH_3OH)^-$. In all the spectra small peaks for empty carcerand were observed. Either guests escaped at the high temperature involved, or the carceplexes underwent bond-breaking processes which allowed the guests to escape.[8]

7.3.3 Guest Rotations in Carceplexes

The ^1H NMR spectra of **9–15** provide invaluable and interesting information about the structures and dynamics of the carceplexes. All proton signals of the guest are moved 1–4 p.p.m. upfield from their normal positions. All guest protons are subject to the large shielding effects of the eight aryl groups which compose much of the globe-like part of the host. The host alone has a longitudinal C_4 axis and four shorter equatorial C_2 axes, as well as five mirror planes. The equatorial σ_h plane passes through the four S atoms, whereas the four σ_v polar planes are defined by either the two poles and two *anti* S atoms, or by two *anti* intrahemispheric OCH_2O groups and the two poles. The multiplicities of the signals due to the inward pointing H^a and upward pointing H^b protons (see host of **9–12**) provide indicator systems for possible constraints of the rotational degrees of freedom of guest relative to host. If rotations of guests around both the long polar and short equatorial axes are fast on the ^1H NMR time scale, the multiplicities of the H^a and H^b signals are the same as that expected for the host taken alone. If rotations of non-like-ended guests (*e.g.* $CH_3COCH_2CH_3$ in **11**) around the long axis are fast, but around the short ones are slow on the ^1H NMR time scale, the H^a and H^b signals of the northern and of the southern hemispheres might show different chemical shifts. In the unlikely event that rotations of the like-ended guests (*e.g.* $CH_3CH_2COCH_2CH_3$) should be slow around both the long and short axes, then the H^a and H^b protons located in the eastern and western hemispheres might be different from one another. If such guests are non-like-ended and not rotating, then the H^a and H^b protons in the northern and southern as well as in the eastern and western hemispheres should have different environments, and therefore different signals.[8]

The ^1H NMR spectrum of $H \cdot CH_3COCH_2CH_3$ provided two sets of signals for the H^a and H^b protons, one for the northern and one for the southern hemispheres, which shows that rotations of the guest around the short equatorial axes are slow on the ^1H NMR time scale. The spectrum of $H \cdot CH_3CH_2COCH_2CH_3$ (**11**) gave only one kind of H^a and H^b signals, indicating that rotation around the long polar axis was fast. The spectra of $H \cdot CH_3CH_2OH$ (**9**),

H·(CH$_3$)$_2$NCHO (**10**), H·CH$_3$OH·CH$_3$OH (**13**) and H·CH$_3$CN (**15**) likewise gave only one type of signals for H^a and H^b, because these guests rotated around all axes rapidly on the ^{1}H NMR time scale.[8]

7.4 Carceplexes with OCH$_2$O Connecting Groups

We were led to a second kind of carcerand through examination of CPK models. Substitution of four OCH$_2$O groups for the four CH$_2$SCH$_2$ groups arranged about the equator in the host of **13–15** provided the generalized carcerand structure **35**, which is portrayed in both a polar and an equatorial view in **35a** and **35b**. The interior cavity of the host of **13–15** is slightly larger than that of **35** due to the greater length of the sulfur bonds in the former compared to the bonds to oxygen in the latter. The model of **35** in its equatorial region appeared more conformationally comfortable than that of the host of **13–15**, particularly when the unshared electron pairs of the two OCH$_2$O oxygens are oriented away from the hollow interior of **35**, *anti* to the enforced orientations of the unshared electron pairs of its two *ortho* oxygens. Accordingly, the CH$_2$ of the OCH$_2$O interhemispheric-spanning groups face away from the cavity.[10,11]

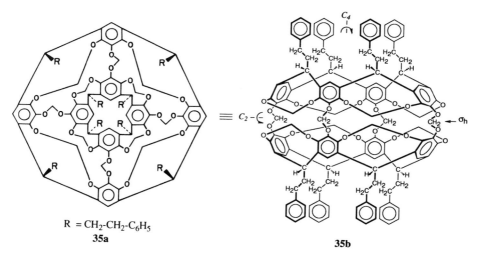

R = CH$_2$-CH$_2$-C$_6$H$_5$
35a

35b

7.4.1 Syntheses

The syntheses of carceplexes **35·G** were carried out as follows. Tetrabromide **27** was metalated in tetrahydrofuran at −78 °C with *t*-BuLi, the tetralithium product formed was quenched with (CH$_3$O)$_3$B to form the arylboronic tetraester, which without isolation was oxidized with H$_2$O$_2$–NaOH to produce tetrol **36** (53%) and triol **37** (23%), as formulated in Chart 7.3. The critical shell closures of two moles of **36** to give carceplexes **38**, **39**, and **40** were carried out

---

[11] J.C. Sherman and D.J. Cram, *J. Am. Chem. Soc.*, 1989, **111**, 4527.

under high dilution conditions in purified dry solvent-Cs_2CO_3–CH_2BrCl at 60–100 °C. In each shell closure, one mole of solvent was incarcerated to provide the following excellent yields of chromatographically pure carceplexes: for **38** which contains $(CH_3)_2NCOCH_3$, 54%; for **39** containing $(CH_3)_2NCHO$, 49%; and for **40** containing $(CH_3)_2SO$, 61%. The sum of elemental analyses for C, H, O, S, and N for each carceplex ranged from 99.73 to 99.90%, which indicated the complexes to be one-to-one. Any carceplex formed with Cs^+ as guest would have remained at the top of the chromatograph column. Three different kinds of mass spectral determinations gave ions of mass indicative of carceplexes each containing one molecule of incarcerated solvent.[10–12]

In CPK models, $CH_2(CH_2CH_2)_2NCHO$ is too large to be incorporated into the cavity of carcerand **35**. In an attempt to synthesize empty host, the shell closure was conducted in this solvent, but only polymeric product was obtained. Another shell closure was conducted in $CH_2(CH_2CH_2)_2NCHO$ which was 0.5 mol % in $(CH_3)_2NCOCH_3$ to give only a 10% yield of **38**. When

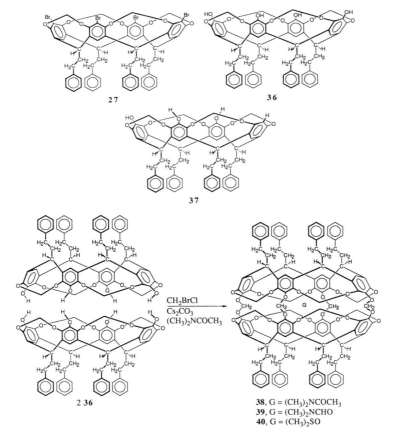

Chart 7.3

[12]D.J. Cram, *Nature (London)*, 1992, **356**, 29.

a 1:1 molar ratio of $(CH_3)_2NCOCH_3$ and $(CH_3)_2NCHO$ was used as solvent, a 27% collective yield of 5:1 molar ratio of **38** to **39** was produced, showing that structural recognition occurred in the reaction favoring the larger of the two potential guests. These results were interpreted as follows.[10,11]

The nucleophilic substitution reactions of the shell closures and the competing polymerizations involve phenoxides which are solvated by solvent. When the solvating molecules are too large to fit into the cavity of the product-determining transition state, only polymerization occurs, because the reacting groups are held too far apart for shell-closing reactions by the partially enclosed solvent that fills the cavitand parts. That the cavitand parts should contain empty space is unlikely on entropic grounds. Organic liquids are about 30% empty space which is broken into small volumes because they inefficiently contact one another due to their shapes. The entropy of dilution of these small volumes would have to be overcome to gather them together to create large enough empty volumes to produce empty carcerand in the product-determining transition states of the shell closures. This transition state must occur after one bridge is in place and during the reaction that completes the second or third bridge. Model examination indicates that once two *anti* bridges are in place, a solvating molecule can not easily escape the interior of the shell, since hydrogen-bonded interhemispheric hydroxyls and the intrahemispheric methylenes block the way. The volume of this transition state is larger than that of the final carcerand, and apparently $(CH_3)_2NCOCH_3$ has more complementary contacts with the shell parts than the smaller $(CH_3)_2NCHO$. In effect, the shell closures are templated by a single solvent molecule solvating the two cavitand parts at the same time, and thus orienting them for the shell closure.[10,11]

7.4.2 Carceplex Crystal Structure

The crystal structure of carceplex **38**·5CHCl$_3$ was determined, and Chart 7.4 provides both side and end stereoviews of the structure in which the interstitial CHCl$_3$ molecules are omitted. The most important feature of the crystal structure is the unequivocal presence of $(CH_3)_2NCOCH_3$ incarcerated in the host, sterically inhibited from departing the interior without breaking bonds.

A very interesting additional characteristic of the crystal structure is that the northern hemisphere is rotated around the long polar axis with respect to the southern hemisphere by about 25°. This axis is not quite linear because the pair of planes 'a' and pair of planes 'b' of diagram **43** (see Chart 7.5) are not quite parallel. The two 'a' planes deviate from being parallel by 5.2° and the two 'b' planes by a value of 4.2°. This small tilt and rotation are visible in diagram **44** of Chart 7.5, in which the four interhemispheric OCH$_2$O bridging groups are drawn. In **44**, the four oxygens of the northern hemisphere are connected to one another to form a *near square*, and the corresponding oxygens of the southern hemisphere form a second near square. Notice that the oxygen's unshared electron pairs and the two hydrogens of the attached CH$_2$ groups of **44** generally face along the surface of the shell or outward, but not inward. These

Carcerands and Carceplexes

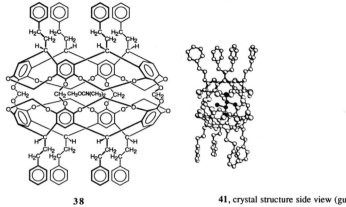

38

41, crystal structure side view (guest atoms dark)

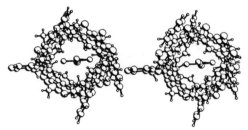

42, crystal structure end view

Chart 7.4

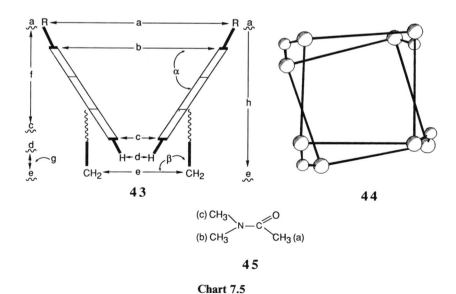

43 **44**

$$\underset{(b)\ CH_3}{\overset{(c)\ CH_3}{\diagdown}} N - C \overset{O}{\underset{CH_3\ (a)}{\diagup}}$$

45

Chart 7.5

rotations, tilts, and spatial dispositions accommodate the incarcerated guest's volume and shape, and also allow the rims of the two hemispheres to pack against each other efficiently to provide close $CH_2O\cdots H_2C$ contacts, and to minimize the cavity volume. The rotation and tilt also destroy the five planes of symmetry of the carcerand portion of the complex to produce a chiral shell. The D_4 symmetry with a C_4 long polar axis and the four C_2 short equatorial axes are left approximately intact. The cavity itself is shaped like a US football, whose long axis is 10.9 Å (distance between two 'd' planes of **43**) and whose short equatorial axes are 6.2 Å (limiting distance between opposite bridging oxygens, or 9 Å minus 2.8 Å).[10]

All hydrogens are omitted from the guest in both the side **41** and end **42** views of the crystal structures of **38** (Chart 7.4). The six non-hydrogen atoms of the $(CH_3)_2NCOCH_3$ guest **45** in **41** or **42** are nearly coplanar. The end view **42** of this guest resembles an airplane floating in a wind tunnel. The two ends at the poles of the host have the largest pores in the shell, each being defined roughly as a circle of four aryl hydrogens with a diameter of 4.2 Å ('d' in **43**), which leaves a hole of about 3.2 Å in diameter. Methyl group 'a' in **45** attached to the carbonyl group is close to being centered in the temperate zone of the northern hemisphere of **41**, whereas the 'b' methyl group is in the tropic zone slightly north of the equator. The carbonyl oxygen occupies the tropic zone slightly south of the equator. The long axis of guest **45** that passes through the 'a' and 'c' methyls is close to being coincident with the long polar axis of the cavity. The $(CH_3)_2NCOCH_3$ molecules in **41** are largely ordered except for the $(CH_3)_2N$ and $COCH_3$ groups exchanging their positions. Thus in the crystal, each guest in effect communicates its rotational position around its long axis to guests in neighboring carceplex molecules. The means of this communication is as interesting as it is obscure.[10]

7.4.3 Guest Movements in Carceplex

Chart 7.6 provides drawings **46–48** of MM2-calculated minimum energy structures for the respective carceplexes **38**, **39**, and **40**, in which eight methyl groups replace the pendant $CH_2CH_2C_6H_5$ groups; the oxygens are shaded. No twist forms were calculated. The placement of the $(CH_3)_2NCOCH_3$ guest in **46** and in crystal structure **41** are very similar, both corresponding closely to what is obtained in a CPK model of **38**.[10] Notice, however, that in **46–48**, all of the unshared electron pairs of the OCH_2O bridges face outward and the CH_2 groups inward, the opposite of what is found in all of our crystal structures of the carceplexes and hemicarceplexes containing OZO moieties.

$$\underset{\mathbf{49}}{\underset{\text{(4.49) } H_3C}{\overset{O}{\overset{\|}{C}}}\underset{\underset{CH_3 \text{ (1.98)}}{|}}{N}-CH_3 \text{ (4.40)}} \qquad \underset{\mathbf{50}}{\underset{\text{(3.77) } H}{\overset{O}{\overset{\|}{C}}}\underset{\underset{CH_3 \text{ (4.00)}}{|}}{N}-CH_3 \text{ (2.90)}}$$

$$\underset{\mathbf{51}}{\text{(3.86) } H_3C \overset{\overset{O}{\|}}{\underset{}{S}} CH_3 \text{ (3.86)}}$$

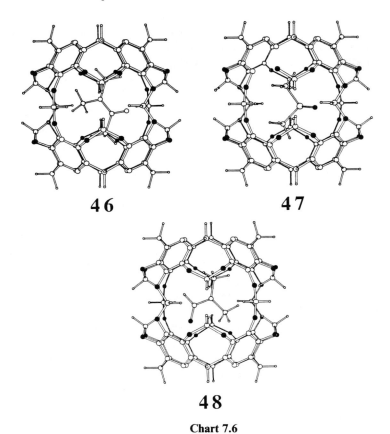

Chart 7.6

All of the protons in the ^{1}H NMR spectra of carceplexes **38–40** in CDCl$_3$ solution at 25 °C were assigned. The signals of the guest protons were moved dramatically upfield, which is consistent with their close proximity to the shielding zones of the aryls of the host. The Δδ shift values in p.p.m. due to incarceration are noted parenthetically next to the appropriate protons in **49**, **50**, and **51**. These large shifts due to aryl shielding provide a splendid indicator for not only incarceration, but also for general locations of the protons within the cavity. Those protons with the higher Δδ values are located closest to the polar regions, and those with the lower are more equatorially positioned.[10] The cavity-facing H$_c$ and shell-facing H$_s$ protons of the *intrahemispheric* OCH$_2$O bridges of the host were the most sensitive to the guest structure. Thus H$_c$ and H$_s$ of **38** with (CH$_3$)$_2$NCOCH$_3$ as guest appear as doublets in multiplets at δ 4.64 and 6.08, respectively, whereas H$_c$ and H$_s$ of **39** with (CH$_3$)$_2$NCHO and of **40** with (CH$_3$)$_2$SO as guests both occur as doublets at δ 4.50 and 6.12. These results indicate that the largest guest (CH$_3$)$_2$NCOCH$_3$ is too closely held by the host to be able to rotate freely around the short equatorial C_2 axes, but can rotate around the long C_4 axis. As a result the (CH$_3$)$_2$N and COCH$_3$ parts affect differently the two hemispheres they occupy, particularly the chemical shifts of the H_c and H_s protons of the two hemispheres. In contrast, the simpler ^{1}H NMR spectra of host containing $(CH_3)_2NCHO$ or $(CH_3)_2SO$ indicate these guest molecules rotate rapidly on the NMR time scale about all axes to give averaged signals. Molecular model examinations are consistent with these interpretations.[10]

7.5 Inner Phase Effects on Physical Properties of Guests

7.5.1 Molecular Communication Through the Shell

A question addressed in this work is how much the shell of carceplexes insulates its guests from the effects of molecules or surfaces external to the host's shell. A possible way of answering this question was to determine if incarcerated guests were subject to the phenomenon of aromatic solvent-induced shifts (ASIS) of proton resonances of solutes with respect to their shifts in non-aromatic solvents. The ^{1}H NMR spectrum of $(CH_3)_2NCOCH_3$ in $C_6D_5NO_2$ was compared to that of the same amide in $CDCl_3$. The ^{1}H NMR spectrum of this same amide incarcerated as in **38** was also taken in the two solvents. The δ of $(CH_3)_2CO$ served as an external reference in these experiments. A modest ASIS *upfield* shift for each of the three kinds of protons for $(CH_3)_2NCOCH_3$ dissolved directly in the two solvents was observed, which averaged δ 0.07 ± 0.03. Incarcerated $(CH_3)_2NCOCH_3$ dissolved in the two solvents provided a larger ASIS *downfield* shift for each of the three kinds of protons, which averaged δ 0.30 ± 0.06. Thus incarcerated $(CH_3)_2NCOCH_3$ proved to be more sensitive to the magnetic properties of the aromatic solvent than non-incarcerated amide.[10]

This effect is explained as follows. The shell of the carcerand is composed largely of eight benzenes, each substituted by three adjacent oxygens in the shell's equatorial region and by two alkyl groups and a hydrogen in the polar region. This arrangement makes the equatorial region rich in the electronegative atoms and the polar in relatively electropositive atoms. The guest cannot interchange its northern and southern ends, and its parts are subject to the polar gradients of the shell. The $C_6D_5NO_2$ as solvent aligns itself on the surface of the carceplex to complement its dipole with those of the shell by a π–π interaction, an effect transmitted to the π-system of the amide guest. Thus, the solvent 'communicates' with the guest through solvent-induced dipolar effects in the shell's π-system. Directly induced dipolar effects between amide dissolved in $C_6D_5NO_2$ are more randomized due to movements of solvent and solute with respect to one another, which are rapid on the ^{1}H NMR time scale. Thus, the carceplex provides a more organized interaction between amide and solvent leading to a larger ASIS effect, which is reversed in direction through mediation by the shell's dipolar effects.[10]

Another form of communication between the guest of a carceplex and external phases is evident in the fact that carceplex **38** containing $(CH_3)_2$N-$COCH_3$ has an R_f value of 0.5, while **39** containing $(CH_3)_2NCHO$ and **40** containing $(CH_3)_2SO$ have an R_f value of 0.6 on 0.5-mm silica gel plates with 3:1 chloroform–hexane as the mobile phase. We interpret the effect of guest on absorption properties of their carceplexes to guest-induced dipole–dipole interactions between the shell and the silica gel surface. The fact that the directions of the dipoles are more averaged by rapid movement with respect to their shell for $(CH_3)_2NCHO$ and $(CH_3)_2SO$ than with $(CH_3)_2NCOCH_3$ as guest accounts for the observed differences in chromatographic behavior.[10]

7.5.2 Possible New Type of Diastereoisomerism

Incarceration of the amides modifies their infrared (IR) spectra. The IR spectrum (KBr pellet) of **38** which contains $(CH_3)_2NCOCH_3$ exhibits two carbonyl stretch bands of equal intensity at 1648 and 1665 cm^{-1}, lying between the normal carbonyl band for this guest as a liquid at 1640 and as a gas at 1695 cm^{-1}. Similarly, the IR for $(CH_3)_2NCHO$ in **39** has a single carbonyl stretch band at 1700 cm^{-1}, lying between the normal carbonyl for this guest as a liquid at 1675 and as a gas at 1715 cm^{-1}. This observation is in harmony with our characterizing (in Section 7.4) the inner phase of carceplexes as being between those of a liquid, a gas, and the confining walls, depending on the relative amounts of occupied and empty space.[10]

The existence of two carbonyl bands of equal intensity for **38** and only one for **39**, we interpret as follows. The crystal structure of **38** reveals that the northern hemisphere is rotated about 25° relative to the southern hemisphere, which introduces a chiral helical element into the complex. Possibly the hybridization at N in $(CH_3)_2NCOCH_3$ is not sp^2 but $sp^{2.1}$ or $sp^{2.2}$, which makes N slightly pyramidal and therefore chiral. Combining the chiral element of the host with that of the guest provides two diastereomers, each with its own carbonyl stretching frequency. Although these diastereomers should interconvert very rapidly, they are likely to do so slowly on the IR time scale ($\sim 10^{-13}$ s). In **39** the shell offers little hindrance to rotations of $(CH_3)_2NCHO$, and therefore the carceplex probably exists as a mixture of many diastereomeric species with little differences in their environments, and therefore in their carbonyl stretching frequencies.[10]

7.6 Comparisons of Carceplexes, Spheraplexes, Cryptaplexes, Caviplexes, Zeolites, and Clathrates

Carceplexes are composed of host and guest components that cannot separate from one another without the breaking of covalent bonds. Their existence does not depend upon host–guest attractions nor on other than gross size complementarity. Rather their existence depends upon physical envelopment of guests during shell closures leading to carceplexes. Guests reported thus far include metal cations, halide anions, argon, Freon 114 ($ClCF_2CF_2Cl$), amides, nitriles, sulfoxides, alcohols, ethers, and ketones. Dissociation rate constants are in effect infinitely small.

Spheraplexes and cryptaplexes exist only because of strong host–guest attractions of a pole–dipole type. Spherands are hollow hosts that are preorganized for having a stereoelectronically complementary relationship with appropriate guests. Cryptands are not hollow and undergo conformational reorganizations during complexation. Both spheraplexes and cryptaplexes exhibit substantial activation free energies for dissociation of highly complementary complexing partners. Caviplexes exist because of weak dipole–dipole attractions between complementary complexing partners, the cavitand parts being rigid enough not to collapse or to fill their own cavities by conformational reorganizations. Activation free energies for dissociation of caviplexes in solution are very low. Zeolites are insoluble polymeric inorganic solids containing enforced channels into which a variety of complementary guests can *enter* and *depart* depending chiefly on size complementarity and temperature. The zeolites can be thermally emptied of their volatile guests and still maintain their structural integrity. Their study is limited to the solid state. Clathrates are crystalline solids of organic compounds containing interstitial solvent (guest) molecules. Channel clathrates are subject to guest exchange between clathrate solids and liquid or gas phases containing potentially exchangeable guests. Clathrate organization depends on intermolecular lattice forces, and when clathrates are dissolved that organization disappears.

Carceplex interiors are unique enough to justify their being called a new phase of matter for the following reasons. (1) These complexes are closed molecular cells of definable internal volumes that maintain their structural integrity as solids, dissolved solutes, or even as short-lived gases (subject to DC–MS). (2) Their guests cannot escape except by covalent bond breaking processes. (3) Carceplexes possess discrete molecular properties such as molecular weight, volume, molecular formula, and chromatographic retention time, and are subject to crystal structure determination. (4) Carceplex interiors are definable mixtures of free space and guest-occupied space, the relative amounts of which can be designed. (5) The only limitations on guests that can be incarcerated appear to be those of molecular shape and volume. The guest must possess a molecular volume less than that of the inner phase of the carcerand interior. (6) More than one guest and even more than one kind of guest can occupy the inner phase simultaneously, the only limitation being that of the available space. (7) The properties of the guest can grossly modify the properties of the carceplex. (8) The solubility properties of carceplexes are subject to control by pendant-group manipulation. (9) The inner phases of carceplexes are unique places where reactions might be carried out.

CHAPTER 8

Hemicarcerands and Constrictive Binding

> *'Presume that carcerands containing portals connecting the inner and outer phases can be made, and that means of driving guests in and out of the carcerands can be devised. To what uses might they be put? If high structural recognition is observed and the in and out rates are high, they might be attached to solid supports useful in separation science. If an indicator system is incorporated in the host that signals host occupation by a guest, hosts might find use in analytical science. If medicinals can serve as guests, timed release of drugs might be developed. If the structures of the eight pendant groups are manipulated so that the systems incorporate themselves into natural membranes, the hosts might serve as delivery systems of guests to these membranes. If $M(acac)_3$ salts of radioactive metals are incorporated into non-metabolizable hosts, complexes of these salts might be useful in radiation diagnostics or therapy. If catalysts such as 'floating porphyrin' can be made to serve as guest along with oxygen and selected hydrocarbons, a protected catalyst can be prepared for selected oxidations. What fun!'*

Liquids are inefficient in their ability to occupy space and are composed of 5–40% vacuum and 95–60% molecular volume. The empty space accounts for the mobile properties of molecules in the liquid state associated with fluidity. The empty space reflects the lack of complementarity in the packing of molecules, is distributed in many small volumes throughout the liquid, and accommodates the translational, rotational, and vibrational degrees of freedom of its molecules. Our inability to synthesize empty carcerands in the liquid phase is at least partially due to the entropically opposed consolidation of many small vacuums into the large one of molecular dimensions inside the carcerand. How might empty hollow molecules be prepared and their solution chemistry be examined?

One strategy to overcome the entropic problem is to carry out shell closures in the gas phase under high vacuum. This approach is not practical for compounds of molecular weights of 2000–3000 unless the starting materials and products are stable to extremely high temperatures. We adopted a second strategy, which involves making carceplexes containing portals large enough for prisoner molecules to escape their cages at high temperatures, but to remain incarcerated at temperatures which allow their isolation, purification, and characterization under ordinary working conditions. We refer to such hosts as hemicarcerands, and to their complexes as hemicarceplexes.

We have devised two approaches to synthesizing hemicarceplexes. One strategy is to omit one of the small building blocks used to construct a carcerand, providing one portal which connects the interior with the medium in which the complex is dissolved. The dimensions of the omitted building block determine those of the portal, which is likely to expand at high temperatures and contract at low ones. In a second approach, the four equatorial linking groups which connect two rigid hemispherical cavitands are long, and flexible enough to co-operatively adopt unstable conformations which generate temporary portals large enough for appropriately sized molecules to depart or enter host interiors at appropriate temperatures. At high temperatures in high boiling

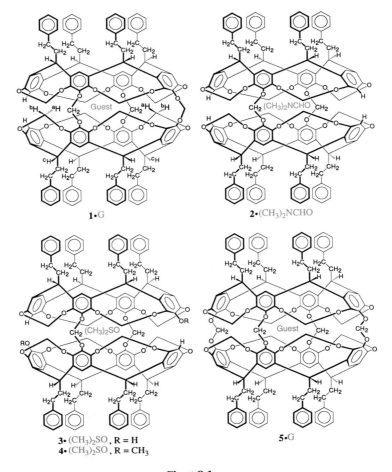

Chart 8.1

solvents whose molecules are too large to enter either type of portal, hemicarceplexes of either kind should liberate their more volatile guest prisoners to form free hemicarcerands by removal of the freed guest from the medium by evaporation. The empty hemicarcerand thus becomes available for study of its ability to incarcerate any number of possible guests.

8.1 Hemicarcerand Containing a Single Portal

The target hemicarceplexes **1·G** and **2·(CH$_3$)$_2$NCHO** are closely related to carceplexes **3·(CH$_3$)$_2$SO**, **4·(CH$_3$)$_2$SO**, and **5·G**, all of which are formulated in Chart 8.1.[1] Carceplex **3·(CH$_3$)$_2$SO** was an anticipated by-product in the synthesis of **1·(CH$_3$)$_2$SO**, and without characterization was converted to fully characterized **4·(CH$_3$)$_2$NCHO**, whereas parent system **5·G** was discussed in Section 7.1.

8.1.1 Synthesis

Octol **6** (see Section 7.2) was converted to a mixture of diols **7** and **8** and the desired triol **9** (9%) by the method outlined in Chart 8.2. A second synthesis

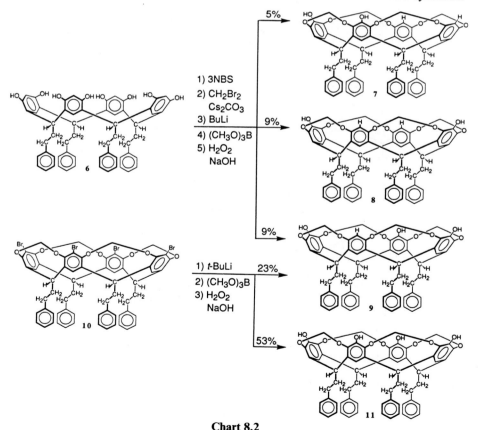

Chart 8.2

involved tetrabromide **10** as starting material which led to **9** (23%) and tetrol **11** used in the synthesis of **5·G**. The phenolic mixtures were easily separated chromatographically, and diol **8** and triol **9** were submitted to shell closures resembling those used to convert **11** to **5·G** (Section 7.1). Triol **9** and CH_2ClBr–Cs_2CO_3 in $(CH_3)_2NCHO$, $(CH_3)_2NCOCH_3$, or $(CH_3)_2SO$ led to **1**·$(CH_3)_2NCHO$ (20%), **1**·$(CH_3)_2NCOCH_3$ (42%), and **1**·$(CH_3)_2SO$ (51%). Carceplex **3**·$(CH_3)_2SO$ (5%) was isolated as a by-product in the synthesis of **1**·$(CH_3)_2SO$, but was converted by CH_3I–NaH–$(CH_2)_4O$ at reflux to **4**·$(CH_3)_2SO$ (>80%) for full characterization. Diol **8** when subjected to the same shell closure conditions in $(CH_3)_2NCHO$ gave **2**·$(CH_3)_2NCHO$ (22%). The high yields obtained in these shell closures coupled with the absence of empty hemicarcerand as product indicates that the reactions were templated by a mole of solvent which solvated the polar transition states leading to these complexes. Any hemicarcerand formed containing Cs^+ as guest would have remained at the top of the chromatograph column used to purify these hemicarceplexes.

When **1**·$(CH_3)_2NCHO$ or **1**·$(CH_3)_2NCOCH_3$ in mesitylene solution was heated to 165 °C, free **1** and the respective guests were produced, whereas **1**·$(CH_3)_2SO$ was stable to this treatment even after extended periods of time. Decomplexation of **1**·$(CH_3)_2SO$ required heating its solution in 1,2,4-$C_6H_3Cl_3$ at 214 °C for 48 hours.

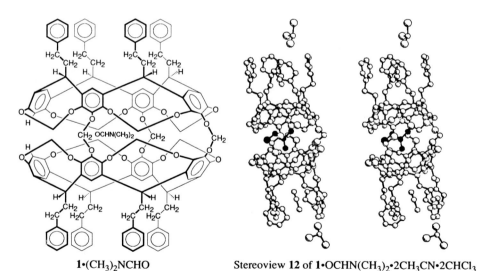

1·$(CH_3)_2NCHO$ Stereoview **12** of **1**·$OCHN(CH_3)_2$·$2CH_3CN$·$2CHCl_3$

8.1.2 Crystal Structure

A sample of **1**·$(CH_3)_2NCHO$ was recrystallized from CH_3CN–$CHCl_3$ to give **1**·$(CH_3)_2NCHO$·$2CH_3CN$·$2CHCl_3$, whose crystal structure (R = 0.168) is portrayed in stereoview **12**. The hemicarceplex has C_2 symmetry, the C_2 axis passing through the oxygen and nitrogen of the guest. The $(CH_3)_2NCHO$ guest is fully incarcerated, with its carbonyl group pointing toward the portal. The long axis of the cavity is 13.4 Å, and is the distance between the planes of the four aryl hydrogens of the northern and southern hemispheres. The short axis is 7.2 Å, and is defined as the distance between the two interhemispheric oxygens on the two bridges flanking the shell hole. Each CH_3CN is packed in a cavity composed of four $CH_2CH_2C_6H_5$ groups, with its nitrogen atom directed inward. Each packet of $(CH_2CH_2C_6H_5)_4 \cdot NCCH_3$ groups is capped with a $CHCl_3$ molecule, which is disordered about an inversion center. The five heavy atoms of $(CH_3)_2NCHO$ are coplanar, with one methyl occupying part of the northern hemisphere's cavity, and the other methyl group's axis being roughly coincident with the long hemicarceplex axis, its hydrogens occupying much of the hemisphere's cavity. As in the crystal structure of **5**·$(CH_3)_2NCOCH_3$, the whole northern hemisphere is rotated around the long axis of the whole complex by about 13° with respect to the southern hemisphere. This rotation diminishes the cavity and portal sizes, and allows the OCH_2O groups in the northern hemisphere to contact the oxygens of the OCH_2O groups in the southern hemisphere and *vice versa*. Without this rotation, CPK model examination shows that four sets of $O_2CH_2\cdots H_2CO_2$ groups abut one another, one set bifurcating the portal.[1]

8.1.3 Characterization

Satisfactory elemental analyses for each element present in **1**·$(CH_3)_2NCHO$, **1**·$(CH_3)_2NCOCH_3$, **1**·$(CH_3)_2SO$, and free **1** were obtained whose sums lay between 99.78 and 100.08%, and for each element except oxygen for **2**·$(CH_3)_2NCHO$ and **4**·$(CH_3)_2SO$. These same complexes gave positive-mode FAB mass spectra, all but one of which gave major signals for the hemicarceplexes, the hemicarcerand, or both. In the cases of **1**·$(CH_3)_2NCHO$ and **1**·$(CH_3)_2NCOCH_3$, **1** gave the parent ion. The spectrum of **1**·$(CH_3)_2NCOCH_3$ also gave an 18% signal for the carceplex. Complex **1**·$(CH_3)_2SO$ gave **1** as the parent ion, but a weak molecular ion for the carceplex. Free **1** gave its molecular ion as the parent ion. Attempts to obtain a FAB-MS of **2**·$(CH_3)_2NCHO$ failed, probably because of its insolubility in the matrix. Complex **3**·$(CH_3)_2SO$ gave a 70% intensity for **3** and one of 20% intensity for **3**·$(CH_3)_2SO$, whereas **4**·$(CH_3)_2SO$ provided a molecular ion for only free **4** (90%). The FTIR spectra in $CDCl_3$ solutions of **1**·$(CH_3)_2NCHO$ and **1**·$(CH_3)_2NCOCH_3$ gave carbonyl bands at 1681 and 1644 cm^{-1}, respectively, whereas that for **1**·$(CH_3)_2NCHO$ as a KBr pellet was found at 1682 cm^{-1}.

In the ^{1}H NMR spectrum in $CDCl_2CDCl_2$ at 22 °C of **2**·$(CH_3)_2NCHO$, the proton signals of both NCH_3 groups are sharp, are located upfield of 0 p.p.m., and integrate for three protons each, similar to those of **5**·$(CH_3)_2NCHO$.[2] In the spectrum of **3**·$(CH_3)_2SO$ under the same conditions several host signals were slightly broad, but all signals were sharp at 80 °C, and no decomplexation occurred at this temperature. The spectrum of **4**·$(CH_3)_2SO$ in $CDCl_2CDCl_2$ at 52 °C gave a six proton singlet at −1.26 p.p.m. for the guest's methyl groups

[1] D.J. Cram, M.E. Tanner, and C.B. Knobler, *J. Am. Chem. Soc.*, 1991, **113**, 7717.

and a singlet at 3.93 p.p.m. for the host's methoxy protons. Molecular model examination of $3 \cdot (CH_3)_2SO$ and $4 \cdot (CH_3)_2SO$ show them to be nearly as rigid as $5 \cdot (CH_3)_2SO$, with no portals available for escape by guest without covalent bonds being broken.

The R_f values on thin layer chromatographic plates of silica gel with 15% hexane: 85% $CHCl_3$ as the mobile phase, decreased in the following order: **1** > $1 \cdot (CH_2)_2SO$ > $1 \cdot (CH_3)_2COCH_3$ > $1 \cdot (CH_3)_2NCHO$. As with the carceplexes $5 \cdot G$ (Section 7.4), the guest communicates with the surface of the silica gel through the shell of the host probably by induced dipolar effects.

The ^1H NMR spectra of the four hemicarceplexes provides conclusive evidence for incarceration of guests within the shell. The guest signals were all shifted upfield by 2–4 p.p.m. due to the proximity of their hydrogens to the shielding faces of the eight aryl groups that line the cavity. In each complex, proton integration confirmed the one-to-one stoichiometry for the host and guest. The chemical shifts of the guests were all within 0.2 p.p.m. of those reported for the corresponding carceplexes $5 \cdot G$.[2] At least one guest proton signal was upfield of that of $(CH_3)_4Si$ at 0.00. The host's signal for the inward-facing intrahemispheric bridging groups was broadened due to partial restriction of guest rotations. The same proton signals for empty **1** were sharp at 22 °C. In the ^1H NMR spectrum of $1 \cdot (CH_3)_2NCOCH_3$ at 60 °C, the number of signals of the host's shell protons are doubled. This phenomenon is attributed to the inability of the $(CH_3)_2NCOCH_3$ molecules to rotate around the short axes of the host, which makes the northern and southern hemispheres slightly different, as was observed for $5 \cdot (CH_3)_2NCOCH_3$, even at temperatures as high as 175 °C.[2]

When the $CDCl_3$ spectral solutions for the three carceplexes of **1** were cooled to -53 °C, the ^1H NMR spectra of the host showed further loss of symmetry, presumably due to slowed motions of the guests relative to the host. The guest signal of $1 \cdot (CH_3)_2SO$ was unaffected by cooling unlike that of $5 \cdot (CH_3)_2SO$. Cooling of the $CDCl_3$ solution of $1 \cdot (CH_3)_2NCHO$ resulted in the splitting of the guest's three signals into two sets of unequal proportions to provide a 5:1 ratio of isomers at -53 °C. We interpret this result as reflecting two different guest orientations of minimum energies that possess a 13 kcal mol^{-1} energy barrier at -13 °C for interconversion. A similar duality of structure was observed for $1 \cdot (CH_3)_2NCOCH_3$ at -53 °C, whose peak intensities indicated the two species to be present in a 25:1 ratio. Molecular model examination of $1 \cdot (CH_3)_2NCHO$ and $1 \cdot (CH_3)_2NCOCH_3$ suggest that the more stable structures resemble that observed in crystal structure **12** (Section 8.1.2) in which the carbonyl oxygen points toward the hole, one NCH_3 is in the northern and the other NCH_3 in the southern hemispherical cavity.

8.1.4 Decomplexation at High Temperatures and Complexation at Ambient Temperatures

Decomplexation kinetics of $1 \cdot (CH_3)_2NCHO$, $1 \cdot (CH_3)_2NCOCH_3$, and $1 \cdot (CH_3)_2SO$ were followed by ^1H NMR spectral changes in 1,2,4-$Cl_3C_6H_3$ and

[2] J.C. Sherman, C.B. Knobler, and D.J. Cram, *J. Am. Chem. Soc.*, 1991, **113**, 2167.

exhibited good first order behavior. At 140 °C, the half-lives $t_{1/2}$ of the complexes were: $1 \cdot (CH_3)_2NCHO$, 14 hours; $1 \cdot (CH_3)_2NCOCH_3$, 34 hours; $1 \cdot (CH_3)_2SO$, too long to measure. At 195 °C, $1 \cdot (CH_3)_2SO$ gave $t_{1/2} = 24$ hours. An examination of 1 indicates the portal to be tablet shaped, somewhat complementary to the two planar amide guests. Neither a model of $(CH_3)_2NCHO$ nor $(CH_3)_2NCOCH_3$ could be either introduced or forced out of the model of 1 without breaking bonds, although the smaller size of the former nearly allowed its model to be forced inside. The tetrahedral character of $(CH_3)_2SO$ made its shape less complementary to that of the portal, which accounts for the much higher temperature required to decomplex $1 \cdot (CH_3)_2SO$.

Molecular models indicate that $CDCl_3$ is much too large to be incarcerated in 1, but that models of ubiquitous small molecules such as N_2, O_2, CO_2, and H_2O can readily enter and depart 1. Dissolution of 1 in $CDCl_3$ is the equivalent of introducing holes in the solvent that might be filled by such small dissolved molecules. Solutions of 1 in untreated $CDCl_3$ gave an ^1H NMR spectrum that exhibits four separate sets of host signals that arise from $1 \cdot N_2$, $1 \cdot O_2$, $1 \cdot H_2O$, and 1, which exchange guests slowly on the ^1H NMR, but rapidly on the human time scale at ambient temperature. The host's proton signals most sensitive to the presence of guests are those of the eight inward-facing intrahemispheric OCH_2O bridges labeled ^aH in the formula of $1 \cdot G$, which in free 1 occur at 3.93 and 4.09 p.p.m. in $CDCl_3$ at 22 °C. These protons are moved upfield by 2.2 p.p.m. from their diastereotopically related counterparts (^bH) which face away from the cavity. Usually the presence of a guest in 1 deshields the ^aH protons and moves them downfield relative to those of free 1. The methine (^cH) protons are insensitive to diamagnetic guests, but are broadened and moved downfield from their positions in the ^1H NMR spectrum of 1 (4.80 and 4.90 p.p.m.) to that of $1 \cdot O_2$ at 5.24 and 5.38 p.p.m. By ^1H NMR measurements the association constants at 22 °C (K_a) for 1 binding N_2 and O_2 were found to be ~ 180 M^{-1} ($\Delta G^0 \sim -3$ kcal mol^{-1}) and ~ 44 M^{-1} ($\Delta G^0 = -2.2$ kcal mol^{-1}) respectively. The peak broadening and chemical shift changes that occur when O_2 enters free hemicarcerand have been found to be a good diagnostic tool for the existence of free hemicarcerands.

The diameter of a xenon atom in a clathrate crystal is 4.43 Å.[3] The portal of a CPK model of 1 could be forced open enough to accept a 4.4 Å sphere diameter model. When a solution of 1 in $CDCl_3$ was saturated with Xe gas, the Xe formed $1 \cdot Xe$ at 22 °C, the half-life for the reaction $Xe + 1 \rightarrow 1 \cdot Xe$ being about 1.5 hours. The chemical shift of ^{129}Xe in its ^{129}Xe NMR spectrum of $1 \cdot Xe$ was -101 p.p.m. compared to free Xe in $CDCl_3$. Addition of CH_3CN to the $CDCl_3$ solution precipitated the 1:1 complex. The K_a for 1 binding Xe was estimated to be ~ 200 M^{-1} in $CDCl_3$ at 22 °C.[1]

In concert with expectations based on molecular model examination, 1 was found capable of incarcerating either one or two molecules of CH_3CN. When dissolved in a 30:1 (v:v) $CHCl_3:CH_3CN$ mixture and equilibrated for 48 hours at 25 °C, 1 gave $1 \cdot CH_3CN$ which was precipitated with pentane to give a complex whose elemental analysis and ^1H NMR spectrum showed it to be

[3] F. Lee, E. Gabe, J.S. Tse, and J.A. Ripmeester, *J. Am. Chem. Soc.*, 1988, **110**, 6014.

one-to-one. This hemicarceplex had a decomplexation half-life of ~ 13 hours in CD_2Cl_2 at 22 °C. When a 1 mM solution of **1** in a 5:2, $CHCl_3$:CH_3CN mixture was allowed to evaporate over a 3-day period, white needles separated whose composition was found by ^1H NMR to be 50% **1·CH_3CN** and 50% **1·$2CH_3CN$**. The ^aH protons of the host of **1·CH_3CN** were found at 4.19 and 4.30 p.p.m., whereas those of **1·$2CH_3CN$** were at 4.71 and 4.75 p.p.m. The guest's methyl protons in **1·CH_3CN** occurred at -2.42 p.p.m., whereas in **1·$2CH_3CN$** they were found as a singlet at -2.15 p.p.m. The relative positions and multiplicity of these signals suggest that the two CH_3CN molecules are aligned on a common axis with their methyl groups facing one another, as in $N{\equiv}CCH_3{\cdots}CH_3C{\equiv}N$. This arrangement locates the CH_3 groups further from the shielding aryl groups of the host than alternative arrangements, and is consistent with the general egg shape of the cavity. In $CDCl_3$ solution at 22 °C, **1·$2CH_3CN$** converts to **1·CH_3CN** with a half-life of 26 min. We attribute the relatively high stability of **1·CH_3CN** compared to **1·$2CH_3CN$** as reflecting the high entropic cost of collecting three molecules that compose **1·$2CH_3CN$**, coupled with the loss of much of the translational degrees of freedom of the guests in passing from **1·CH_3CN** to **1·$2CH_3CN$**.

Additional examples of hemicarceplexes that form at ambient temperature are **1·CH_2Cl_2**, **1·CH_2Br_2**, and **1·CS_2**, the first two by dissolution in guest as solvent, the last by dissolution in 6% CS_2 in $CHCl_3$ (v:v). The complexes were precipitated by addition of pentane, and each complex was shown to be one-to-one by elemental analyses. Both **1·CH_2Br_2** and **1·CS_2** had half-lives for decomplexation greater than 400 hours when dissolved in CD_2Cl_2.

8.1.5 Complexation at Elevated Temperatures

When heated in the presence of massive excesses of appropriately sized guests G, hemicarcerand **1** gave **1·G** in a complexation driven by mass law. Complexes were precipitated with pentane and were stable at ambient temperature to recrystallization, chromatography, and ordinary laboratory manipulation, and were shown by elemental analyses and ^1H NMR spectra to be one-to-one. By these means, complexes of **1** with pyridine, benzene, THF, Et_2NH, $CH_3(CH_2)_3NH_2$, and α-pyrone were prepared and characterized.

Table 8.1 lists the chemical shifts (δ values, 500 MHz in $CDCl_3$ at 22 °C) in the ^1H NMR spectra of free guests and guests incarcerated by **1**, as well as the Δδ difference between them. Without exception, incarceration results in movement of the guest's proton signals upfield by amounts for $-\Delta\delta$ values that vary between a high of 4.54 p.p.m. for ^cH of pyridine to a low of 1.44 for ^aH of pyridine. These large Δδ upfield shifts reflect the fact that most of the interior of **1** is lined with the shielding faces of eight benzene rings. Models show this shielding effect could be greatest in the northern and southern temperate zones and least in the equatorial region of the hemicarcerand.

The magnitudes of the $-\Delta\delta$ values in Table 8.1 for the various protons provide structural information as to the average positioning of the various guest parts inside the cavity of **1**. For example, ^cH of pyridine obviously is

Table 8.1 Chemical shifts (p.p.m.) of guest signals in the 500 MHz ^{1}H NMR spectra of G and of **1·G** in CDCl$_3$ at 22 °C

Guest	Proton	δ of G in **1·G**	δ of free G	−Δδ
^{a}H$_3$C–N(–^{b}H$_3$C)–C(=O)–^{c}H	^{a}H	−0.43	2.86	3.99
	^{b}H	0.22	2.94	2.73
	^{c}H	4.04	7.99	3.95
^{a}H$_3$C–N(–^{b}H$_3$C)–C(=O)–CH$_3^c$	^{a}H	−1.43	2.94	4.37
	^{b}H	0.97	3.02	2.05
	^{c}H	−2.24	2.09	4.33
(CH$_3$)$_2$SO		−1.20	2.61	3.81
H$_2$O		−1.87	1.55	3.42
CH$_3$CN		−2.42	2.00	4.42
2 CH$_3$CN		−2.15	2.00	4.15
CH$_2$Cl$_2$		2.64	5.30	2.66
CH$_2$Br$_2$		≈2.5	4.95	≈2.5
^{a}H$_2$, ^{b}H$_2$ (tetrahydrofuran)	^{a}H	−2.2	1.85	3.07
	^{b}H	−0.23	3.75	3.98
pyridine (^{a}H, ^{b}H, ^{c}H)	^{a}H	6.24	7.68	1.44
	^{b}H	2.84	7.30	4.46
	^{c}H	4.08	8.62	4.54
C$_6$H$_6$		3.87	7.36	3.49
CH$_3$–N(–CH$_2$–CH$_2^b$)(–CH$_3^a$)–H^c	^{a}H	−2.94	1.08	4.02
	^{b}H	−0.48	2.65	3.13
	^{c}H	−1.48		
CH$_3^a$–^{b}H$_2$C–CH$_2^c$–^{d}H$_2$C–NH$_2^e$	^{a}H	−3.39	0.90	4.29
	^{b}H	−1.24	1.33	2.57
	^{c}H	−0.94	1.41	2.35
	^{d}H	−1.13	2.27	3.80
	^{e}H	−3.18		

located in the northern or southern temperate zones of **1** while ^{a}H of pyridine is close to the equator. The −Δδ values of the protons of CH$_3$(CH$_2$)$_3$NH$_2$ located toward the two ends of the molecule are greater than those for protons located in the middle. Molecular model examination of **1**·CH$_3$(CH$_2$)$_3$NH$_2$ shows that the long axis of the guest and the long axis of the host must be close to being coincident, no other arrangements being available. The cavity in this complex is more completely and uniformly filled than the others.

8.2 Hemicarcerand Containing Four Potential Portals

In this section we describe the synthesis, crystal structure, and binding properties of hemicarcerand system **13**, containing the same two polar caps and eight solubilizing $C_6H_5CH_2CH_2$ appendages as **1–5**, but with four semi-mobile equatorial spacer groups, 1,2-$OCH_2C_6H_4CH_2O$, connecting the two polar caps. Unlike **1**, CPK models of **13** indicate that when the polar caps are aligned over one another as in the drawing of **13**, the portals are open and large enough to allow models of molecules as large as *p*-xylene to enter and exit the inner phase of the hemicarcerand. In the model of **13**, if one polar cap is rotated with respect to the other around the polar axis, the portals close, the inner volume decreases, and many new atom-to-atom close contacts are formed.

We undertook the study of **13** to answer these questions. (1) To what extent is **13** capable of incarcerating guest molecules as a function of temperature, guest size, and shape? (2) Is there high molecular recognition shown in guest incarceration? (3) Do the portals close by rotation of the polar caps with respect to one another? (4) What is the mechanism of guest substitution in this four portal system? (5) To what extent does incarceration restrict the molecular rotations of incarcerated guests? (6) What are the relationships between the structures of guests and free energies of complexation? (7) How does the partition of free energies between enthalpic and entropic parameters relate to guest structure? (8) How do the activation free energies for entrance and egress of guests relate to guest structures? This type of host is new, and these kinds of questions have never been addressed before.[4]

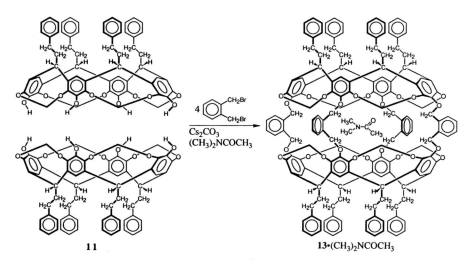

8.2.1 Synthesis

Two moles of tetrol **11** were shell closed by their reactions with four moles of 1,2-$(BrCH_2)_2C_6H_4$ in $(CH_3)_2NCOCH_3$–Cs_2CO_3 to give **13**·$(CH_3)_2NCOCH_3$

[4] D.J. Cram, M.T. Blanda, K. Paek, and C.B. Knobler, *J. Am. Chem. Soc.*, 1992, **114**, 7765.

(23% yield), which was easily purified by chromatography and crystallization. This hemicarceplex is stable to ordinary laboratory manipulations at temperatures below 70 °C, and indefinitely at ambient temperature and below. The complex was characterized by elemental analyses, ^{1}H NMR spectrum, and FAB-MS. The free host **13** was obtained by heating **13**·$(CH_3)_2NCOCH_3$ at 160 °C for 24 hours in 1,4-$(CH_3)_2CHC_6H_4CH(CH_3)_2$, a solvent much too large to enter the interior of **13**. Free host **13** was fully characterized. Its FAB-MS gave a 10% peak (**13**, 100%) corresponding to **13**·$4H_2O$. When O_2 or N_2 was bubbled through a solution of **13** in $CDCl_2CDCl_2$, these gases rapidly gave **13**·O_2 and **13**·N_2, respectively, mixed with **13**. The ^{1}H NMR spectrum of **13** is very sensitive to the presence of these guests in its interior. In particular, the δ value of the eight inward-facing OCH_2O protons moved upfield by 0.12 p.p.m. in the solution containing O_2 and by 0.27 p.p.m. in the solution containing N_2. The coalescence temperatures of **13**·O_2 and **13**·N_2 appear to be below ambient temperatures. The two complexes are rapidly interconverted by simply passing a stream of the new gas into a solution of the one to be displaced. Thus the entrance and egress of O_2 and N_2 to and from the inner phase of **13** at 25 °C is rapid on the human, but slow on the ^{1}H NMR time scale. As noted with **1**, this phenomenon provides a useful diagnostic test for empty host.[4]

8.2.2 Crystal Structure

Crystallization of **13**·$(CH_3)_2NCOCH_3$ from $(CH_3)_2NCOCH_3$–$CHCl_3$–$O(CH_2CH_2)_2O$ gave crystals which contained as many as eighteen disordered solvent molecules per host, one incarcerated, the others being present as solvates. This disorder inhibited refinement past $R = 0.30$, but the structure was good enough to provide interesting conclusions. Chart 8.3 contains a stereodrawing **14** of the carcerand part of the crystal structure of **13**·(solvate)$_n$, along with **15**, a computer-drawn structure of **13** minus its eight $C_6H_5CH_2CH_2$ groups. This drawing approximates a CPK model based on the crystal structure coordinates. This structure illustrates how the polar twist closes the portals and increases the number of close contacts between atoms of the rims of the polar caps and those of the equatorial spacer groups.

The four 1,2-$OCH_2C_6H_4CH_2O$ units located in the equatorial region of **14** possess a conformation in which the northern polar cap is rotated about 21° with respect to the southern polar cap about a C_2 polar molecular axis. Each set of the four $ArOCH_2Ar$ oxygens attached to each polar cap forms a near-square plane ($\pm$ 0.04 Å, least squares). The two squares are ~2.4 Å apart and are close to being parallel to one another, as shown by the fact that the angle formed between the normals to these near planes is only 0.6°. The distance between the two oxygen atoms found in each xylyloxy bridge varies between 2.76 and 2.95 Å. The unshared electron pairs of all eight oxygens atoms face the inner phase of the host, and their attached 1,2-xylyl groups face outward. The unshared electron pairs of all of the other oxygens (those of the eight OCH_2O groups) face outward in an enforced arrangement, thus producing eight sets of out-in-out partially compensating dipoles associated with the eight sets of

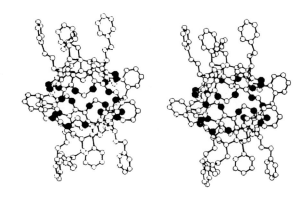

Stereoview **14** of **13**·solvate minus guest

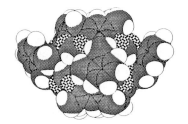

15, energy-minimized structure of **13** minus $8C_6H_5CH_2CH_2$ groups

16

Chart 8.3

three-proximate oxygens. The aryls of the xylyl groups are arranged in a way that provides the host with a C_2 rather than a C_4 polar axis. Their arrangement, the two squares defined by the eight $ArOCH_2Ar$ groups, and the twist of the two polar caps are visualized in partial structure **16** of Chart 8.3.

8.2.3 Guest Variation in Hemicarceplexes

New complexes of **13** were formed either directly from **13**·$(CH_3)_2NCOCH_3$ or from preformed empty **13**. In the former method, **13**·$(CH_3)_2NCOCH_3$ was heated at 80–200 °C in the new guest as solvent or co-solvent, the guest substitution reaction being driven to completion by mass action. Usually aromatic or halogenated solvents such as $C_6H_5CH_3$ or $CHCl_2CHCl_2$ dissolved **13**·$(CH_3)_2NCOCH_3$, whereas more polar solvents such as $(CH_2)_4O$, $CH_3(CH_2)_3OH$, $CH_3COCH_2CH_3$, $CH_3CH_2O_2CCH_3$, or CH_3CN did not. Of the five equimolar solvent mixtures of $CH_3C_6H_5$ with these five more polar solvents examined, only the first led to a mixture of complexes, 85:15, **13**·$(CH_2)_4O$–**13**·$C_6H_5CH_3$. In the other four examples, only complexes of the smaller, more polar guests were formed under conditions of kinetic control of products.

We probed the geometric constraints of the portal through which guests had to pass during entry or egress from the inner phase of **13** by heating either free **13** or **13·(CH$_3$)$_2$NCOCH$_3$** in solvent guests of increasing size and differing shapes. The largest molecules that formed detectable complexes were C$_6$H$_5$CH$_2$CH$_3$, 1,4-(CH$_3$)$_2$C$_6$H$_4$, and CHCl$_2$CHCl$_2$. No complexes were detected with the following potential guests: 1,2-(CH$_3$)$_2$C$_6$H$_4$, 1,3-(CH$_3$)$_2$C$_6$H$_4$, (CH$_3$)$_3$CC$_6$H$_5$, 1,4-CH$_3$C$_6$H$_4$CH$_2$CH$_3$, 1,4-CH$_3$CH$_2$C$_6$H$_4$CH$_2$CH$_3$, 1,4-CH$_3$C$_6$H$_4$CH(CH$_3$)$_2$, 1,3,5-(CH$_3$)$_3$C$_6$H$_3$, or 1,4-(CH$_3$)$_2$CHC$_6$H$_4$CH(CH$_3$)$_2$. A sharp discontinuity depending on the shape of the guest is evident in the fact that, whereas p-xylene and ethylbenzene enter the inner phase of **13**, o-xylene and m-xylene do not. These results conform to expectations based on CPK model examinations. The complexes of **13** that are stable at ambient temperatures were fully characterized.

8.2.4 Rotations of Guests Relative to Host

The rates of decomplexation of all guests except O$_2$ and N$_2$ were on the order of 10^{-6} s^{-1} at 25 °C. Thus ^{1}H NMR spectra proved valuable for determining the presence of guest in host, particularly since spectra of both species changed upon complexation. The signals of the inward-turned protons of the host's OCH$_2$O protons varied from δ 3.82 for **13·CH$_3$CN** to δ 4.08 for empty **13** to δ 4.16 for **13·CH$_3$C$_6$H$_5$** (CDCl$_3$ as solvent, 25 °C, 500 MHz spectra). Two different kinds of protons for the northern and southern hemispheres were observed at 25 °C only when the guest was CH$_3$CH$_2$O$_2$CCH$_3$ or C$_6$H$_5$CH$_2$CH$_3$. The two kinds of signals indicate that the long axes of these two guests are aligned with the long north–south axis of the host and that these guests do not rotate end-to-end rapidly on the ^{1}H NMR time scale. The presence of only one kind of host signal in the other non-like-ended complexes suggests that these guests probably rotate end-to-end rapidly under these conditions.

Three of the complexes were subjected to temperature-dependent spectral studies. At −80 °C, the spectrum of **13·CH$_3$CN** showed little change, indicating that even at this temperature the guest rotations around all axes were rapid on the spectral time scale. The spectrum of **13·C$_6$H$_5$CH$_3$** changed upon cooling of the CDCl$_3$ solution to provide a coalescence temperature T_c of −9 °C (Δv = 60 Hz), giving $\Delta G^{\ddagger}$ = 13 kcal mol^{-1} for end-to-end rotation of the guest relative to the host. The spectral signals of **13·CH$_3$CH$_2$O$_2$CCH$_3$** coalesced at 100 °C (Δv = 60 Hz) to provide $\Delta G^{\ddagger}$ = 18 kcal mol^{-1} for a similar rotation. Molecular model examinations of **13·C$_6$H$_5$CH$_3$** and **13·CH$_3$CH$_2$O$_2$CCH$_3$** indicate these rotations must occur when the host is in an untwisted conformation in which the guests are much less closely held. Probably part of the activation free energy for end-to-end rotations of these guests must involve overcoming the attractions of the bridging atoms for one another in their twisted conformation. Smaller and more compact guests such as CH$_3$CN or CHCl$_3$ can probably rotate without the host untwisting, whereas 1,4-CH$_3$C$_6$H$_4$CH$_3$ and CHCl$_2$CHCl$_2$ are probably incapable of end-to-end rotations at ordinary working temperatures.

8.2.5 Proton Magnetic Resonance Spectra of Guests in Hemicarceplex

The proton signals of guests incarcerated in **13** are shifted upfield by 0 to 4.4 p.p.m. relative to those of free guests ($\Delta\delta$ values) in $CDCl_3$ at 25 °C (500 MHz spectra). This upfield shift from their normal positions is due to guest incarceration in a magnetically anisotropic environment composed largely of the faces of benzene rings in the polar caps and OCH_2O and OCH_2Ar groups in the respective temperate and tropic zones of their globe-shaped containers. The time-averaged positions of the various guest protons suggested by CPK model examination are roughly borne out by the magnitudes of the $\Delta\delta$ values coupled with the expectation that in terms of shielding power, polar caps > temperate zone > tropical zone. The longest and most tightly held rigid guest is 1,4-$CH_3C_6H_4CH_3$, whose CH_3 groups certainly occupy the polar caps ($\Delta\delta = 4.17$), and whose ArH protons occupy the tropic zone ($\Delta\delta = 1.02$). The terminal protons of $CH_3CH_2C_6H_5$ (the CH_3 and p-H) are pressed into the polar caps, as shown by their respective $\Delta\delta$ values of 4.09 and 4.10, while the o-H resides in the tropic zone ($\Delta\delta = 1.25$), and the other protons in the temperate zone ($\Delta\delta = 2.73$ for CH_2 and 4.10 for m-H).[4]

As expected from the general parallel alignment of the long axes of host and guest in the complexes, the terminal protons of the aliphatic guests have higher $\Delta\delta$ values than the internal protons. For example the CH_3 and OH groups of the guest in **13**·$CH_3CH_2CH_2CH_2OH$ provide $\Delta\delta$ values of 3.85 and 4.39, respectively, whereas the other three kinds of protons give values between 1.75 and 2.65 p.p.m. Similar patterns are observed in the ^1H NMR spectra of the other carceplexes.[4]

8.2.6 Mechanism of Guest Substitution of Hemicarceplexes

When **13**·$(CH_3)_2NCOCH_3$ was heated at 100 °C with a potential new deuterated guest as solvent, the ^1H NMR spectrum of the host over time displayed the original spectrum plus the growth of the spectra of transient empty **13** and of final product, the new hemicarceplex containing a molecule of solvent. Table 8.2 lists the first-order-rate constants k_1, s^{-1}, for disappearance of **13**·$(CH_3)_2$

Table 8.2 *Pseudo first-order rate constants for disappearance of **13**·$(CH_3)_2N$-$COCH_3$ in deuterated solvents at 100 °C*

Solvent	Maximum % of free **13**	k_1 (sec^{-1} × 10^4)
C_6D_5Br	3	93
C_6D_5Cl	4	83
1,2-$(CD_3)_2C_6D_4$	100	8.5
1,4-$CD_3)_2C_6D_4$	8	3.5
$C_6D_5CD_3$	<1	2.0
$CDCl_2CDCl_2$	45	0.62

NCOCH$_3$ as well as the maximum percentage of free **13**, that temporarily accumulated. With 1,2-(CD$_3$)$_2$C$_6$D$_4$, only free carcerand **13** was produced, since molecules of this solvent cannot pass through the portal of **13**. The maximum of **13** observed with the other solvents was 45% with CDCl$_2$CDCl$_2$.

These results indicate that substitution occurs by a two-step mechanism, the first being decomplexation to give empty hemicarcerand, the second involving its complexation with a solvent molecule. The low concentration of initial guest produced compared to that of the solvent, a factor of ~4000, should render the dissociation effectively irreversible. Equation 8.1 relates the rate constants to the mechanism.

$$\mathbf{13}\cdot(CH_3)_2NCOCH_3 \xrightarrow{k_1} \mathbf{13} + (CH_3)_2NCOCH_3 \xrightarrow[\text{solvent}]{k_2} \mathbf{13}\cdot\text{solvent} \quad (8.1)$$

The rate constants k_1 for decomplexation vary by the large factor of 150, and decrease with changes in solvent in the order listed in Table 8.2. This substantial difference in rate must be associated with the ability of the solvents to differentially solvate the transition states for decomplexation. Molecular model examination of possible transition states for decomplexation indicates the complex must be in its extended, untwisted conformation in which the cross sections of three possible portals are adjusted so that the cross section of the portal employed for passage of the guest can be maximized. The different shapes and polarizabilities of the solvents must control the extent to which they can solvate the ground and the transition states involved in these decomplexations, both for host and guest parts.

8.2.7 Dependence of Decomplexation Rates on Guest Structures

The first-order-rate constants k_1 for decomplexation of ten different hemicarceplexes of host **13** dissolved in CDCl$_2$CDCl$_2$ were determined at 100 °C. The total spread in passing from the highest rate constant observed, $t_{1/2}$ 38 min. for CH$_3$CN, to the lowest observed, $t_{1/2}$ 409 min. for CH$_3$CH$_2$O$_2$CCH$_3$, was only a factor of 10. The guests arranged in decreasing order of their rate constants for decomplexation are: CH$_3$CN > (CH$_3$)$_2$NCHO > (CH$_3$)$_2$NCOCH$_3$ > (CH$_2$)$_4$O > C$_6$H$_5$CH$_3$ > CH$_3$(CH$_2$)$_3$OH > CH$_3$CH$_2$COCH$_3$ > CHCl$_2$CHCl$_2$ > CH$_3$CH$_2$O$_2$CCH$_3$. This order reflects such variables as guest size, shape, conformational flexibility, as well as the electronic character of the atoms involved.

Interestingly, the solvent effects on decomplexation rates are a power of 10 more important than the guest structural effects, at least over the range of guests examined. We interpret this finding as reflecting the fact that host reorganization from a twisted to an untwisted conformation involves the loss of many close contacts between atoms on the bowl rims and those of the four bridging o-xylyl units. Essentially the same activation energy cost has to be paid before guests as small as CH$_3$CN and as large as CHCl$_2$CHCl$_2$ can escape incarceration. The high sensitivity of the decomplexation rate to the solvent structure indicates that in the transition state for complexation–decomplex-

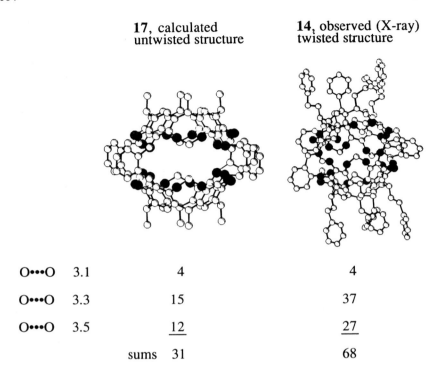

		17, calculated untwisted structure	14, observed (X-ray) twisted structure
O•••O	3.1	4	4
O•••O	3.3	15	37
O•••O	3.5	<u>12</u>	<u>27</u>
	sums	31	68

Δ = 37 close contacts associated with wrapping

Chart 8.4

ation, host–host interhemispheric and host–guest attractions are being exchanged for host–solvent and guest–solvent attractions. We conclude the transition-state structure more closely resembles the dissociated than the associated state of the hemicarceplex.

Chart 8.4 provides views of **13** (minus the eight $C_6H_5CH_2CH_2$ feet) in its calculated untwisted form, drawing **17**, and of the observed crystal structure **14**, which is twisted. Listed below the structures are the kinds and numbers of close contacts associated with each. The close contacts of the twisted structure **14** exceed those of the untwisted structure **17** by 37, of which 22 are O----C and 15 are C----C types. In other work, we found a rough correlation between the number of intermolecular close contacts of a similar variety generated in complexes in $CDCl_3$ formed by four-fold 'lock and key' dimers,[5] whose binding free energies in $CDCl_3$ in some cases were > 9 kcal mol^{-1} (see Section 6.4).

8.2.8 Constrictive and Intrinsic Binding

Through the use of 500 MHz ^1H NMR spectral changes, the rate constants for

---

[5] D.J. Cram, H.J. Choi, J.A. Bryant, and C.B. Knobler, *J. Am. Chem. Soc.*, 1992, **114**, 7748.

Table 8.3 Thermodynamic and kinetic parameters for complexation of **13** and decomplexation of **13·G** in $1,2\text{-}(CD_3)_2C_6D_4$ at 100 °C

Guest	$(CH_3)_2NCOCH_3$	$CH_3CH_2O_2CCH_3$	$CH_3CH_2COCH_3$	$C_6H_5CH_3$
ΔG^0 (kcal mol^{-1})	−3.7	−3.8	−5.3	−3.4
ΔH (kcal mol^{-1})	−1.5	−3.1	−2.5	+2.2
ΔS (cal K^{-1} mol^{-1})	+6	+2	+7.5	+15
complexation				
$\Delta G^\ddagger$ (kcal mol^{-1})	23.5	25.9	24.8	25.8
$\Delta H^\ddagger$ (kcal mol^{-1})	16.3	19.2	17.7	20.7
$\Delta S^\ddagger$ (cal K^{-1} mol^{-1})	−19.4	−17.9	−19.0	−13.6
decomplexation				
$\Delta G^\ddagger$ (kcal mol^{-1})	27.2	29.8	30.1	29.2
$\Delta H^\ddagger$ (kcal mol^{-1})	20.5	22.2	20.8	14.3
$\Delta S^\ddagger$ (cal K^{-1} mol^{-1})	−18.3	−23.1	−25.3	−40.0

complexation of **13** with G and decomplexation of **13·G** were determined at three or four temperatures in the 50–130° C range in $1,2\text{-}(CD_3)_2C_6D_4$ as solvent. The guests G involved were $(CH_3)_2NCOCH_3$, $CH_3CH_2O_2CCH_3$, $CH_3CH_2COCH_3$, and $C_6H_5CH_3$. Table 8.3 records the activation free energy, enthalpy, and entropy values for association and dissociation calculated from the rate constants at 100° C.

We define *intrinsic binding* as the free energy ΔG^0 of complexation of a guest by a host, and *constrictive binding* as the free energy of activation $\Delta G^\ddagger_{assoc}$ for that complexation, which is equal to $\Delta G^\ddagger_{dissoc} - (-\Delta G^0)$. In other words constrictive binding refers to the free energy which must be provided to reach the transition state for complexation–decomplexation from the complexed state minus the intrinsic binding free energy of the complexing partners. Since $\Delta G_{constrictive}$ is part of an activation energy term, it is always positive.

Interestingly, the constrictive binding free energies $\Delta G^\ddagger_{assoc}$ that must be overcome for **13·G** to dissociate vary little with guest structure changes for $(CH_3)_2NCOCH_3$, 23.5 ; $CH_3CH_2O_2CCH_3$, 25.9; $CH_3CH_2COCH_3$, 24.8; and $C_6H_5CH_3$, 25.8 kcal mol^{-1}. We conclude that most of the constrictive binding is associated with the host reorganizing from a highly twisted structure to an untwisted structure in which the cross section of one portal is maximized and many close contacts are lost. Thus the energetics of constrictive binding are much more controlled by the structure of the common host **13** than by the structures of the four different guests. As guests become progressively larger, possibly with $CHCl_2CHCl_2$, this generalization probably will not apply.

The ΔG^0 values for intrinsic binding at 100 °C vary with guest structure as follows: $C_6H_5CH_3$, −3.4; $(CH_3)_2NCOCH_3$, −3.7; $CH_3CH_2O_2CCH_3$, −3.9; $CH_3CH_2COCH_3$, −5.3 kcal mol^{-1}. The intrinsic binding free energies correlate *inversely* with guest size in terms of the number of heavy atoms in the molecule. The last three guests contain C=O dipoles whose positive carbon atoms are electronically complementary to the sixteen inward-turned unshared electron pairs of the host's eight ArOCH$_2$ oxygens. Also these three guests all make enthalpic contributions to the binding free energies that are negative ranging from ΔH of -2.1 to -3.1 kcal mol^{-1}. Their corresponding entropic contributions to the intrinsic binding free energies are also negative, ranging from $-T\Delta S$ of -0.75 to -2.8 kcal mol^{-1}. Thus, the intrinsic binding free energies for the three guests are *both* enthalpically and entropically driven.

In contrast, the ΔG^0 value for **13** binding C$_6$H$_5$CH$_3$ is -3.4 kcal mol^{-1}, being opposed by $\Delta H = +2.2$ and driven by $-T\Delta S = -5.6$ kcal mol^{-1}. We attribute the unfavorable enthalpy of binding of C$_6$H$_5$CH$_3$ by **13** to be due to the lack of complementarity in shape between the extensive flat surfaces of C$_6$H$_5$CH$_3$ and the curved surfaces of the interior of **13**. Accordingly, **13**·C$_6$H$_5$CH$_3$ has few host–guest close contacts. In contrast, the flat surfaces of the 1,2-(CD$_3$)$_2$C$_6$D$_4$ solvent should provide many close contacts with C$_6$H$_5$CH$_3$ in its dissociated state.

8.2.9 Driving Forces for Intrinsic and Constrictive Binding

The entropies of binding all four guests of Table 8.3 have positive signs for the formation of **13·G**, which is highly unusual for processes in non-polar solvents when two molecules are collected and their movements restricted relative to one another. Positive entropies as high as 6 cal mol^{-1} K^{-1} have been observed by Collet for a cryptophane complex with CH$_2$Cl$_2$,[6] and we have observed entropies as high as 40 cal mol^{-1} K^{-1} for dimerization of certain velcrands in CDCl$_3$[5] (Section 6.5). We interpret the present case as reflecting two additive factors. Solvated guest releases somewhat organized solvent molecules that become part of the medium, thereby driving the complexation process. This solvolytic driving force for complexation has often been observed in hydroxylic solvents, mainly water, but seldom in hydrocarbon solvents.

The second factor reflects the entropy of dilution of empty space.[5,7] An empty host such as **13** possesses an inner space of substantial size. When a guest such as C$_6$H$_5$CH$_3$ enters the interior of **13**, much of the empty space is displaced by the guest, and driven by entropic dilution it becomes dispersed into many smaller spaces throughout the bulk solvent phase. Thus entropic driving forces for complexation can be observed even in hydrocarbon solvents.

Table 8.3 lists the $\Delta H\ddagger$ and $\Delta S\ddagger$ for association of empty **13** with four different guests to form the transition states for complexation. The respective enthalpic $\Delta H\ddagger$ (kcal mol^{-1}) and entropic $-T\Delta S\ddagger$ (kcal mol^{-1}, 100 °C) contributions to $\Delta G\ddagger_{assoc}$ for the four complexes are: 16.3 *vs.* 7.2 for (CH$_3$)$_2$NCOCH$_3$; 19.2 *vs.* 6.7 for CH$_3$CH$_2$O$_2$CCH$_3$; 17.7 *vs.* 7.1 for CH$_3$CH$_2$COCH$_3$; and 20.7 *vs.* 5.1 for C$_6$H$_5$CH$_3$. Thus for all four guests $\Delta H\ddagger$ and $-T\Delta S\ddagger$ components for the constrictive binding are positive. The enthalpic contribution varies from 69 to 80% of the total constrictive binding while entropic contributions vary from 31 to 20% with changes in the guest structure. The entropic contribution to this transition state barrier is lowest for C$_6$H$_5$CH$_3$ as guest, the only one free of

---

[6] A. Collet, *Angew. Chem., Int. Ed. Engl.*, 1989, **28**, 1246.
[7] D.J. Cram, K.D. Stewart, I. Goldberg, and K.N. Trueblood, *J. Am. Chem. Soc.*, 1985, **107**, 2574.

conformational mobility which likely would be frozen out in the transition state for complexation.

8.3 Chiral Recognition by Hemicarcerands

Collet *et al.* reported the first example of stereoselectivity for a hemicarcerand enantiomer binding the enantiomers of a racemic guest. The cyclotriveratrylene-based cryptophane **18** served as host and CHFClBr as guest. At 53 °C there exists a difference in thermodynamic stability of $\Delta\Delta G^0 = 260$ cal mol^{-1} between the two diastereomeric complexes.[8]

18

(*R*)-19

(*R*$_4$)-**20**•$CHCl_3$

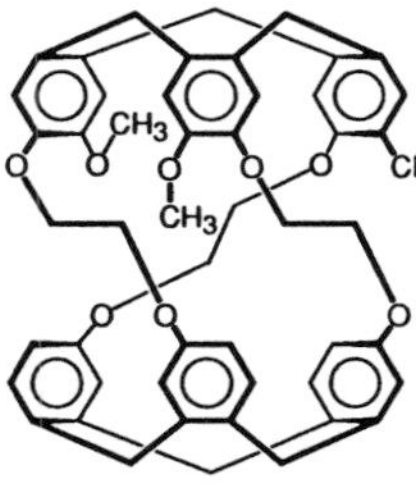

Host (R_4)-**20** was used in our studies of both kinetic and thermodynamic chiral recognition in incarceration. The compound was prepared by shell closure of two moles of bowl-shaped tetrol **11** with four moles of *R*-**19** binaphthyl dibromide in $(CH_3)_2NCOCH_3$–Cs_2CO_3 at 40 °C. During isolation the initially formed (R_4)-**20**·$(CH_3)_2NCOCH_3$ underwent guest exchange with $CHCl_3$ to give (R_4)-**20**·$CHCl_3$ (12%). Similarly, (S_4)-**20**·$CHCl_3$ (13%) was prepared from **11** and (*S*)-**19**. When **20**·$CHCl_3$ enantiomers in neat solvents were heated at the temperatures indicated, guest exchange occurred to give **20**·G as one-to-one complexes (^{1}H NMR integrations): with G = 1,4-$(CH_3)_2C_6H_4$, 100 °C, 18 hours; $CH_3CHICH_2CH_3$, 70 °C, 4 hours; $CH_3CHOHCH_2CH_3$ in 1,3-$(CH_3)_2$-5-$(CH_3)_3CC_6H_3$, 90 °C, 24 hours; and $(CH_3)_2CHCH_2Br$, 90 °C, 4 hours.[9]

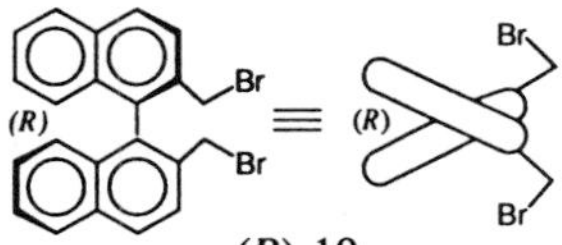

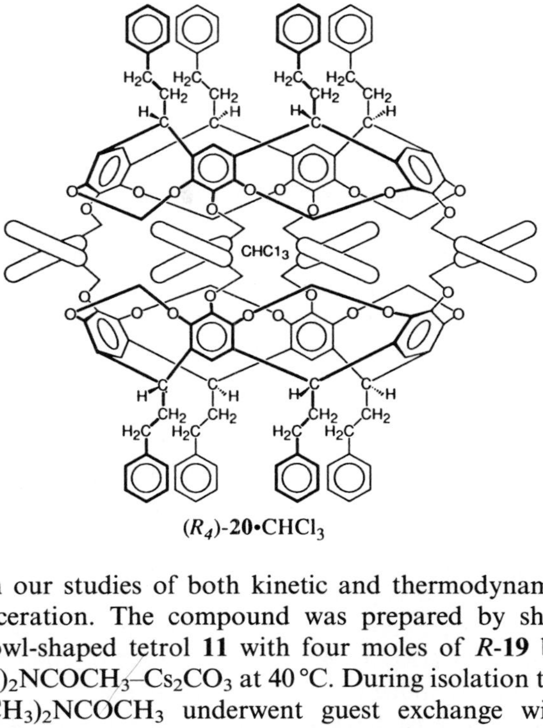

These hemicarceplexes liberated their guests when dissolved in $CDCl_3$ at 23 °C with $t_{1/2}$ values (^{1}H NMR spectral changes) as follows: **20**·$CH_3CHICH_2CH_3$, ~50 000 hours; **20**·1,4-$(CH_3)_2C_6H_4$, 4 hours;

[8] J. Canceill, L. Lacombe, and A. Collet, *J. Am. Chem. Soc.*, 1985, **107**, 6993.

$20 \cdot CH_3CHOHCH_2CH_3$, 2 hours; and $20 \cdot (CH_3)_2CHCH_2Br$, 0.33 hours. The rates for decomplexation decrease by a factor of more than 10^4 with the small change in guest structure in going from $CH_3CHOHCH_2CH_3$ to $CH_3CHICH_2CH_3$. An even more striking example of structural recognition is the failure of $20 \cdot CHCl_3$ to produce $20 \cdot CH_3CH_2C_6H_5$ when heated in $CH_3CH_2C_6H_5$ for 20 hours at 140 °C, whereas $20 \cdot 1,4\text{-}(CH_3)_2C_6H_4$ was readily formed under milder conditions.[9]

In CPK models of (R_4)-**20** or (S_4)-**20**, the size of the four chiral portals varies with the size of the dihedral angles between the two naphthalene rings in the four binaphthyl units. When the dihedral angles are minimized, the two hemisphere rims contact one another in many places, the cavity and portal sizes are minimized, and the hemicarcerand appears to be unstrained. When the angle is maximal, both the portals and the cavity are large, but many close contacts are lost and bond angle strains appear. Interestingly, a model of $1,4\text{-}(CH_3)_2C_6H_4$ nicely fits the interior of a model of (R_4)-**20** when the dihedral binaphthyl angles are minimal, and probably in its most stable conformation, whereas $CH_3CH_2C_6H_5$ does not.[9]

For study of chiral recognition in guest release, diastereomeric one-to-one hemicarceplexes were each prepared by heating in the dark (R_4)-**20** or (S_4)-**20** dissolved in enantiomerically pure $(S)\text{-}CH_3CH_2CH(CH_3)CH_2Br$ to give respectively $(R_4)\text{-}\mathbf{20}\cdot(S)\text{-}CH_3CH_2CH(CH_3)CH_2Br$ and $(S_4)\text{-}\mathbf{20}\cdot(S)\text{-}CH_3CH_2CH(CH_3)CH_2Br$. The first-order-rate constants for guest release from the diastereomeric complexes were determined in $CDCl_3$ at 23 °C by ^1H NMR spectral changes to give $k_{R_4S} : k_{S_4S} = 7$.[9]

In a second study, (S_4)-**20** was heated at 100 °C dissolved in racemic $CH_3CHBrCH_2CH_2Br$ until a diastereomeric equilibrium had been established (no further ^1H NMR spectral changes) in which one diastereomeric complex dominated by a factor that ranged from 1.5:1 to 2:1 in three runs. This ratio shows that the two diastereomers differed by ~ 300 cal mol^{-1} in free energy at 100 °C. The dissociation rate constants of the two isomers were then determined at 23 °C in $CDCl_3$ to provide $k_{fast} : k_{slow} = 5$. The *less* thermodynamically stable isomer gave the *faster* rate. Similarly, from $(S_4)\text{-}\mathbf{20}\cdot CHCl_3$ and racemic $CH_3CH_2CHBrCH_2Br$ was prepared an equilibrated diastereomeric mixture whose components were present in the ratio of 2:1. The dissociation rate constants were determined at 23 °C in $CDCl_3$ to give $k_{fast} : k_{slow} = 9$. In contrast to the former complex, the *more* stable diastereomer gave the *faster* rate.[9]

At 23 °C in $CDCl_3$ the $\Delta\Delta G^\ddagger$ values for the diastereomeric complexes dissociating were as follows: $20 \cdot CH_3CH_2CH(CH_3)CH_2Br$, 1.1; $20 \cdot CH_3CHBrCH_2CH_2Br$, 1.0; and $20 \cdot CH_3CH_2CHBrCH_2Br$, 1.3 kcal mol^{-1}. The $\Delta\Delta G^0$ values at 100 °C for the two latter diastereomers are ~ 300 cal mol^{-1}. Usually $\Delta\Delta G^0$ values for diastereomeric complexes decrease with increasing temperature. If in our study $\Delta\Delta G^0$ remained at ~ 300 cal mol^{-1} at 23 °C, the $\Delta\Delta G^\ddagger$ value for the complexation *diastereomeric transition states* would be 1.6 kcal mol^{-1} for $20 \cdot CH_3CH_2CHBrCH_2Br$ ($k_{fast} : k_{slow} = 15$) and 0.7 kcal mol^{-1} for $20 \cdot CH_3CHBrCH_2CH_2Br$ ($k_{fast} : k_{slow} \sim 3$).

[9] J.K. Judice and D.J. Cram, *J. Am. Chem. Soc.*, 1991, **113**, 2790.

Differences in steric repulsions and dipole–dipole alignments in the diastereomeric transition states are probably responsible for the chiral selectivity in decomplexation. For each of the three chiral guests examined, the host discriminated between CH_3 vs. Br or CH_3CH_2 vs. $BrCH_2$ groups attached to the chiral center. The sizes of covalently bound CH_3 and Br calculated from their volumes and surface areas differ only by 5–10%. Probably differences in host–guest dipole–dipole alignments in the diastereomeric transition states play the bigger role.

CHAPTER 9

Varieties of Hemicarcerands

'The design and synthesis of new organic compounds can be at one and the same time a beguiling and bewildering activity. Usually new compounds are designed and made because of hoped for physical, chemical, or biological properties they are likely to possess. Loose correlations of properties with chemical structure are used to guide a molecular designer, who reasons by analogy, both with respect to selecting the target compound and deciding how that compound might be prepared from available materials. With diligence and luck the target compound is made, and its properties are just close enough to those hoped for to tempt the investigator into synthesizing a series of analogues with the hope of improving the properties and uncovering closer correlations between structure and activity. By this iterative procedure this art–science of organic chemistry evolves.'

The previous chapter described three hemicarcerands designed to incarcerate organic compounds of molecular weights that range from about 40 to about 110. This chapter deals with hosts designed to complex by incarceration guests of higher and lower molecular weights, particularly the former.

9.1 An Octalactone as a Hemicarcerand

Octalactone $2 \cdot CH_2Cl_2$ was synthesized by the reaction of four moles of tetrol 1^1 with four moles of 1,3-$C_6H_4(COCl)_2$ in Cs_2CO_3–$(CH_3)_2NCOCH_3$ in 5% yield (Chart 9.1). Exchange of the initial guest $(CH_3)_2NCOCH_3$ with CH_2Cl_2 occurred during purification of the product by silica gel chromatography with CH_2Cl_2 as the mobile phase. The octalactone was fully characterized as its one-to-one complex with CH_2Cl_2, which, when heated at 110 °C in $CHCl_2CH$-Cl_2 for 12 hours, was converted to $2 \cdot CHCl_2CHCl_2$ (90%, fully characterized). This complex, when recrystallized from $C_6H_5NO_2$, gave solvate $2 \cdot CHCl_2CH$-$Cl_2 \cdot 2C_6H_5NO_2$ as crystals suitable for crystal structure determination. The stereoview of the result is shown in Chart 9.1.[2]

In this crystal structure, the $CHCl_2CHCl_2$ guest resides comfortably in the cavity with its long axis roughly coincident with the long axis of the host. One

---

[1] L.A. Tunstad, J.A. Tucker, E. Dalcanale, J. Weiser, J.A. Bryant, J.C. Sherman, R.C. Helgeson, C.B. Knobler, and D.J. Cram, *J. Org. Chem.*, 1989, **54**, 1305.
[2] M.L.C.Quan, C.B. Knobler, and D.J. Cram, *J. Chem. Soc., Chem. Commun.*, 1991, 660.

Varieties of Hemicarcerands

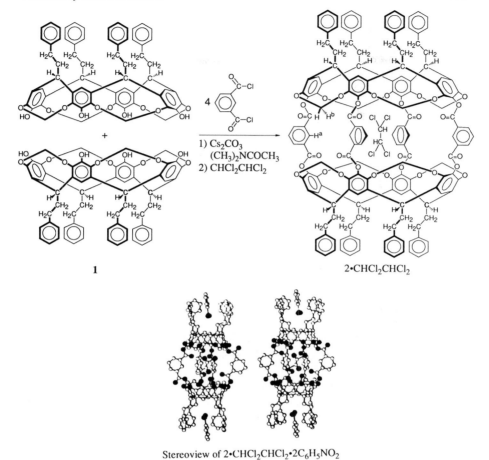

Chart 9.1

molecule of $C_6H_5NO_2$ is inserted between the four $C_6H_5CH_2CH_2$ groups appended to the southern and a second between the four $C_6H_5CH_2CH_2$ groups of the northern hemisphere of the globe-shaped container, in each case with the NO_2 group oriented inward. Although all 8 of the lactone groups possess the more stable *anti* configuration (two attached aryls *trans* to one another), each of the four bislactone bridges has one carbonyl pointing inward toward the cavity and the other pointing outward away from the cavity. Each hemisphere contains two inward-pointing and two outward-pointing carbonyls distributed around the polar axis in an *in–in–out–out* pattern.[2]

In the ^1H NMR spectra of **2·G**, the δ values of ^aH are guest-sensitive, while those of ^bH are both conformation- and guest-sensitive. Comparison of the chemical shifts of these protons in **2·CDCl$_3$** with those for **2·CHCl$_2$CHCl$_2$** led to the conclusion that the latter possessed approximately the same structure in the crystalline and solution (in CDCl$_3$) states at 25 °C. At higher temperatures, the lactone conformations of **2**·CHCl$_2$CHCl$_2$ dissolved in CDCl$_2$CDCl$_2$ equilibrate to provide a time-averaged ^{1}H NMR spectrum with T_c for ^{b}H of 80 °C, and a $\Delta G\ddagger$ value for the transition of about 18 kcal mol^{-1}.^{2}

Hemicarceplex **2**·CHCl$_2$CHCl$_2$ is stable indefinitely at 25 °C as a solid or in solution, but slowly decomplexes at 100–134 °C. The first-order rate constant at 100 °C for decomplexation of **2**·CHCl$_2$CHCl$_2$ in CDCl$_2$CDCl$_2$ to give **2**·CDCl$_2$CDCl$_2$ provided a $t_{1/2}$ value of 18 hours. A study of the rate as a function of four temperatures provided $E_a \cong 25$ kcal mol^{-1}, much of which must be constrictive binding (Section 8.2.8).^{2}

A survey employing ^{1}H NMR and thin-layer chromatographic criteria for complexation was conducted for **2**·2CH$_2$Cl$_2$ when dissolved in potential guests as solvents. Of those examined, CH$_3$CON(CH$_2$CH$_2$)$_2$O and 1,2-Cl$_2$C$_6$H$_4$, formed characterizable complexes at 100–125 °C, whereas 1,4-Cl$_2$C$_6$H$_4$, 1,4-(CH$_3$)$_2$C$_6$H$_4$, CHBr$_2$CHBr$_2$, 4-CH$_3$C$_6$H$_4$CN, C$_6$H$_5$NO$_2$, and Et$_3$PO failed to produce isolable complexes.^{2} The low yield of the shell closure coupled with the complications in the ^{1}H NMR spectra of the hemicarceplexes of **2** discouraged further investigation of the system.

9.2 An Octaimine as a Hemicarcerand

As a CPK molecular model, structure **5·G** possesses a preorganized cavity roughly complementary to a model of [3.3]paracyclophane, although the four portals of **5** were too small to allow admission of that model to its interior without the breaking of covalent bonds. Furthermore, potential guests B–R of Chart 9.2 in models appear to be firmly held by the model of **5**, and yet **5** has portals large enough to potentially allow admission of these large guests without serious strain. However, a model of A is much too large and structurally awkward a compound to enter a model of **5**. Thus a study of **5** was undertaken, and triphosphonamide A was adopted as a solvent when needed to dissolve and bring into contact host and potential guests at high temperatures.^{3}

9.2.1 Synthesis and Characterization

A simple synthesis of **5** was developed in which readily available tetrabromide **3**^{4} was metalated in (CH$_2$)$_4$O with BuLi (-78 °C, 2 minutes) and O(CH$_2$CH$_2$)$_2$NCHO was added to produce tetraaldehyde **4** (75%). A 2:1 molar ratio of 1,3-diaminobenzene to **4** in dry pyridine was stirred at 65 °C for 4 days to give, after chromatography (silica, CH$_2$Cl$_2$), a 45% shell-closure yield of **5**·CH$_2$Cl$_2$, which was fully characterized. The sum of the elemental analyses of empty **5** was 100.13%, and all of the protons in the 500 MHz ^{1}H NMR spectrum of **5** in CDCl$_3$ (except those of the pendant C$_6$H$_5$CH$_2$CH$_2$ groups) were assigned by a homonuclear correlation COSY experiment. Application of NOE experiments to ^{a}H, ^{b}H, and ^{c}H protons of **5·G** in CDCl$_3$ demonstrated that all eight configurations about the C=N bonds have their two attached aryl

---
[3] M.L.C. Quan and D.J. Cram, *J. Am. Chem. Soc.*, 1991, **113**, 2754.
[4] J.C. Sherman and D.J. Cram, *J. Am. Chem. Soc.*, 1989, **111**, 4527.

Varieties of Hemicarcerands

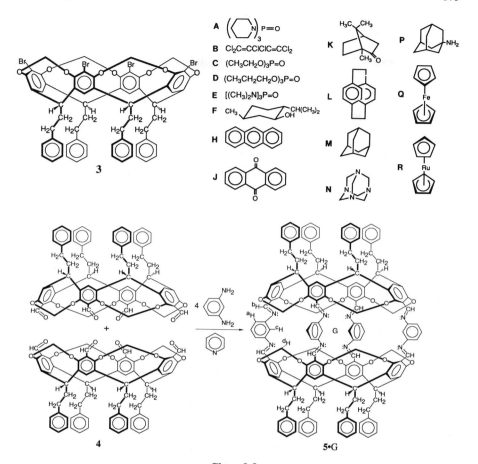

Chart 9.2

groups *anti* to one another. This configuration conforms to expectation based on a molecular model of **5** that possesses a C_4 axis, and with the **5·**ferrocene crystal structure (Section 9.2.4).

9.2.2 Complexation

Although **5** was prepared in pyridine, this guest was exchanged with CH_2Cl_2 during the isolation, and CH_2Cl_2 was easily removed by heating the complex to 150 °C under vacuum. Molecules as small as pyridine and CH_2Cl_2 in molecular models of **5** appear able to enter and exit the interior of **5** without detectable activation energies, as expected. Medium-sized compact molecules were introduced into the interior of **5** either by melting a host–guest mixture of solids many hours at temperatures of 80–168 °C, or by heating them in concentrated solutions in A, a solvent too large to enter **5**. Under the latter conditions,

[3.3]paracyclophane, 9,10-dimethoxyanthracene, 9,10-dimethylanthracene, dibenzenechromium, carboranes (1,7-$C_2B_{10}H_{12}$ and 1,10-$C_2B_8H_{10}$) and tetrabutyl phosphate failed to form complexes detectable by ^{1}H NMR.

The fourteen guests which formed hemicarceplexes that were easily isolated, purified, and characterized are formulated in Chart 9.2 and are identified by the letters B through R. All complexes were stable in solution at ambient temperatures. All of them gave distinctive ^{1}H NMR spectra in $CDCl_3$, and thin-layer chromatograph R_f values which ranged from 0.16 for 5·hexamethylphosphoramide (5·E) to 0.60 for 5·ruthenocene (5·R) on silica gel (3% EtOAc in CH_2Cl_2). All complexes except those of hexamethylphosphoramide (5·E), anthracene (5·H), and camphor (5·K) gave good elemental analyses and FAB-MS molecular ions. The guests played major roles in imparting physical properties to the complexes, although no host–guest covalent bonds connected the complexing partners. As expected, as the guests made larger contributions to the masses of the complexes, their roles in determining the physical properties became more important.[3]

The ^{1}H NMR signals of the guests were moved upfield upon complexation by as much as $\Delta\delta = 4.12$ p.p.m. for the 2,3-protons of anthraquinone or as little as $\Delta\delta = 0.42$ p.p.m. for ruthenocene. Unlike the equatorial regions of the complexes, the polar regions of the host's cavity are lined with the shielding faces of aryl rings. Comparisons of the magnitudes of upfield shifts of various protons provided structural conclusions consistent with those based on CPK model examination of many hemicarceplexes. For example, in 5·anthracene, 5·anthraquinone, 5·[2.2]paracyclophane, and 5·menthol, the protons terminating the long axes of the guests are shifted much further upfield than those located toward the middle of the guest, indicating that the long axes of host and guest in these complexes are roughly coincident. When N,N-dideuterioamantidine was substituted for amantidine (P) in preparing 5·amantidine, the NH_2 signals at -2.40 p.p.m. (2.78 p.p.m. upfield of their non-incarcerated position) in the complex disappeared. Thus the NH_2 group must be located in a polar cap of 5·P.[3]

The inward-facing host protons ^{c}H and ^{d}H of 5 gave equally informative ^{1}H NMR chemical shifts reflecting guest characters and locations. Thus ^{c}H moves upfield by 1.03 and ^{d}H downfield by 0.09 p.p.m. upon complexation with [2.2]paracyclophane. We conclude that the equatorial ^{c}H proton is shielded by the aryl faces of the guests, whereas the tropical ^{d}H region of the host is slightly deshielded by the aryl edges of the guest.[3]

9.2.3 Decomplexation

The $t_{1/2}$ for guest departure by 13 of the complexes were measured in $CDCl_2CDCl_2$ and/or $CDCl_3$ by following ^{1}H NMR signal changes with time at temperatures of 25–134 °C. Values ranged from a low of 3.2 hours at 25 °C for 5·hexachlorobutadiene in $CDCl_3$ to a high of 19.6 hours for 5·ferrocene in $CDCl_2CDCl_2$ at 112 °C. The kinetic stability order for guest loss by the complexes was as follows: ferrocene > [2.2]paracyclophane > adamantane > ruthenocene > amantidine > hexamethylenetetramine > camphor > anthra-

Varieties of Hemicarcerands

quinone ~ tripropyl phosphate > anthracene > menthol > triethyl phosphate ~ hexachlorobutadiene. Variable-temperature ^{1}H NMR kinetic measurements provided activation energies E_a for decomplexation of ~ 19 and ~ 28 kcal mol^{-1} for **5·ruthenocene** and **5·adamantane**, respectively. Notice that adamantane $C_{10}H_{18}$, which is rigid and nearly spherical, is much more difficult to liberate than flexible tripropyl phosphate $C_9H_{21}O_4P$.[3]

9.2.4 Crystal Structure

Chart 9.3 depicts **5·ferrocene**, **6**, a side stereoview of **5·ferrocene**'s crystal structure, **7a**, an end view of the crystal structure, and **7b**, an end view including only the guest and the octaimine bridges whose two sets of four terminal carbons are connected by heavy lines that form two nearly parallel squares. This beautiful crystal structure corresponds well to that predicted from CPK model examination. Aside from the ferrocene guest and appendages, the host possesses an approximate C_4 axis which is roughly coincident with the long axis of the ferrocene guest. The configurations about all eight C=N bonds have their two attached aryl groups *anti* to one another, which is particularly visible in stereoview **6**. The northern and southern hemispheres are almost directly over one another in **6**, the twist angle being about 0°. Interestingly, the ferrocene guest possesses a staggered (crystallographically required) conformation. View **7a** suggests the 1,3-diiminobenzene connecting groups are arranged like paddles in a paddle wheel around the outside of the central cavity of **5**. The planes of these groups are roughly perpendicular to the best planes of their attached aryl-carbon atoms. The average dihedral angle between the best planes of the benzene connecting groups and those of their attached CH=N: groups is 43° (37–52°).[5]

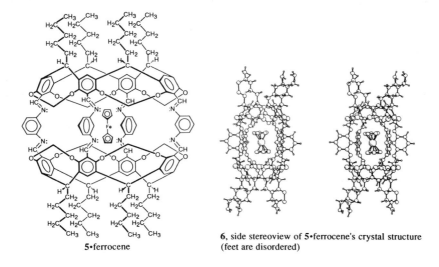

5·ferrocene

6, side stereoview of 5·ferrocene's crystal structure (feet are disordered)

Chart 9.3

[5] C.B. Knobler and D.J. Cram, unpublished results.

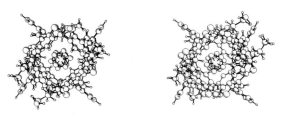

7a, end stereoview of **5**•ferrocene

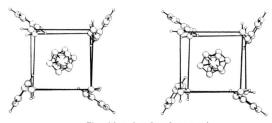

7b, abbreviated end stereoview

Chart 9.3 (*continued*)

9.3 An Octaamide as a Hemicarcerand

The equatorial spacer groups of octalactone **2** and octaimine **5** are both seven atoms in length with a *meta*-substituted benzene in the middle. The synthesis, crystal structure, and binding properties of octaamides **8** and **9** were also undertaken (see Chart 9.4). This potential hemicarcerand differs structurally from **2** and **5** only in the sense that eight CONH groups are substituted for the eight CO_2 groups of **2** or the eight CH=N groups of **5**. The octaamide was expected to be more hydrolytically stable than either of the other two hemicarcerands. We chose to make two hemicarcerands: **8**, because compounds with $C_6H_5CH_2CH_2$ appendages give simpler ^{1}H NMR spectra in the upfield region and better crystal structures; and **9**, because those with $CH_3(CH_2)_4$ 'feet' are more soluble.[6]

9.3.1 Syntheses and Characterization

The synthesis of **8** was accomplished by the reaction sequence $3^4 \rightarrow 12 \rightarrow 13 \rightarrow 8$, and of **9** by the sequence $10^7 \rightarrow 14 \rightarrow 9$. The shell closures $13 \rightarrow 8$ and $14 \rightarrow 9$, with 1,3-$(H_2N)_2C_6H_4$ as reactant, were both carried out in Et_3N-$(CH_2)_4O$ at 55 °C under high dilution conditions to give **8** and **9** in 7% yield. The former, **8**, gave elemental analyses showing it to be free of guest, whereas the latter, **9**, indicated one mole of CH_2Cl_2 to be present, probably as guest. The ^{1}H NMR and ^{13}C NMR spectra of both hemicarcerands in $CDCl_3$ exhibited effective C_{4d}

---

[6] H.-J. Choi, D. Bühring, M.L.C. Quan, C.B. Knobler, and D.J. Cram, *J. Chem. Soc., Chem. Commun.*, 1992, 1733.
[7] J.A. Bryant, M.T. Blanda, M. Vincenti, and D.J. Cram, *J. Am. Chem. Soc.*, 1991, **113**, 2167.

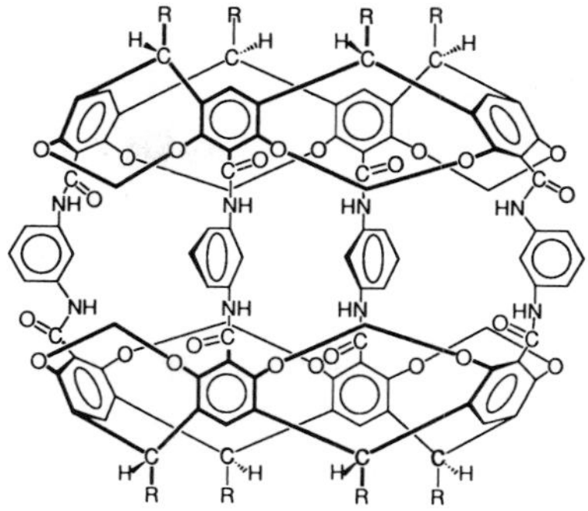

Varieties of Hemicarcerands

8, R = CH$_2$CH$_2$C$_6$H$_5$
9, R = (CH$_2$)$_4$CH$_3$

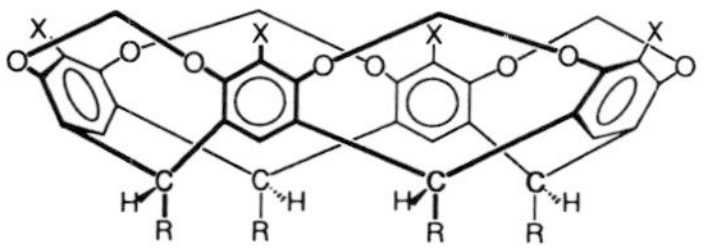

3, X = Br, R = CH$_2$CH$_2$C$_6$H$_5$
10, X = Br, R = (CH$_2$)$_4$CH$_3$
11, X = CO$_2$Me, R = CH$_2$CH$_2$C$_6$H$_5$
12, X = CO$_2$Li, R = CH$_2$CH$_2$C$_6$H$_5$
13, X = COCl, R = CH$_2$CH$_2$C$_6$H$_5$
14, X = COCl, R = (CH$_2$)$_4$CH$_3$

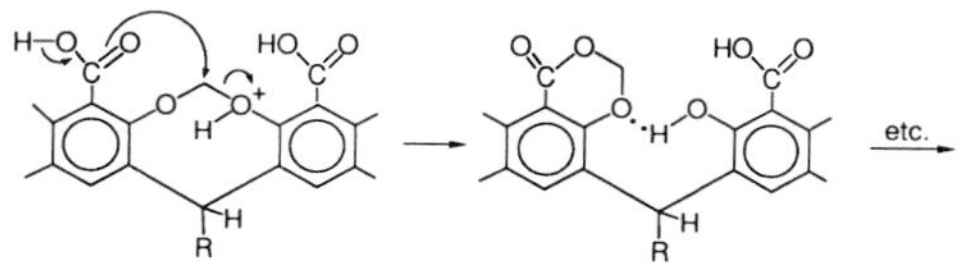

Chart 9.4

symmetry undoubtedly because of conformational equilibrations and solvent entrance and departure of the cavity being rapid on the NMR time scale (see Chart 9.4).

The base-catalyzed hydrolysis of tetraester **11** gave a salt solution, acidification of which resulted in a four-fold, acid-catalyzed rearrangement of the tetracarboxylic acid to form the phenolic lactone acetal **19** (18% yield). The compound was characterized through its crystal structure, a stereoview of which is found in Chart 9.5. An examination of CPK models of the tetracarboxylic acid indicates that the carboxyl groups are well located to act as neighboring groups that participate in exchanging four eight-membered for four six-membered rings to give **19**. See **15** for a suggested mechanism.

9.3.2 Crystal Structure

The crystal structure of **8·6H$_2$O·10C$_6$H$_5$NO$_2$** proved to be very interesting. The six H$_2$O molecules occupy the cavity as shown in stereoview **16** in Chart 9.5. Each of the four nitrogen atoms are within 3.0 Å of one of the oxygens of the H$_2$O molecules. Four inward-pointing carbonyl oxygens of the host are within 2.8–3.0 Å of the H$_2$O-oxygen atoms. Although not located, the hydrogen atoms of the guest H$_2$O molecules undoubtedly form a network of host-to-guest and

line drawings   stereoviews

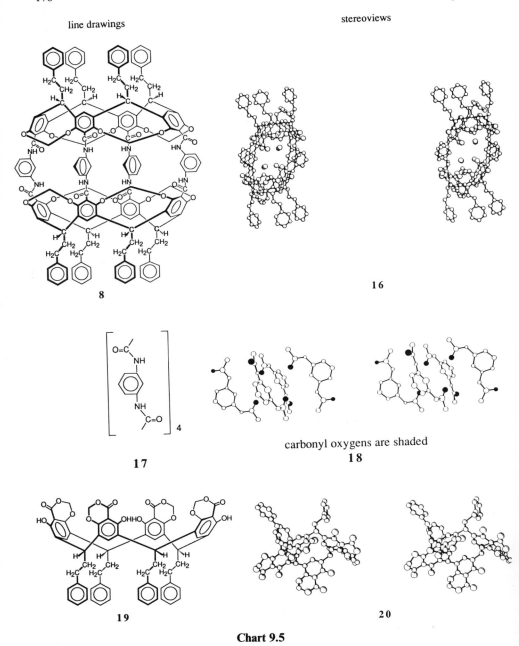

carbonyl oxygens are shaded

Chart 9.5

guest-to-host hydrogen bonds. The host itself is centrosymmetric, but lacks a C_4 axis due to displacement of the northern and southern hemispheres by about 1.9 Å, which is visible in stereoview **16**.[6]

Chart 9.5 also contains stereoview **18** of the four $CONHC_6H_4NHCO$ units

Varieties of Hemicarcerands

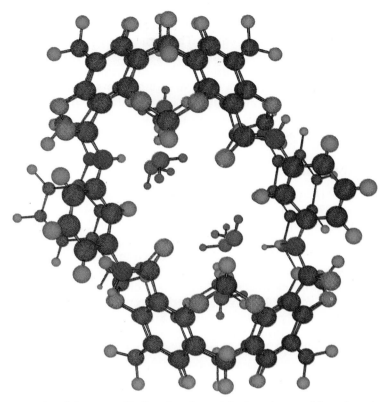

Figure 9.1 *Colored drawing of ball-and-stick model of octalactam **8** based on the coordinates of crystal structure of **8·6H$_2$O**, but with eight CH$_3$ groups replacing the eight C$_6$H$_5$CH$_2$CH$_2$ feet. The color code is: C, gray; H, blue; N, purple; O, red*

that bridge the two polar caps. Each of the eight amide groups is *anti-planar*, and close to being coplanar with their centrally attached benzenes. Each of the four bridges contains one inward- and one outward-pointing carbonyl arranged so that the inward-pointing carbonyl oxygens are distant from one another. The distance from the carbonyl oxygen of the northern hemisphere at about 11 o'clock in **18** to the opposite carbonyl oxygen in the southern hemisphere at about 5 o'clock is about 6.80 Å. The four carbonyl oxygens of the northern and the four carbonyl oxygens of the southern hemispheres are each within 0.04 Å of being coplanar, the distance between the least-squares planes being 4.2 Å. Each hemispheric set of four nitrogen atoms is also nearly coplanar within 0.04 Å, and the two nitrogen planes are about 4.8 Å distant. The crystal structure **16** was refined to give an R value of 0.11.

The octaamide pictured in color in Figure 9.1 shows how the six water guest molecules probably occupy the interior of **8·6H$_2$O**. Each of the two hemispheres of **8** contains a water, with each of its two hydrogens pointing toward the face of an aryl. The four remaining waters occupy the parts of the cavity defined by the $(m\text{-CONHC}_6\text{H}_4\text{NHCO})_4$ intrahemispheric bridging groups. The four oxygens are arranged to form a rough square held in place by four N—H····OH$_2$ and four CO····HOH hydrogen bonds. These guest-to-bridge hydrogen bonds seem to dominate the structure rather than HOH····OH$_2$ hydrogen bonds. Also obvious in Figure 9.1 is the displacement of the two hemispheres from having a common axis, and how the (H$_2$O)$_6$ guest structure adapts to this displacement.[6]

Crystal structure **20** of tetralactone **19** was refined to $R = 0.12$. Although the hydrogens were not located, the intramolecular C=O····HOC oxygen-to-oxygen distances are between 2.5 and 2.6, suggesting hydrogen bonding.

9.3.3 Complexation

Attempts to obtain stable complexes of octaamide **8** involved heating to 120–170 °C for 27 hours either solutions of **8** in guest, or solutions of host and guest dissolved in tripiperidylphosphoramide at 120–160 °C. The solutions were cooled and a large amount of hexane was added. The precipitate was washed, dried, and its 500 MHz ^{1}H NMR spectrum taken in CDCl$_3$. Twelve potential guests were examined: aspirin, adamantane, 1,4-diisopropylbenzene, azulene, bromobenzene, 3-bromotoluene, 4-methylanisole, tetrachloroethane, 1,4-dibromobenzene, menthol, *N,N'*-tetramethylterephthalic diamide, and 1,4-diacetoxybenzene. Only the last complex, **8**·1,4-CH$_3$CO$_2$C$_6$H$_4$O$_2$CCH$_3$, was isolated as a stable one-to-one complex. Its CH$_3$ protons were moved upfield from their normal $\delta = 2.3$ signals to $\delta = -1.0$ when incarcerated. Although most of the other potential guests probably formed complexes reversibly, they were unstable to the conditions of isolation. Apparently 1,4-diacetoxybenzene uniquely combines intrinsic and constrictive binding to octaamide **8** which provide a slow enough decomplexation rate to allow manipulation of the hemicarceplex at ambient temperature.[6]

9.4 A Near Hemicarcerand Based on [1.1.1]Orthocyclophane Units

A molecular model examination of the hypothetical carcerand **21** provided the inspiration for host structures based on unit **22** whose syntheses and properties are described in Chapters 7 and 8 and Sections 9.1–9.3. Besides containing in its shell six tetrabenzene substructures **22**, structure **21** also contains eight tribenzene substructures **23**, which provided the origins of the conceptions of the hosts described in this and the next sections (see Chart 9.6). The demonstration that the acid-catalyzed condensation of veratrole (1,2-dimethoxybenzene) with CH$_2$O gave cyclotriveratralene[8] **24** in a good yield suggested that analogous reactions might be used to synthesize shallow bowls which, when triply bridged, might provide orthocyclophanic hemicarcerands such as chiral system **25** or *meso* system **26** with acetylenic bridges.[9] We were encouraged in the latter part

---

[8] A.S. Lindsey, *J. Chem. Soc.*, 1965, 1685.
[9] D.J. Cram, M.E. Tanner, S.J. Keipert, and C.B. Knobler, *J. Am. Chem. Soc.*, 1991, **113**, 8909.

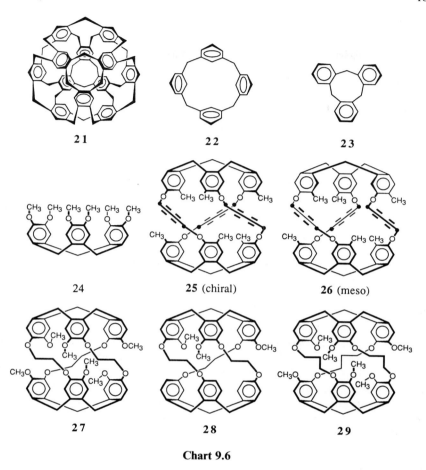

Chart 9.6

of this endeavor by A. Collet's elegant syntheses and studies of cryptophanes **27–29**.[10]

9.4.1 Syntheses

Our syntheses employed as starting material 4-bromo-3-methoxybenzyl alcohol[11] **30**, treatment of which with P_2O_5 and Et_2O gave the three-fold condensation product **31** in 40% yield. The bromine atom in **30** acts as a group that protects the 4-position from alkylation by benzyl cations during the conversion of **30** to **31** on the one hand, and provides synthetic flexibility for further reactions of **31** on the other. In the synthesis leading to **25** and **26**, **31** was lithiated and the product quenched with CH_3I to give racemic chiral compound **32** in 83% yield. The three anisyl units of **32** were cleaved with BBr_3

---

[10] J. Canceill, A. Collet, G. Gottarelli, and P. Palmeri, *J. Am. Chem. Soc.*, 1987, **109**, 6454.
[11] M. Elliot, N.F. James, and B.P. Pearson, *J. Sci. Food Agric.*, 1967, **18**, 325.

in CH_2Cl_2 to give the trisphenol **33** (86%), which when alkylated with $HC{\equiv}CCH_2Br$ produced the tris-acetylenic compound **34** (88%).

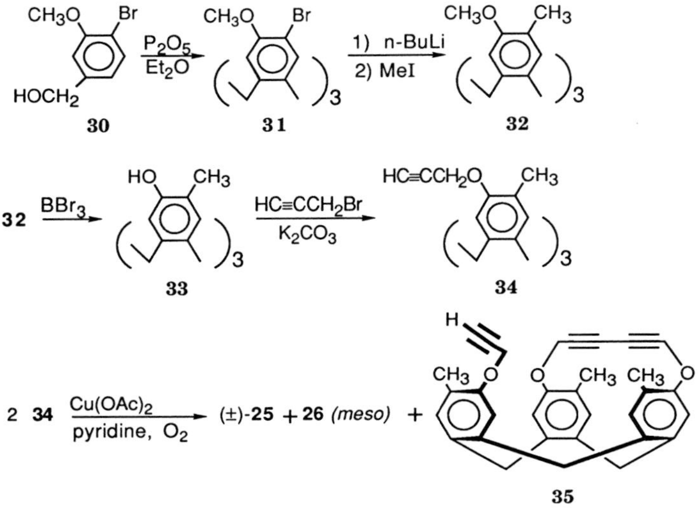

[Scheme showing compounds 30 → 31 (P$_2$O$_5$, Et$_2$O) → 32 (1) n-BuLi, 2) MeI); 32 → 33 (BBr$_3$) → 34 (HC≡CCH$_2$Br, K$_2$CO$_3$); 2 × 34 (Cu(OAc)$_2$, pyridine, O$_2$) → (±)-25 + 26 (meso) + 35]

The shell closures were accomplished by coupling the three terminal acetylenes of one molecule of **34** with those of a second molecule of **34** by oxidation with anhydrous $Cu(OAc)_2$ in dry pyridine at 60 °C.[9] The conditions of the reaction were crucial to the production of even the low yields of 4% of **25** and 2% of **26** obtained. At 85 °C, the reaction gave **35** as a trace by-product. Diastereomers **25** and **26** were easily separated chromatographically. In thin layer chromatography on silica gel with 1:1 CCl_4:CH_2Cl_2, **25** had an R_f value of 0.49, whereas **26** had an R_f of 0.29, which probably reflects the difference in propensity of each host for binding solvent.

9.4.2 Crystal Structures

The identities of the isomeric hosts (±)**25** and *meso* **26** were established through their crystal structures,[9] side stereoviews of which are recorded in Chart 9.7. The crystals of the former were grown from CH_2Cl_2, and those of the latter from n-C_6H_{14}-$CHCl_3$-EtOH. The interior of **25** racemate was occupied by two molecules of disordered CH_2Cl_2, and that of **26** by one molecule of $CHCl_3$. These guests are omitted in the views of the crystal structure, whose oxygens are darkened to help the eye superimpose the structural pairs on one another.

Notice that the northern hemispheres of **25** and **26** are rotated around the polar axis with respect to the southern hemispheres, in **25** by 120°, and in **26** by 60°. In CPK models, this molecular twisting in **25** closes the three large portals to impart a compact spherical shape to the host. This twist minimizes the volume of the cavity which is composed of two polar caps and an equatorial region lined by six CH_3 groups and three $OCH_2C{\equiv}C{-}C{\equiv}CCH_2O$ moieties

Varieties of Hemicarcerands

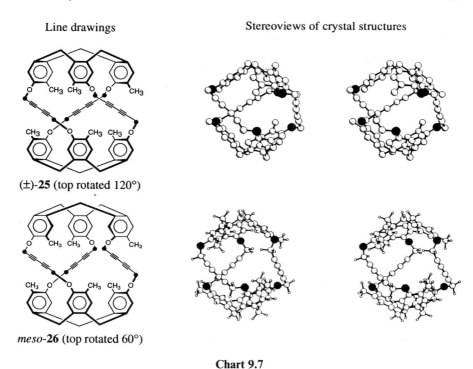

Chart 9.7

that lie along the rims of the polar caps. The D_3 symmetry of both the open and closed conformations of **25** is beautifully visible in both the crystal structure and models of the host. The open structure appears strain free in models, and possesses a large enough cavity to accommodate models of molecules as large as Fe(acac)$_3$ with 22 heavy atoms. The closed structure is highly complementary to spherical molecules such as CHCl$_3$ (four heavy atoms).

In contrast to the crystal structure and CPK models of **25**, the 60° rotation of the northern relative to the southern hemisphere found in **26** is accompanied by a sideways displacement of one hemisphere with respect to the other, so they do not share a common axis. The resulting cavity is approximately ellipsoidal in shape. The shell surface contains many small holes in the tropic and temperate zones of the globe, and the cavity size is larger than that of **25**. In other words, when the northern and southern polar caps possess the same configuration as in **25**, they pack together rim-to-rim in a more highly complementary fashion than in **26**, whose northern and southern polar caps are enantiomeric to one another.

9.4.3 Characterization

In the solid state, **25** strongly complexed solvents such as CH$_2$Cl$_2$, C$_6$H$_5$CH$_3$, and CHCl$_3$ to give one-to-one complexes which were stable to heat under high vacuum at 130 °C. These guests were easily washed out with CH$_3$CN, and the

25·CH₃CN formed free **25** at 100 °C under vacuum for 24 hours. In contrast, **26**·CH₂Cl₂ gave up its guest when heated to 60 °C for 100 hours. Both hosts provided good elemental analyses and molecular ions in their FAB-MS.[9]

9.4.4 Complexation

Hexachloroacetone in CPK models can enter the cavity of models of **25** only with difficulty in its most extended form, whereas models of CHCl₂CHCl₂ and C₆H₆ enter and depart the cavity of **25** with ease. Accordingly, (CCl₃)₂CO was used as a non-complexing solvent for studies of **25** binding certain guests. At −20 °C in this solvent, the exchanges of bound guest with the following unbound guests were slow on the ¹H NMR time scale, and $-\Delta G^0$ values (kcal mol⁻¹) for (±)-**25** binding these guests were as follows: CHCl₃, 5.7; (CH₃)₃COH, 4.9; CH₂Cl₂, 4.8; propylene oxide, 4.4; cubane, 4.2. In contrast, C₆H₆ and CHCl₂CHCl₂ entered and departed (±)-**25** rapidly on the ¹H NMR time scale at 21 °C. Estimates of their $-\Delta G^0$ values from ¹H NMR titration data are 4.1 kcal mol⁻¹ for C₆H₆ and 2 kcal mol⁻¹ for CHCl₂CHCl₂.

Examination of CPK models of the complexes provides the following correlation of guest–host structure with binding. The surface of CHCl₃ and the interior surface of compact (±)-**25** contact one another over a greater area than any other of the guests without disturbing rim-to-rim contacts of the host, or placing constraints on rotations of guest with respect to host. In **25**·(CH₃)₃COH the location of the OH group appears limited to the polar region. A model of **25**·CH₂Cl₂ resembles a toy rattle, the guest surface being substantially less than that of the inner surface of the host. The shape of propylene oxide provides even fewer contact points between host and guest in **25**·propylene oxide. Cubane is too large to fit into the cavity of the compact form of **25**, so some rim-to-rim contacts must be lost in **25**·cubane that are present in free **25**. Also, the eight protruding hydrogens of cubane inhibit any shell-to-carbon contacts in **25**·cubane. The tablet shape and long circumference of benzene limits its location to the equatorial region in **25**·C₆H₆, with its plane perpendicular to the C_3 axis of the host. The host has to untwist slightly to accommodate benzene, losing some of its rim-to-rim contacts and providing little shared surface area for host and guest, which accounts for its relatively low binding. Although a **25**·CHCl₂CHCl₂ model provides a large common surface area for host and guest, most of the rim-to-rim contacts are lost because of the large volume of the guest. We conclude that the rim-to-rim contacts are present in free **25**, and loss of these by complexation lowers the binding free energy accordingly.

Collet[12] in earlier work reported the synthesis from **24** of **27** and **29** (Chart 9.5), the former in its enantiomeric form. Complex **29**·CHCl₃ also exists in its most compact, twisted form (60° rotation, crystal structure), as is illustrated in drawing **29**. From Collet's ΔH and ΔS values for **29** binding its most complementary guest in CDCl₂CDCl₂ as solvent, we calculate that, at −20 °C, $-\Delta G^0 \stackrel{\bullet}{=} 4.2$ kcal mol⁻¹, which compares with **25** binding CDCl₃ at −20 °C in

---

[12] A. Collet, *Tetrahedron*, 1987, **43**, 5725.

$(CCl_3)_2CO$ of $-\Delta G^0 = 5.7$ kcal mol^{-1}. In CPK models, the compact forms of **25**, **27**, and **29** resemble one another in their shapes and internal volumes.

In $(CD_3)_2CO$ at $-20\,°C$, the 500 MHz ^{1}H NMR spectra[9] of guests bound by acetylenic host **25** show substantial upfield proton chemical shifts $\Delta\delta$ values as follows: cubane, 2.79; CHCl$_3$, 4.24; $(CH_3)_3$COH, 3.04; CH_2Cl$_2$, 3.99; propylene oxide CH_3, 3.23, CH_2, 2.88–3.82, CH, 4.00; C$_6$H$_6 \approx 2.5$. Because both **25** and propylene oxide are chiral, they form diastereomeric complexes both easily visible in the ^{1}H NMR spectrum and present in equal amounts. The large upfield shifts are due to the shielding ring currents in the six benzene rings whose faces line much of the cavity of **25**.

The crystal structure of $(\pm)$-**25** portrays a large number of rim-to-rim close contacts (van der Waals plus 0.2 Å or less) due to the interhemispheric twist of the molecule. These atom-to-atom close contacts number 88, which are missing in **25** in its most extended form.

9.5 Rigidly Hollow Hosts That Encapsulate Small Guests

Lacking among hosts previously described are those which are rigidly hollow and contain small portals complementary to two or three heavy-atom molecular guests. Here we describe the synthesis, crystal structures, and binding properties of hosts **36** and **37**.[13] These hosts bear a structural resemblance to **38** of Collet and Lehn, which is conformationally flexible enough to fill much of its own cavity, and has much larger portals than **36** and **37**. Host **38** has been shown to bind $CH_3NH_3^+$ in halogenated solvents.[14]

36, X = OTs
37, X = H

38

9.5.1 Synthesis

The syntheses of **36** and **37** followed the reaction sequences **39**[15] → **40** → **41**, **31** + **41** → **42** → **43** → **36** → **37**. Good yields were obtained up to the critical shell-closing step **43** + p-CH$_3$C$_6$H$_4$SO$_2$NH$_2$ → **36**, which went in only 10%

---

[13] M.E. Tanner, C.B. Knobler, and D.J. Cram, *J. Org. Chem.*, 1992, **57**, 40.
[14] J. Canceill, A. Collet, J. Gabard, F. Kotzyba-Hibert, and J-M. Lehn, *Helv. Chim. Acta*, 1982, **65**, 1894.
[15] S.A. Sherrod, R.L. da Costa, R.A. Barnes, and V. Boekelheide, *J. Am. Chem. Soc.*, 1974, **96**, 1565.

yield. Reduction of tritosylamide **36** with sodium anthracenide in tetrahydrofuran gave **37** (65%).[13]

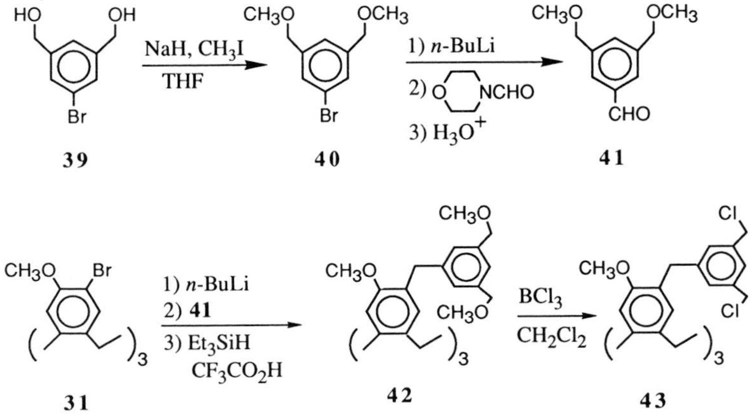

[Scheme: compound **39** (HO, OH, Br substituted arene) → NaH, CH$_3$I, THF → **40** (CH$_3$O, OCH$_3$, Br) → 1) n-BuLi; 2) O(CH$_2$CH$_2$)$_2$NCHO (morpholine formamide); 3) H$_3$O$^+$ → **41** (CH$_3$O, OCH$_3$, CHO)]

[Scheme: compound **31** (CH$_3$O, Br, trimer) → 1) n-BuLi; 2) **41**; 3) Et$_3$SiH, CF$_3$CO$_2$H → **42** → BCl$_3$, CH$_2$Cl$_2$ → **43**]

9.5.2 Crystal Structures

Chart 9.8 depicts line structures and the corresponding crystal structures of **36**·2CH$_3$CN (**44**), **37**·2CH$_3$OH (**45**), and **37**·CH$_2$Cl$_2$ (**46**). Two molecules of CH$_3$CN per molecule of host were present in **44**, one located in the host's cavity (partial occupancy) and oriented perpendicularly to the host's C_3 axis. The second CH$_3$CN molecule was present as a solvate external to the cavity. Variation in the orientation of the three tosyl groups of **44** destroyed the intrinsic C_3 symmetry of the host. Compound **37** crystallized with two molecules of CH$_3$OH per molecule of host, one molecule of CH$_3$OH occupying one of three equivalent positions inside the cavity. A second molecule of CH$_3$OH is located outside the cavity in **45** in a position remote from the observer, and is not shown in **45**. Both **44** and **45** are side stereoviews. When crystals of **37** were grown from CH$_2$Cl$_2$, **37**·CH$_2$Cl$_2$ was produced, whose cavity is empty in its crystal structure **46**, a bottom stereoview, shown on Chart 9.8. In this view the observer is looking down the C_3 molecular axis of the crystal with the [1.1.1]orthocyclophane unit nearest the eye. The CH$_2$Cl$_2$ present as a solvate is positioned remote from the observer and provides by disorder a C_3 pattern about the C_3 axis of the host. Pseudo-hydrogen bonding of CH$_2$Cl$_2$ to strong acceptors has been observed previously with highly preorganized systems[16] but is absent in **46**.

The important feature of the two crystal structures **45** and **46** is that the host structures **37** are virtually identical, even though **45** contains a guest and **46** is empty. In both structures the unshared electron pairs and hydrogens of the three nitrogens are directed away from the cavity. *Thus the host's organization is not dependent on occupancy of its cavity.*[13]

---

[16] K.M. Doxsee, M. Feigel, K.D. Stewart, J.W. Canary, C.B. Knobler, and D.J. Cram, *J. Am. Chem. Soc.*, 1987, **109**, 3098.

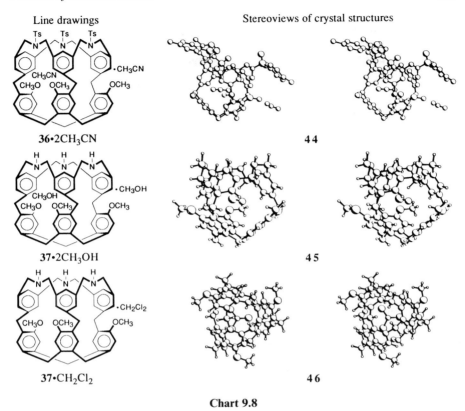

Chart 9.8

9.5.3 Complexation of Tritosylamide Host

Examination of CPK molecular models of hosts **36** and **37** indicates their cavities to be complementary to the ubiquitous small molecules such as O_2, N_2, H_2O, and CO_2, but to be too small to accommodate $CDCl_3$ or CH_2Cl_2 as guests. Thus dissolution of these hosts in such solvents is tantamount to dissolving holes in liquids. Comparison of the 500 MHz ^1H NMR spectra of $CDCl_3$ solutions of **36** in the absence and presence of O_2, N_2, H_2O, and CO_2 provided the following conclusions. (1) The host's proton signals in solutions free of these potential guests changed very little with temperature from 21 to $-40\,°C$. (2) When present, N_2 enters the cavity of **36** exhibiting a binding constant K_a of about $10^2\,M^{-1}$ at temperatures from -3 to $-40\,°C$, the complexation being entropically disfavored. The coalescence temperature is close to $10\,°C$, and the barrier to decomplexation is about 14 kcal mol^{-1} at this temperature. (3) When present, O_2 enters the cavity at $21\,°C$ and greatly broadens all of the host's proton signals except those of the tosyl groups that are remote from the cavity. From -4 to $-40\,°C$, a new set of broad signals appears indicating **36**·O_2 is formed and decomplexes with an activation free energy of about 14 kcal mol^{-1}, and has a K_a value of about $10^2\,M^{-1}$. (4)

Molecules of H_2O when present enter **36**, but the coalescence temperature for water complexation was below $-40\,°C$ and could not be reached due to freezing. (5) The host is almost completely complexed in a CO_2-saturated solution in $CDCl_3$ at $-28\,°C$, and decomplexation at this temperature is slow on the ^1H NMR time scale.[13]

Decomplexation of **36·CH_3OH** in $CDCl_3$ at $22\,°C$ is slow on the ^1H NMR time scale. Incarceration of CH_3OH results in movements of the two sets of guest proton signals upfield by $\Delta\delta = 5.48$ (CH_3) and $\Delta\delta = 5.64$ p.p.m. (OH). Molecular model examination of **36·CH_3OH** shows the CH_3 and OH groups are pressed into the shielding regions of their surrounding aryl groups. At $22\,°C$, the K_a value for **36** binding CH_3OH is $10\,M^{-1}$, and $-\Delta G^0 = 1.4$ kcal mol^{-1}, $\Delta H = -6.6$ kcal mol^{-1}, and $\Delta S = -18$ cal mol^{-1} K^{-1}, the binding being enthalpy driven. Host **36** also binds CH_3CN in $CDCl_3$, but much more weakly. At $-28\,°C$ the protons of complexed CH_3CN were moved upfield by $\Delta\delta = 5.33$.[13]

9.5.4 Complexation of Triamine Host

In $CDCl_3$ at $-28\,°C$, guest exchange between **37·N_2** and dissolved N_2 became slow on the ^1H NMR time scale, and triamine **37** bound N_2 a little more strongly than did tritosylamide **36**. Triamine **37** showed about the same binding behavior toward O_2 and H_2O in $CDCl_3$ as did **36**. Triamine **37** binds CH_3OH in $CDCl_3$ at $22\,°C$ somewhat more strongly than **36**, with $K_a = 47\,M^{-1}$, $-\Delta G^0 = 2.3$ kcal mol^{-1}, $\Delta H \cong -8.5$ kcal mol^{-1}, and $\Delta S = -21$ cal mol^{-1} K^{-1}. The two sets of proton signals of the guest in **37·CH_3OH** are moved upfield by $\Delta\delta = 5.15$ for CH_3, but the OH signal was not observed in the spectrum, indicating that the OH group is rapidly exchanging its proton with external CH_3OH. Such exchange might occur by a proton-relay mechanism involving the three amino groups, or by direct hydrogen bonding of guest CH_3OH with external CH_3OH through the rather wide portals of the host. The binding by **37** of CH_3CN in $CDCl_3$ at $22\,°C$ is considerably stonger than by **36**, the K_a value for formation of **37·CH_3CN** being in the range $1-10\,M^{-1}$. The $\Delta\delta$ value for bound CH_3CN in **37·CH_3CN** is 4.81 p.p.m. The crystal structure of **36·CH_3CN** indicates the sulfonamide nitrogen is sp^2 hybridized, and that this effect coupled with the steric requirements of the Ts groups in **36** imposes conformational constraints on the $ArCH_2NTsCH_2Ar$ moieties, which draws the aryl groups closer together and decreases the cavity size of **36** compared to that of **37**.[13]

Host **37** weakly binds CH_3CH_2OH in $CDCl_3$ with $K_a < 5\,M^{-1}$. In the presence of a large excess of CH_3CH_2OH in $CDCl_3$, the bound guest's ^1H NMR spectral signals are upfield of those of the standard $(CH_3)_4Si$. The guest's CH_2 protons are diastereotropic due to their residence in the cavity of a chiral host. Their $\Delta\delta$ values are 4.60 and 4.84 p.p.m. The CH_3 group signal shows $\Delta\delta = 4.91$ p.p.m. in **37·CH_3CH_2OH**. The OH protons are exchanging too fast to be observed. The decomplexation rate for **37·CH_3CH_2OH** at $22\,°C$ in $CDCl_3$ gave $t_{1/2} = 40$ minutes. Thus although **37** is a very weak binder of EtOH at $22\,°C$ the decomplexation rate is on the human time scale, indicating that the complexation rate is also slow. A CPK molecular model of **37·EtOH** can be easily assembled, but the guest can be forced out through one of the portals only with some difficulty. Thus **37·EtOH** once formed appears to be held together not only by the usual contact forces but also by *weak constrictive binding* which must be overcome for dissociation to occur. This behavior of **37·EtOH** contrasts with that of **37·CH$_3$OH**, which at 22 °C is formed and dissociates instantaneously on the human time scale, and yet is more stable that **37·EtOH**. Thus **37** shows structural recognition of both a thermodynamic and kinetic variety.[13]

9.6 A Host With a Large Cavity

A useful way of considering the size of portals in the hemicarcerands is to count the number of atoms composing the rings through which potential guests must pass to become incarcerated. Lactone **2**, imine **5**, and lactams **8** and **9** all possess four 28-membered ring portals, whereas the polyacetylene host **25** possesses three 32-membered ring portals. Tosylamide host **36** and its amine **37** both

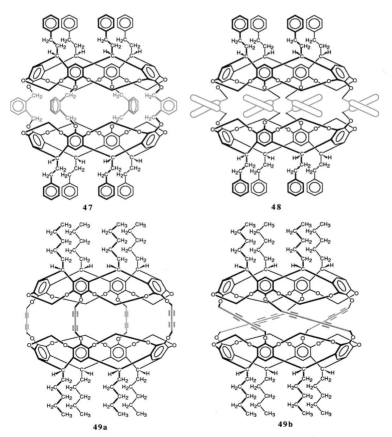

[17] D.J. Cram, R. Jaeger, and K. Deshayes, *J. Am. Chem. Soc.*, 1993, **115**, 10111.

contain three 19-membered ring entryways, the tetra *o*-xylyl host **47** (Section 8.2) has four 26-membered rings and tetrabinaphthyl host **48** (Section 8.3) possesses four 30-membered, but highly constraining, rings. In this section, we discuss the octaacetylenic host **49**, whose four 30-membered ring entryways exhibit the linearity demanded by its CH$_2$—C≡C—C≡C—CH$_2$ components.[17] These long hemisphere-linking components contain a minimum of hydrogens to block entry and exit of guests, and allow the host to assume conformation **49a** which has a polar axis longer than that of any system yet reported. In molecular models, **49a** can twist to conformation **49b**, which decreases the cavity size and closes the portals.

9.6.1 Synthesis

The shell-closing reaction leading to **49** involved oxidative coupling of the tetraacetylenic compound **50** with pyridine–O$_2$–Cu(OAc)$_2$ and went in 5–8% yields. Notice that **49** and **50** contain pentyl feet, rather than the more often used 2-phenylethyl feet of **47** and **48**. Generally, pentyl feet provide the hemicarceplexes with greater solubility than the 2-phenylethyl feet, and even **49** has a limited solubility in organic media. Host **49**, isolated after chromatographic purification (CH$_2$Cl$_2$–silica gel) was free of guest, as expected. Molecular models of common solvents such as CH$_2$Cl$_2$, CHCl$_3$, C$_6$H$_6$, C$_6$H$_5$CH$_3$, C$_5$H$_5$N, and (CH$_2$)$_4$O can easily be inserted into models of **49**. Reduction of the eight carbon–carbon triple bonds of **49** with H$_2$–PdC–C$_6$H$_6$ produced **51** (91% yield), whose complexing properties have not been examined.[17]

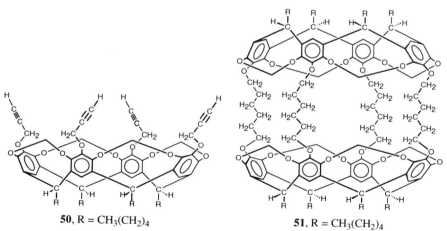

50, R = CH$_3$(CH$_2$)$_4$ **51**, R = CH$_3$(CH$_2$)$_4$

9.6.2 Complexation

Molecular model examination of **49** indicated 1,3,5-tri-*tert*-butylbenzene (TTB) is too large to pass through the portals of **49** and enter the cavity of the host. Experimentally, molten TTB was used as a solvent for introduction of certain guests into **49**, since this host when heated for 7 days at 140 °C under argon failed to complex TTB.

Varieties of Hemicarcerands

Table 9.1 Complexation conditions, yields, and decomplexation half-lives in $CDCl_3$ at 25 °C for **49**·guests

Guest Structure	Label	Solvent[a]	Time (h)	Complexation Temp. (°C)	% complexed[b]	Decomplexation $t_{1/2}$ (h)[c]
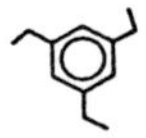	A	guest	144	80	33	960
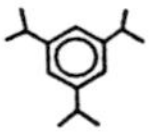	B	guest	72	90	90	1608
	C	guest	168	80	60	13.5
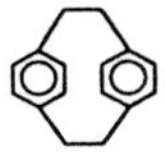	D	TBB[d]	48[e]	80	66	5
	E	TBB[d]	168	80	44	0.5
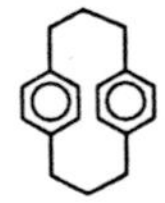	F	TBB[d]	168	80	82	13
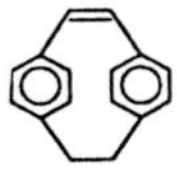	G	TBB[d]	48	120[f]	90	8
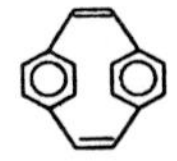	H	TBB[d]	48	80	90	11
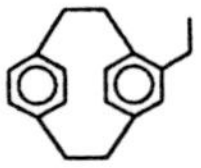	J	TBB[d]	48	120[f]	48	216
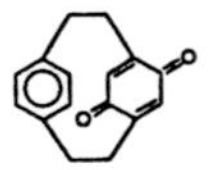	K	TBB[d]	48	140[f]	80	24
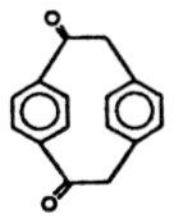	L	TBB[d]	72	80	61	1
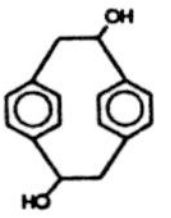	M	TBB[d]	48	90	57	4

[a]Melts at 80–140 °C. [b]Mixtures of empty host and 1:1 complexes. [c]Calculated from first-order rate constants for decomplexation obtained from ^{1}H NMR spectral changes with time. [d]1,3,5-[(CH$_3$)$_3$C]$_3$C$_6$H$_3$. [e]A large amount of excess insoluble guest was present. [f]Lower temperatures and times gave lower percentages complexed.

When a solution of *p*-xylene, 1,4-diethylbenzene, 1,4-diisopropylbenzene, 1,4-di-*tert*-butylbenzene, *tert*-butylbenzene, 1,3,5-trimethylbenzene, hexamethylphosphoramide, adamantane, α-pinene, 1-aminoadamantane (amantidine), or 2-ketoadamantane with **49** was heated, cooled, and the host and potential complex isolated, only free host was recovered. We conclude that these guests enter the cavity of **49**, but the guests escape during the isolations, which were carried out at ambient temperatures.

Another group of guests failed to enter the cavity of **49** at high temperatures because their size and/or shape precluded their formation of complexes. These compounds were [3.4]paracyclophane, ferrocene, ruthenocene, tripiperidylphosphine oxide, 1,3-dicarbomethoxyadamantane, 4,12-dinitro[2.2]paracyclophane, 1,2,3,6,7,8-hexahydropyrene, 1,3-diphenylacetone, *trans*-stilbene, triphenylmethane, 2,6-dimethoxynaphthalene, 9,10-anthraquinone, and cortisone.

Table 9.1 records the structures of the guests that entered **49** at 80–140 °C either with molten guest as solvent, or in TTB as solvent, and whose hemicarceplexes survived isolation. Guests A–M of Table 9.1 range in their nonhydrogen atoms from twelve for 1,3,5-triethylbenzene (A) and 1,3-dimethyladamantane (C) to eighteen for [3.3]paracyclophane (F), 4-ethyl[2.2]paracyclophane (J), and the three disubstituted [2.2]paracyclophanes (K–M). These guests are either rigid or have very few degrees of freedom of rotation of their parts. In CPK molecular model experiments, these guests are either very difficult to force into the cavity of **49**, or *cannot quite* be pushed into the cavity without breaking the bonds of the host. Once inside the host model, the guest models do not inhibit the twisting of the two hemispheres of the host relative to one another, as in **49b**, thereby increasing the number of host-to-guest and host-to-host contacts. Not surprisingly, the wider guests were qualitatively more difficult to introduce into the cavity than the narrower ones.[17]

The presence of an additional single methylene in the bridge of [3.3]paracyclophane (F) was enough to destroy its ability to enter **49**, since [3.4]paracyclophane failed to form a complex. A less dramatic example of structural recognition was the ability of 1,3,5-triethylbenzene (A) to be incarcerated at high temperatures and be retained upon cooling, but 1,3,5-trimethylbenzene was small enough to escape during isolation.[17]

9.6.3 Correlation of Decomplexation Rates with Guest Structures

Table 9.1 records the half-lives at 25 °C for decomplexation of the twelve hemicarceplexes of **49** in $CDCl_3$, which decreased from $t_{1/2}$ = 1608 hours for B and 960 hours for A to 5 for D and 0.5 hour for E. Guest shape rather than the number of non-hydrogen atoms dominates the order. For example, the $t_{1/2}$ = 960 hours for 1,3,5-triethylbenzene (A) containing twelve non-hydrogen atoms is about three powers of ten greater than that for 2,9-diketo[2.2]paracyclophane (L) with $t_{1/2}$ = 1 hour and containing eighteen non-hydrogen atoms. The 30-membered ring through which a guest must pass to enter the cavity of **49** in its open form (**49a**) is roughly rectangular in shape, being much longer in its axial compared to its equatorial dimension. Thus 1,4-disubstituted benzene derivatives possessing rectangular cross sections easily enter and exit the cavity. Only when they become extended in the third dimension as in the paracyclophanes is constrictive binding visible in the exit half-lives. In contrast, the 1,3,5-trisubstituted benzenes possess cross sections that are more square, one of whose dimensions is non-complementary to the narrow rectangular portal when the substituents are ethyls as in A, or isopropyls as in B. Model examinations indicate that for A or B to enter the portal of **49a** and not pass out the opposite portal, the guest on entering must rotate in such a way as to align its long axis with the long axis of the host, a motion resisted by the walls of the cavity. Of the paracyclophanes, resistance to this entry-rotation motion is pronounced only with 4-ethyl[2.2]paracyclophane (J), whose $t_{1/2}$ = 216 hours.

Resistance to the screw-like motion required for exit (or entry, same transition states) of guests increases in the order for **49**·Guest when the guest is [2.2]paracyclophane (D), $t_{1/2}$ = 5 hours < [3.3]paracyclophane(F), 13 hours, < [3.4]paracyclophane, off scale. The interesting observation that **49**·[2.3]paracyclophane (**49**·E, 0.5 hour) decomplexes faster than either **49**·[2.2]paracyclophane or **49**·[3.3]paracyclophane is explained by the fact that because of the unlike bridge lengths in [2.3]paracyclophane, this guest possesses a preorganized screw-like structure along its long axis, and thus is more complementary to the transition states for exit or entry into the cavity of **49a** than its symmetrical homologues, D and F.[17]

9.7 A Highly Adaptive and Strongly Binding Hemispherand

The wide range of guests incarcerated into host **47** encouraged the study of **52**.[18] Both **47** and **52** contain four six-atom bridges between their two hemispheres,

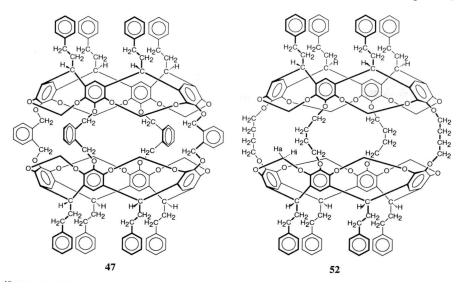

47

52

[18] T.A. Robbins, C.B. Knobler, D.R. Bellew, and D.J. Cram, *J. Am. Chem. Soc.*, 1994, **116**, 111.

and differ only in the sense that the bridges of **52**, $[O(CH_2)_4O]_4$, are more conformationally mobile than the (O-*o*-xylyl-O)$_4$ bridges of **47**. Not only should the 26-membered ring portals of **52** be more adaptive to the shapes of entering guests, but the shells of the resulting hemicarceplexes are more amenable to minimizing their unoccupied internal volumes than those of **47** (Section 8.2).

9.7.1 Synthesis

The synthesis of **52**·$(CH_3)_2NCOCH_3$ was accomplished through the shell closure of two moles of tetraphenol **1** (Section 9.1) with four moles of $TsO(CH_2)_4OTs$ in $(CH_3)_2NCOCH_3$–Cs_2CO_3 at 70 °C in 30–40% yields. The same shell closure conducted in $(CH_3)_2SO$–Cs_2CO_3 gave **52**·$(CH_3)_2SO$ in 15–20% yield. Empty **52** was obtained (94%) by heating **52**·$(CH_3)_2NCOCH_3$ in $(C_6H_5)_2O$ (molecules too large to be incarcerated) for 5 days. Solvent molecules of CH_2Cl_2 went in and out of the inner phase of **52** fast enough at 25 °C to be useful as a chromatographic solvent for **52** and its complexes. The kinetics for decomplexation of **52**·$(CH_3)_2NCOCH_3$ were followed in $C_6D_5NO_2$ at 140–170 °C, and provided an E_a value of 23.5 kcal mol^{-1} for decomplexation in this solvent. At the temperatures and times involved, little $C_6D_5NO_2$ became complexed.[18]

9.7.2 Complexation

Empty **52** was heated up to 5 days at 57–195 °C with the compounds listed in Chart 9.9 either with guest as solvent or with a 100 to 1 molar excess of guest in $(C_6H_5)_2O$. The 29 stable 1:1 complexes listed in Chart 9.8 were formed and characterized. These incarcerated guests fall into Classes A–F with respect to their shapes and sizes. Molecular models of guests A can be pushed into a model of **52**, and in the resulting models of the complexes, the northern and southern hemispheres can be twisted around the polar axis with respect to one another in those complexes containing the smaller guests. Experimentally the ^{1}H NMR spectra of those hemicarceplexes containing non-like-ended guests such as CH_3CH_2I, $(CH_3)_2NCOCH_3$, and $CH_3CH_2O_2CCH_3$ show the presence of only one kind of H$_i$ protons (see **52**), indicating these guests at 25 °C can rotate rapidly on the NMR time scale relative to their host. The guest's methyl group signals are changed by incarceration by values for $\Delta\delta$ that are as high as 4.23 p.p.m. for $CH_3CO_2CH_2CH_3$.[18]

The generally slot-shaped Class B guests in models are easier to push through the slot-shaped portals of **52** than are the more conformationally mobile guests of Class A. The CH_2 protons of these cyclic guests cannot in models of **52**·B penetrate as deeply into the shielding regions of the host as do the CH_3 protons in the **52**·A group, and the $\Delta\delta$ of their ^{1}H NMR signals are lowered accordingly.[18]

All of the aromatic guests of Classes C–F possess slot-shaped cross sections with all their non-hydrogen atoms coplanar except the methyl groups of 1,2- and 1,4-$(CH_3O)_2C_6H_4$. In models constructed with new bonds, all of these

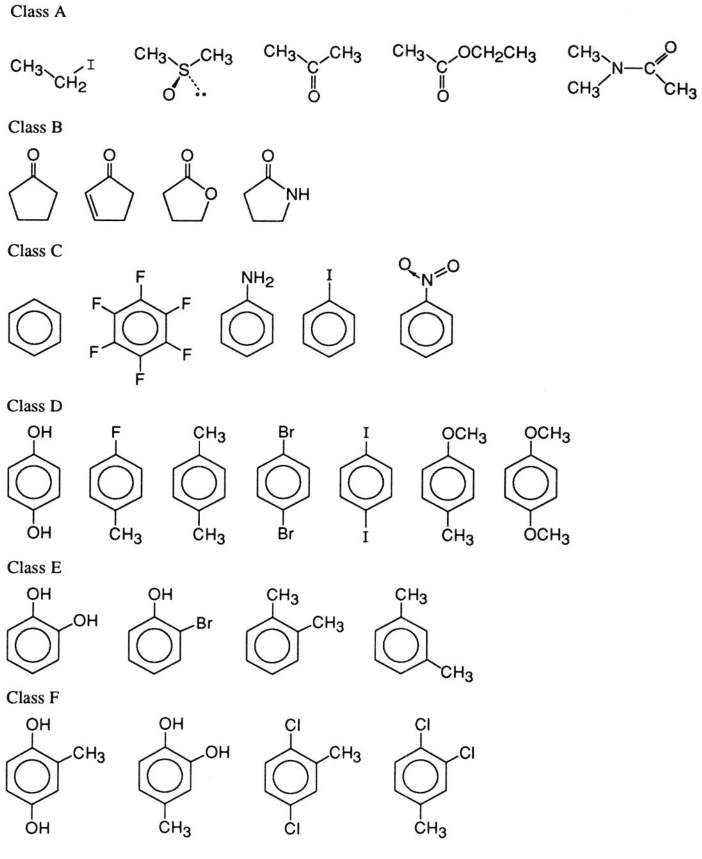

Varieties of Hemicarcerands

Class A

CH₃–CH₂–I, (CH₃)₂S=O, (CH₃)₂C=O, CH₃–C(=O)–OCH₂CH₃, (CH₃)₂N–C(=O)–CH₃

Class B

cyclopentanone, cyclopentenone, γ-butyrolactone, 2-pyrrolidinone

Class C

benzene, hexafluorobenzene (C₆F₆ shown with 5 F), aniline, iodobenzene, nitrobenzene

Class D

1,4-(OH)₂C₆H₄, 1,4-F₂C₆H₄ (1-F, 4-CH₃ — para-fluorotoluene), 1,4-(CH₃)₂C₆H₄, 1,4-Br₂C₆H₄, 1,4-I₂C₆H₄, 4-CH₃-C₆H₄-OCH₃, 1,4-(OCH₃)₂C₆H₄

Class E

1,2-(OH)₂C₆H₄, 2-Br-C₆H₄-OH, 1,2-(CH₃)₂C₆H₄, 1,3-(CH₃)₂C₆H₄

Class F

1,2,4-(OH)₂-CH₃-C₆H₃ (OH, CH₃, OH), 1,2,4-(OH)₂-CH₃-C₆H₃ (OH, OH, CH₃), 2,4-Cl₂-CH₃-C₆H₃, 2,4-Cl₂-CH₃-C₆H₃

Chart 9.9

guests can be pushed into **52** without breaking bonds, and for the larger guests, only with difficult and synchronous conformational changes of the host. To fully twist the hemispheres of the host, the axis of C_6H_6 as guest must assume a north–south pole alignment. As the guests become more highly substituted, this alignment becomes more mandatory, and the twisting of the host's hemispheres is more inhibited, so that in **52**·*p*-$(CH_3)_2C_6H_4$ or **52**·*p*-$I_2C_6H_4$ little twisting is possible. Of the non-like-ended aromatic guests, only C_6H_5I, 4-$CH_3C_6H_4$ OCH_3, 2,4-$Cl_2C_6H_3CH_3$, and 3,4-$Cl_2C_6H_3CH_3$ display two different sets of H_i and H_a protons in their ¹H NMR spectra, which shows only these guests cannot rotate rapidly on the NMR time scale around the short equatorial axes of their hosts at 25 °C. Generally, the size of $\Delta\delta$ values correlate well with how deeply the protons of the guest are forced into the shielding polar caps of the host. Examples are 4.46 p.p.m. for 3,4-$Cl_2C_6H_3CH_3$, 4.40 p.p.m. for 1,4-$(CH_3O)_2C_6H_4$, and 4.10 p.p.m. for 1,4-$NO_2C_6H_4H$. These values roughly correlate with the lengths of the long axes of the guests.[18]

9.7.3 Structural Recognition in Complexation

Qualitative conclusions regarding relative rates of complexation were drawn from the times, temperatures, and inadvertent competition experiments that were conducted. Clearly the small open-chain guests of Class A, with the exception of $(CH_3)_2NCOCH_3$, complex **52** more readily than the cycles of Class B, which in turn complex more easily than the monosubstituted benzenes of Class C. Of the members of this series, $C_6H_5NO_2$ complexed **52** the least readily, being more comparable to the 1,4-disubstituted benzenes of Class E, whose most easily complexed member was 1,4-$(HO)_2C_6H_4$. One or two hydroxyl groups attached to a benzene expedited complexation. Generally, 1,4-disubstituted isomers complexed **51** very much faster than did 1,3-disubstituted isomers.[18]

9.7.4 Crystal Structures

Chart 9.10 contains side stereoviews and the corresponding top partial stereoviews **53–64** of **52**·p-$I_2C_6H_4$·$2C_6H_5NO_2$, **52**·p-$(CH_3)_2C_6H_4$·$2C_6H_5NO_2$, **52**·$O_2NC_6H_5$·$2C_6H_5NO_2$, **52**·o-BrC_6H_4OH·$2C_6H_5NO_2$, **52**·$(CH_3)_2N$-$COCH_3$·$2C_6H_5NO_2$, and **52**·$6H_2O$. In all six top views, the four $(CH_2)_4$ interhemispheric bridging groups lie outside the volumes defined by the attached eight ArO oxygens of the host, which in two sets of four are connected by straight lines to form two squares that are parallel to one another. This arrangement provides an *out–in–out* orientation of unshared electron pairs for the three oxygens attached to adjacent carbons on each of the eight benzene rings of the host. This arrangement provides for the greatest compensation of dipoles and lowest energy. An alternate *in–out–in* arrangement of these oxygens is impossible to construct with CPK models since the methines and one hydrogen of the OCH_2O bridges collide.[18]

In CPK models of **52**, when the $(CH_2)_4$ interhemispheric bridges are all distributed outside their attached oxygens as shown in top views **54–64**, the 26-membered ring portals are generally at their widest in the horizontal dimension, but are shortest in their vertical. We label this conformation $(BO)_4$, meaning *four bridges out*. At the other extreme, the unshared electron pairs of the eight oxygens of the $[O(CH_2)_4O]_4$ bridges face outward forcing the four carbon chains inward. In this conformation, the portals and cavities become longest in the vertical dimension, and shortest in the horizontal. We label this conformation $(BI)_4$, meaning *four bridges in*. Many other combinations and permutations are possible, each of which modifies the shapes of the portals somewhat. These many conformations provide limited flexibility for the used and unused portals to adapt to the variously shaped guests entering the host, and realigning their axes.[18]

Once complexed, molecular models of **52**·Guest in the $(BO)_4$ form show the host has two ways to maximize the area of host–guest contacts and minimize its unused space. In the first of these, the four central C—C bonds of the $(CH_2)_4$ bridges can approach coplanarity with one another as in **61** and **63**, which

Varieties of Hemicarcerands

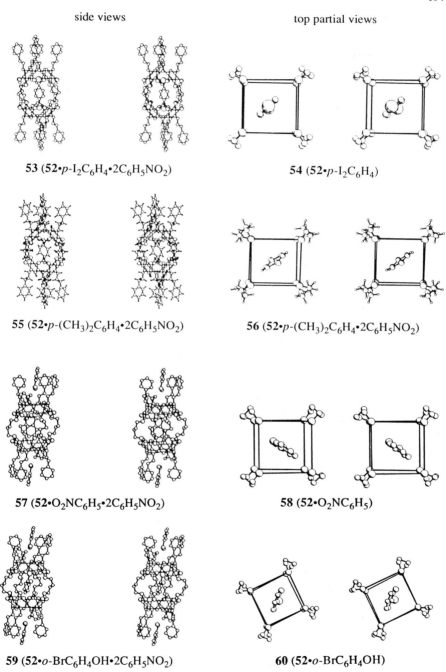

side views — top partial views

53 (**52**·p-I$_2$C$_6$H$_4$·2C$_6$H$_5$NO$_2$) **54** (**52**·p-I$_2$C$_6$H$_4$)

55 (**52**·p-(CH$_3$)$_2$C$_6$H$_4$·2C$_6$H$_5$NO$_2$) **56** (**52**·p-(CH$_3$)$_2$C$_6$H$_4$·2C$_6$H$_5$NO$_2$)

57 (**52**·O$_2$NC$_6$H$_5$·2C$_6$H$_5$NO$_2$) **58** (**52**·O$_2$NC$_6$H$_5$)

59 (**52**·o-BrC$_6$H$_4$OH·2C$_6$H$_5$NO$_2$) **60** (**52**·o-BrC$_6$H$_4$OH)

Chart 9.10 *Crystal structures of* **52**·*guests*

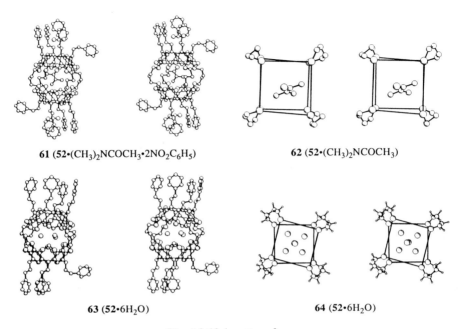

61 (52·(CH$_3$)$_2$NCOCH$_3$·2NO$_2$C$_6$H$_5$)   **62** (52·(CH$_3$)$_2$NCOCH$_3$)

63 (52·6H$_2$O)   **64** (52·6H$_2$O)

Chart 9.10 (*continued*)

minimizes the distances between the oxygen planes (**62** and **64**), and adapts to the relatively short (CH$_3$)$_2$NCOCH$_3$ and (H$_2$O)$_6$ guests. To maximize the distance as in **53** and **55** to accommodate the greater lengths of *p*-I$_2$C$_6$H$_4$ and *p*-(CH$_3$)$_2$C$_6$H$_4$ guests, the four central C—C bonds of the (CH$_2$)$_4$ bridges depart markedly from a coplanar arrangement.[18]

The second way the (BO)$_4$ conformation can adapt to the guest is by twisting the northern hemisphere relative to the southern around the polar axis, a movement that shortens the polar axis of the complex. Of the six structures of Chart 9.10, only **52·6H$_2$O** makes use of this conformational change, which is particularly visible in **64**, whose angle of twist is 15°. The maximum twist angle among the other complexes is only 1°.

In spite of the many different sizes and shapes of the guests in **53–61**, their five crystal structures all belong to the same space group, *P2$_1$/c*. In these five structures, two C$_6$H$_5$NO$_2$ molecules fill the cavitand-like void created by the two sets of four C$_6$H$_5$CH$_2$CH$_2$ appendages, with the nitro groups closest to the globe-shaped trunk of the host. The volumes at 25 °C of the five unit cells range from 6895 Å^3 for **52**·*p*-I$_2$C$_6$H$_4$·2C$_6$H$_5$NO$_2$ (**53**) to 6799 Å^3 for **52**·(CH$_3$)$_2$N-COCH$_3$ (**61**), a change of about 1.4%. Presumably these complexes are isomorphous and can fit into each other's lattices.[18]

The hosts of the six complexes of Chart 9.10 in each case are centrosymmetric, whereas complexes **52**·*p*-I$_2$C$_6$H$_4$·2C$_6$H$_5$NO$_2$ and **52**·*p*-(CH$_3$)$_2$C$_6$H$_4$·2C$_6$H$_5$NO$_2$ as a whole are centrosymmetric. The polar axes of the host and guest are colinear in **52**·*p*-I$_2$C$_6$H$_4$ and **52**·*p*-(CH$_3$)$_2$C$_6$H$_4$. The effective polar length of the cavities approximately equals the distances between the north and south polar carbon planes minus 1.54 Å, the diameter of a covalently bonded carbon. The cavity lengths arranged in decreasing order are as follows (Å): **52**·p-I$_2$C$_6$H$_4$ (10.27); **52**·p-(CH$_3$)$_2$C$_6$H$_4$ (9.89); **52**·o-BrC$_6$H$_4$OH (9.67); **52**·O$_2$NC$_6$H$_5$ (9.60); **52**·(CH$_3$)$_2$NCOCH$_3$ (9.57); and **52**·6H$_2$O (8.14). These distances compare with the long-axis lengths of the guests in CPK models as follows (Å): p-I$_2$C$_6$H$_4$ (10.35); p-(CH$_3$)$_2$C$_6$H$_4$ (8.5); o-BrC$_6$H$_4$OH (8.61); O$_2$NC$_6$H$_5$ (8.20); and (CH$_3$)$_2$NCOCH$_3$ (7.2). Thus the cavity length of **52**·I$_2$C$_6$H$_4$ is 0.09 Å shorter than the guest model length, which is about a 1% departure from a fit in this dimension. The other guests are substantially shorter than the cavity of their host taken from the crystal structures. As expected, the cavities of the twisted complexes **52**·6H$_2$O and **3**·Guest (Section 8.2) are 10 and 20%, respectively, shorter than the untwisted cavities.[18]

The maximum diameters of the cavities are best measured by the diagonals of the two squares of oxygens minus the diameter of a covalently bound oxygen (2.80 Å). The resulting distances (Å) in increasing order are as follows: **52**·I$_2$C$_6$H$_4$ (5.84); **52**·O$_2$NC$_6$H$_5$ (6.12); **52**·(CH$_3$)$_2$NCOCH$_3$ (6.16); **52**·o-BrC$_6$H$_4$OH (6.16); **52**·6H$_2$O (6.18); **52**·p-(CH$_3$)$_2$C$_6$H$_4$ (6.26); and **3**·Guest (6.32). As is intuitively obvious, the longer the cavity of **52** (e.g. **52**·I$_2$C$_6$H$_4$), the narrower it is. Furthermore, the broader its cavity (e.g. **52**·6H$_2$O), the shorter the length. The fact that the spread is small in the broad dimension testifies to the overall rigidity of the cavities.

9.7.5 The Unusual Guest Structure in 52·6H$_2$O

Complex **52**·6H$_2$O formed from **52** after guest had been driven out by heating **52**·(CH$_3$)$_2$NCOCH$_3$ in (C$_6$H$_5$)$_2$O for several days and **52** isolated. Elemental analysis showed the presence of two H$_2$O molecules. When this latter sample was crystallized slowly from o-(CH$_3$)$_2$C$_6$H$_4$ in contact with the atmosphere, **52**·6H$_2$O·4o-(CH$_3$)$_2$C$_6$H$_4$ crystallized. Thus **52** accumulated over time 6 molecules of water, which fitted themselves into the irregular octahedral structure visible in stereoview **64** (Chart 9.10). Chart 9.11 provides a top view of the geometric relationships of the six guest oxygens of the 6H$_2$O guest, and the eight host oxygens that terminate the host's four O(CH$_2$)$_4$O bridges. The distances between the closest oxygens are included in the view of Chart 9.11. The hydrogens were not located in the crystal structure.

Crystal structure **52**·6H$_2$O was readily assembled in CPK models in which the six H$_2$O molecules are hydrogen bonded to one another and to the host as follows. Oxygens of the four equatorial waters form a square held together by one hydrogen bond between each two oxygens, the four remaining hydrogens being hydrogen bonded to the host's eight equatorial oxygens. The remaining two H$_2$O molecules which are located one in each of the polar caps donate hydrogen bonds to the two diagonally-located equatorial oxygens' unshared electron pairs. In this model, the four hydrogen bonds between the oxygens of the guest and host are either bifurcated or averaged. *This arrangement is complementary to the roughly octahedral shape of the inner phase of empty **52**.*

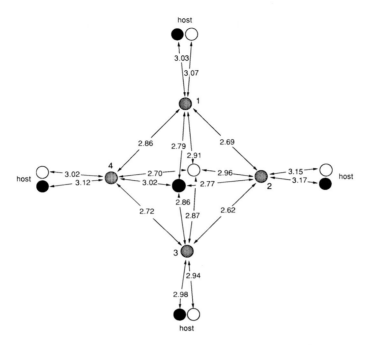

Chart 9.11 *Top view of the arrangement of six guest and eight host oxygen atoms in 52·6H$_2$O crystal structure, with the O···O distances in Å, the dark oxygens above the plane of the page, the cross-hatched oxygens in that plane, and the light oxygens below the plane*

The average close O-to-O distances (Å) in the six H$_2$O guest molecules are as follows: four O$_e$-to-O$_e$, 2.72 ±0.07; and eight O$_e$-to-O$_a$, 2.86 ±0.08. The overall average of O----O close distances in 6H$_2$O is 2.81 Å *vs.* 2.76 Å for ordinary ice.[19] The average of the eight close oxygen distances between host-equatorial and guest-equatorial oxygens is 3.06±0.08 Å. These distances are all within the normal range for oxygen-to-oxygen hydrogen bonding. The average O----O$_e$----O bond angle in which O$_e$ is one of the four equatorial oxygens (numbered 1 through 4 in Chart 9.11) is 107°, close to the 109.5° of ice I.[19] Although the distances and bond angles in 6H$_2$O guest are close to what is observed in ice I, the octahedral oxygen arrangement in 6H$_2$O is not encountered among the I–VIII polymorphs of ice,[19] and therefore is an example of the guest adapting its structure to that of its molecular vessel.

As has become evident from the earlier chapters, the predictions made from CPK molecular model examination have been remarkably well borne out by crystal structure determinations. The greatest test of the utility of these models involves the results reviewed on the host–guest relationships in the hemi-

---

[19] D. Eisenberg and W. Kauzmann, 'The Structure and Properties of Water', Oxford University Press, New York, 1969, p. 85.

carceplexes of Chapter 8 and this chapter. Many of these hosts possess molecular weights well in excess of 2000, and the guests in excess of 200, hundreds of atoms being involved in their complexes. When distances that run in excess of 10 Å match well between model and experiment, the level of faith required to do the next experiments based on models lowers the activation energy for making the effort.

CHAPTER 10

Reactions of Complexed Hosts, of Incarcerated Guests, and Host Protection of Guests from Self Destruction

'At what point should an investigator change the direction of his or her research? Clearly not until certain conditions have been fulfilled. A complete failure of the experiments should be recognized as early as possible. When the research is successful, the first few generations of questions addressed should be answered, enough so that additional questions and the methods of answering them have become obvious. Certainly a package of self-supporting, internally consistent results should be in hand. When the chances of success of a further series of experiments possess a certain predictable level, the challenge, charm, and intrigue of the work starts to dissipate. Another indicator is the point at which surprising results that demand explanation cease to appear. Each investigator can be characterized by his or her tolerance for ambiguity, predictability, failure, success, adventure, misadventure, the abstract, and the concrete. One of the many beautiful things about organic chemical research is that all types of temperaments, talents, inclinations, and propensities of researchers are needed for the field to prosper.'

The inner spaces of carcerands and hemicarcerands are unique. They are submicrophases of molecular dimensions which possess fixed boundaries and designable shapes and sizes. Depending on the spatial and electronic relationships between host and guest, the inner phase can blend some of the properties of the classical liquid, solid, and gas phases. In Chapters 7, 8, and 9, carceplexes and hemicarceplexes were discussed, whose crystal structures demonstrated that whereas hosts are locked in a molecular lattice, their guests were sometimes immobilized as in a crystal, and sometimes mobile as in the gas or liquid phases. In other cases, the guests were disordered only in the sense that two alternative placements of guest with respect to host were observed. For dissolved carceplexes or hemicarceplexes, rotations of guests relative to their hosts about certain axes occurred rapidly, but rotations about other axes were physically blocked for lack of space. While carcerands and hemicarcerands taken as a whole can exist in the gas, liquid, or solid phases, their guests reside in a phase of their own, the inner phase.

Carcerands and hemicarcerands are molecular containers, whose inner phases invite study and elicit a wide range of questions never before asked. What kinds of chemical reactions can be carried out in these inner phases? What kinds of protection from bulk phase reactions can the shells of hosts offer highly reactive guest compounds, which normally dimerize or polymerize? How are the rates of chemical reactions affected by their being carried out in the inner phase? Can the partitioning of guest starting material between products be directed toward more compact products by the confining boundaries of the inner phase? Can highly reactive catalysts be incorporated into hemicarcerands where they are protected from poisoning but catalyze desired but not undesired reactions, based on selective admission of substrates to the inner phase? Can reaction intermediates such as radicals or carbenes be prepared and preserved in the inner phase? Can photochemical, thermal, oxidation and reduction, oxy-Cope and other rearrangements be carried out in the inner phase? Can inner phase chirality induce stereospecificity in inner phase reactions? Can reactions be carried out on carcerands or on carceplex hosts while guests occupy their interiors?

Just a beginning to answers to these questions has been addressed experimentally. This chapter points to results obtained in our research group through 1993, some of which are still fragmentary.

10.1 Energy Barriers to Amide Rotations in the Inner Phase of a Carcerand

Comparisons of the energies of the rotational barrier to amide rotation in the liquid, gas, and inner phases of carcerands provides a means of characterizing these new phases. Accordingly, dynamic ^{1}H NMR experiments were conducted to obtain these barriers for $(CH_3)_2NCOCH_3$ and $(CH_3)_2NCHO$ dissolved in $C_6D_5NO_2$ and inside carceplexes $1 \cdot (CH_3)_2NCOCH_3$ and $1 \cdot (CH_3)_2NCHO$ dissolved in the same solvent. The coalescence of the two CH_3 resonances of $(CH_3)_2NCHO$ inside of $1 \cdot (CH_3)_2NCHO$ dissolved in $C_6D_5NO_2$ occurred at 140 °C and gave an activation free energy of $\Delta G^{\ddagger}_{413 \text{ K}} = 18.9$ kcal mol^{-1}. The coalescence of the two CH_3 resonances of $(CH_3)_2NCHO$ dissolved directly in $C_6D_5NO_2$ occurred at 120 °C and provided $\Delta G^{\ddagger}_{393 \text{ K}} = 20.2$ kcal mol^{-1}. Thus the energy barrier is lower for this incarcerated amide than for the directly dissolved amide. The coalescence of the two CH_3 resonances of $(CH_3)_2N$-$COCH_3$ inside of $1 \cdot (CH_3)_2NCOCH_3$ dissolved in $C_6D_5NO_2$ occurred at 190 °C and gave $\Delta G^{\ddagger}_{463 \text{ K}} = 20.3$ kcal mol^{-1}, whereas for $(CH_3)_2NCOCH_3$ dissolved directly in $C_6D_5NO_2$, the coalescence temperature was 63 °C with $\Delta G^{\ddagger}_{336 \text{ K}} = 18$ kcal mol^{-1}. Thus incarceration of $(CH_3)_2NCOCH_3$ *raises* the rotational barrier relative to that for simple dissolution.[1]

[1] J.C. Sherman, C.B. Knobler, and D.J. Cram, *J. Am. Chem. Soc.*, 1991, **113**, 2194.

Generally C—N rotational barriers of amides decrease in the order polar solvents > non-polar solvents > gas phase. If we fit the current results into this correlation for $(CH_3)_2NCHO$, the barrier decreases in the order polar solvent > carcerand inner phase > vacuum. Molecular model examination of **1**·$(CH_3)_2NCHO$ indicates the carcerand inner phase is a mixture of vacuum and guest occupation of volume that leaves the rotating parts of the $(CH_3)_2NCHO$ less encumbered by the sides of the container than by $C_6D_5NO_2$ solvating molecules. In contrast CPK models of **1**·$(CH_3)_2NCOCH_3$ can barely be assembled because the container sides are compressed against the guest parts. Rotation of the $(CH_3)_2N$ plane out of the $COCH_3$ plane by 90° to generate a transition-state model appears to require more deformation of the container than does the planar guest. Thus the inner phase of carcerands can be designed to be vacuum-like, liquid-like, or even solid-like, depending on the mix of guest occupation and free space.[1]

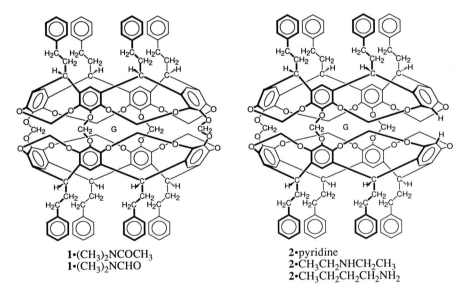

1·$(CH_3)_2NCOCH_3$
1·$(CH_3)_2NCHO$

2·pyridine
2·$CH_3CH_2NHCH_2CH_3$
2·$CH_3CH_2CH_2CH_2NH_2$

10.2 Acidity of Amine Salts in the Inner Phase of a Hemicarcerand

One of the simplest reactions to be carried out in the inner phase is the protonation of amines encapsulated in hemicarcerand **2**.[2] Free pyridine dissolved in $CDCl_3$ was found by ^1H NMR to be readily protonated upon the addition of a large excess of CF_3CO_2D as evidenced by the upfield shift of its protons. Under the same conditions, pyridine incarcerated in **2** gave no evidence of being protonated. Thus incarcerated pyridine is much less basic than free pyridine. Several effects are probably responsible. The rigid host interior has a limited ability to solvate the pyridinium ion, and the shell of the host inhibits the formation of contact ion pairs between the pyridinium and

[2] D.J. Cram, M.E. Tanner, and C.B. Knobler, *J. Am. Chem. Soc.*, 1991, **113**, 7717.

$CF_3CO_2^-$ ions, although shell-separated ion pairs might form. In comparison, individual $CDCl_3$ molecules can arrange themselves to solvate pyridinium ion and accommodate to contact ion-pair formation. It is conceivable but highly unlikely that the unshared electron pair of incarcerated pyridine is so hindered as to be unavailable for protonation.

In contrast to **2·pyridine**, **2·$CH_3CH_2NHCH_2CH_3$** dissolved in $CDCl_3$ was converted instantaneously (^{1}H NMR detected) to its conjugate acid by addition of 100 equivalents of CF_3CO_2D, and the charged guest instantaneously decomplexed to give **2** and $Et_2ND_2^+ \cdot {}^-O_2CCF_3$. Thus **2·$CH_3CH_2NHCH_2CH_3$** is both kinetically and thermodynamically unstable relative to **2** + $Et_2ND_2^+$ in $CDCl_3$. The N: atom of the $CH_3CH_2NHCH_2CH_3$ guest is located in the equatorial region of **2·$CH_3CH_2NHCH_2CH_3$** close to the portal, which undoubtedly facilitates the deuteron transfer. The $CF_3CO_2^-$ counterion generated is envisioned as 'pulling the guest out of the host'. The addition of 100 equivalents of weaker acid CD_3CO_2D to **2·$CH_3CH_2NHCH_2CH_3$** dissolved in $CDCl_3$ resulted only in exchange of NH for ND, but the amount of salt formed as an intermediate was not detected. Thus incarcerated amine undergoes isotopic exchange *via* undetected salt formation much more rapidly than decomplexation occurs.[2]

Hemicarceplex **2·$CH_3CH_2CH_2CH_2NH_2$** dissolved in $CDCl_3$ displayed a third type of behavior when treated with 10 equivalents of CF_3CO_2D. An ^{1}H NMR spectrum of the solution taken immediately provided a new set of guest signals that accounted for two-thirds of the bound amine signals. The downfield shifts of these signals relative to those of **2·$CH_3CH_2CH_2CH_2NH_2$** are consistent with their assignment as the conjugate acid of the amine. The remaining one-third of the amine signals appeared as **2·$CH_3CH_2CH_2CH_2NH_2$**. The new set of *guest* signals was accompanied by a new set of *host* signals. The simultaneous appearance of **2·$CH_3(CH_2)_3NH_2$** and **2·$CH_3(CH_2)_3ND_3^+$** signals in comparable amounts indicated that the pK_a of incarcerated $CH_3(CH_2)_3ND_3^+$ is close to that of CF_3CO_2D in $CDCl_3$. In addition, the results indicate that proton transfers between **2·$CH_3(CH_2)_3ND_3^+$** and **2·$CH_3(CH_2)_3NH_2$** are slow on the ^{1}H NMR time scale. When the above experiment was repeated with the addition of 100 equivalents of CF_3CO_2D, complete salt formation occurred and the ^{1}H NMR spectrum became insensitive to further acid addition. Molecular model examination of **2·$CH_3(CH_2)_3NH_2$** suggests that the isotopic exchange and salt formation reactions occur *via* the hole in the polar cap closest to the NH_2 group.

Acidification of **2·$CH_3(CH_2)_3NH_2$** to form **2·$CH_3(CH_2)_3ND_3^+$** greatly accelerated decomplexation. The 2:1 ratio of acidified to non-acidified complexed amine remained constant with time, while the decomplexation proceeded to completion to give free **2** and $CH_3(CH_2)_3ND_3^+ \cdot {}^-O_2CCF_3$, the reaction half life being about 10 minutes at 22 °C. The addition of a large excess of the weaker acid CD_3CO_2D to a solution of **2·$CH_3(CH_2)_3NH_2$** in $CDCl_3$ at 22 °C resulted only in the exchange of the amine protons for deuterons.[2]

10.3 Hemicarcerand as a Protecting Container

Ordinarily $CH_3(CH_2)_3Li$ reacts instantaneously with CH_2Br_2 in $(CH_2)_4O$ as solvent. The formation of hemicarceplex $2 \cdot CH_2Br_2$ was reported in Section 8.1.3. When a 0.5 mM solution of $2 \cdot CH_2Br_2$ in dry $(CH_2)_4O$ was treated with 300 equivalents of $CH_3(CH_2)_3Li$ for one minute at 25 °C, the ^1H NMR spectrum of the recovered complex showed the CH_2Br_2 to be intact. Thus the reactive CH_2Br_2 was inaccessible to the highly reactive reagent because of the guest protection provided by the host.[2]

In a second experiment, empty **2** dissolved in $CHCl_3$ was mixed with excess CH_2N_2 in $CH_3C_6H_5$ (1 M), and the resulting solution was stirred for 80 minutes at 25 °C. The complex was precipitated by flooding the mixture with hexane, and the precipitate was washed and dried to provide $2 \cdot CH_2N_2$. The guest proton ^1H NMR signal in $CDCl_3$ was observed at δ -0.69, having been moved upfield by incarceration by $\Delta\delta = 3.96$ p.p.m. After standing at 25 °C in $CDCl_3$ for 1 hour, $2 \cdot CH_2N_2$ liberated 15% of its CH_2N_2. After being stored for seven months in the solid state at 25 °C in the dark, $2 \cdot CH_2N_2$ lost 85% of its guest. In effect, **2** acted as a stabilizing container for the notoriously unstable diazomethane.[3]

10.4 A Thermal–Photochemical Reaction Cycle Conducted in the Inner Phase of a Hemicarcerand

Corey et al.[4] reported the photochemical conversion of α-pyrone **3** to the rearranged lactone **4**, which when heated gave the third lactone **5**. Flash

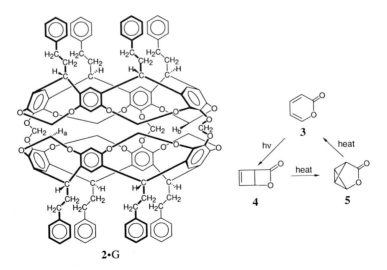

2·G

---

[3] D.J. Cram and M.E. Tanner, unpublished results.
[4] E.J. Corey and J. Streith, *J. Am. Chem. Soc.*, 1964, **86**, 950; E.J. Corey and W.H. Pirkle, *Tetrahedron Lett.*, 1967, 5255.
[5] J. Kreile, N. Munzel, and A. Schweig, *Chem. Phys. Lett.*, 1986, **124**, 140.

vacuum pyrolysis at 600 °C converted **5** back to α-pyrone **3** to complete the three reaction cycle **3** → **4** → **5** → **3**.[5]

This same reaction cycle was carried out in the inner phase of hemicarcerand **2**.[6] Hemicarceplex **2**·α-pyrone (**2·3**) was prepared by heating at reflux a 3:1 (v) C_6H_5Cl-α-pyrone solution of **2**, and precipitating **2·3** by adding hexane to the cooled reaction mixture. The host-to-guest ratio of the complexing partners was found to be 1:1 by elemental analysis and a 500 MHz ^1H NMR spectrum. An FTIR spectrum of **2·3** in $CDCl_3$ exhibited bands at 1733, 1719, 1539, and 1246 cm^{-1} due to incarcerated **3**, which compare with 1737, 1719, 1549, and 1247 cm^{-1} for free **3** dissolved in $CDCl_3$.

Comparison of the UV spectra of free **3** and of **2·3** in $CDCl_3$ shows the absence of any charge-transfer interactions between host and guest. The strong absorbance due to the aromatic rings of the host ends abruptly at 300 nm, leaving a small UV window for the selective photolysis of α-pyrone in **2·3**. Irradiation of 2–4 mM solutions of **2·3** in $CDCl_3$ at 25 °C with a 75-W xenon arc lamp fitted with a filter for λ < 300 nm ($SnCl_2$: aqueous HCl) caused a gradual, quantitative conversion of **2·3** to **2·4** in one hour. The reaction was followed by changes in the 500 MHz ^1H NMR spectrum of the reaction mixture. In the spectrum of **2·3**, the host H^a and H^b signals are broad at 25 °C due to restricted motion of the guest relative to the host on the ^1H NMR time scale, but sharpen somewhat at 60 °C. In the spectrum of **2·4**, these signals are fairly sharp at 25 °C, but are doubled. This duality of signals indicates that in **2·4**, the long axis of guest **4** and that of host **2** are roughly coincident in **2·4**, and that end-to-end rotation of guest relative to host is slow on the ^1H NMR time scale. The chemical shifts of **4** in **2·4** were consonant with those of free **4** in CCl_4 after allowance was made for the shielding effects of the host. The FTIR spectrum of **2·4** in $CDCl_3$ gave a lactone carbonyl stretch as a doublet at 1847 and 1823 cm^{-1}, which compares well with the reported values of 1848 and 1818 cm^{-1} for free **4** in CCl_4.[4] Hemicarceplex **2·4** remained unchanged when stored in the solid state for two weeks, and no sign of decomplexation or rearrangement was observed in a $CDCl_3$ solution at 25 °C over a 24-hour period.[6]

When **2·4** was heated as a solid for 17 hours at 90 °C, 80% of the complex was converted to **2·5**, 13% remained as **2·4**, and 7% had reverted to the α-pyrone complex. The 500 MHz ^1H NMR chemical shifts of **2·5** are consistent with those reported[5] for free **5** in $CDCl_3$ if allowances are made for the shielding effects of the host in the spectrum of **2·5**. The coupling constants for **5** in **2·5** were also in agreement with those reported for **5** in $CDCl_3$.[5] The north–south hemispheric symmetry of the host in **2·5** is indicated in its ^1H NMR signals for H^a and H^b, which were not doubled. The FTIR spectrum of **2·5** in $CDCl_3$ gave a strong peak at 1821 cm^{-1}, not far from the absorptions at 1818 and 1795 cm^{-1} for **5** in $CDCl_3$.[5]

As a final transformation to complete the inner-phase reaction cycle, **2·5** was heated as a solid at 140 °C for 20 hours to give **2·3**, the starting material of the cycle. Within the limits of ^1H NMR spectroscopy, the three transformations that occurred in the inner phase were quantitative.[6]

[6] D.J. Cram, M.E. Tanner, and R. Thomas, *Angew. Chem., Int. Ed. Engl.*, 1991, **30**, 1024.

10.5 Cyclobutadiene Stabilized by Incarceration

Efforts to synthesize and characterize cyclobutadiene **6** have been made for over a century.[7-9] This smallest carbocycle composed of conjugated double bonds has been prepared as a fleeting intermediate by several methods, but has been isolated only in an argon matrix at 8K by the photolysis of α-pyrone.[10-11] Attempts to isolate cyclobutadiene $(CH)_4$ led to the isomeric dimer **7**, which when heated produced cyclooctatetraene **8**. However, Masamune[12] and Maier[13] prepared **9** and **10**, respectively, whose tertiary butyl groups stabilized the cyclobutadiene nucleus by sterically inhibiting its reactions.

We prepared cyclobutadiene in the inner phase of hemicarcerand **2**, where the guest is protected from dimerization by the surrounding shell, and from reactants too large to pass through the portal of **2** at 25 °C.[6] The synthesis of **2·6** was accomplished by irradiating solutions of **2·3** with a 75-W xenon arc lamp for 30 minutes in degassed $CDCl_3$ or $(CD_2)_4O$ solution sealed in a borosilicate glass tube. The 500 MHz ^1H NMR spectra of the solution of $2·(CH)_4$ in $CDCl_3$ gave a sharp singlet for incarcerated cyclobutadiene at δ = 2.27, and in $(CD_2)_4O$, δ = 2.35. The latter signal is Δδ = 3.03 p.p.m. upfield from the ring proton of **9** in $(CD_2)_4O$ due to the shielding effects of the host of $2·(CH)_4$. In comparison, the proton singlet of $2·C_6H_6$ is Δδ = 3.49 p.p.m. upfield of free C_6H_6 dissolved in $CDCl_3$ at 22 °C.[2] The host's inward-pointing H^a and H^b protons (see **2·G**) gave sharp doublets in the ^1H NMR spectrum of $2·(CH)_4$ at δ = 4.27 and 4.36, providing direct evidence that cyclobutadiene possesses a singlet ground state.[6] Triplet $(CH)_4$ would dramatically broaden and shift the host signals, as did triplet O_2 in the spectrum of $2·O_2$.[2]

The sharpness of these H^a and H^b signals shows that in $2·(CH)_4$ the guest is rotating rapidly on the ^1H NMR time scale about all axes with respect to the host. Examination of CPK molecular models of $2·(CH)_4·CO_2$ suggests that

---

[7] M.P. Cava and M.J. Mitchell, 'Cyclobutadiene and Related Compounds', Academic Press, New York, 1967.
[8] T. Bally and S. Masamune, *Tetrahedron*, 1980, **74**, 343.
[9] G. Maier, *Angew. Chem., Int. Ed. Engl.*, 1988, **27**, 309.
[10] O.L. Chapman, C.L. Mcintosh, and J. Pacansky, *J. Am. Chem. Soc.*, 1973, **95**, 614.
[11] C.Y. Lin and A. Cranz, *J. Chem. Soc., Chem. Commun.*, 1972, 1111.
[12] S. Masamune, N. Nakamura, M. Suda, and H. Ona, *J. Am. Chem. Soc.*, 1973, **95**, 8481.
[13] G. Maier, S. Pfriem, U. Schafer, and R. Matusch, *Angew. Chem., Int. Ed. Engl.*, 1978, **17**, 520.

Complexed Hosts, Incarcerated Guests, and Host Protection from Self Destruction

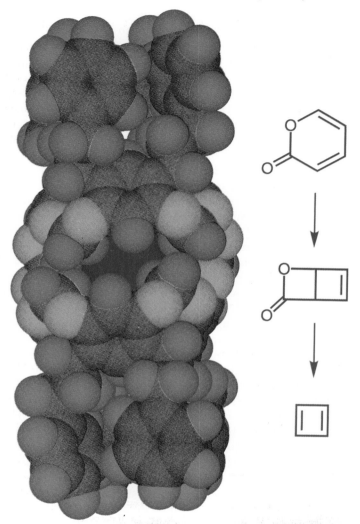

Figure 10.1 *Colored side view of hemicarceplex* **2**·*G, the host of which simulates CPK models based on the coordinates of the crystal structure of* **2**. *G is a sphere which symbolizes α-pyrone, Corey lactone* **4**, *or cyclobutadiene, whose structures appear to the left of* **2**·*G. The color code is: C, gray; H, blue; O, red; guest, dark blue*

shuffling of two such guests in the cavity should be too slow to randomize their positions rapidly enough on the ^{1}H NMR time scale to provide sharp signals for the H^a and H^b.[6] An analogy for the CO_2 ejection from the cavity by the $(CH)_4$ is found in the ejection of only one of two CH_3CN guest molecules by heating a termolecular complex[14] similar to **2**·$(CH)_4$·CO_2.

When the irradiation times of **2**·**3** were varied, both the precursor for and products of **2**·$(CH)_4$ were observed in the ^{1}H NMR spectra of the $CDCl_3$

[14] J.A. Bryant, M.T. Blanda, M. Vincenti, and D.J. Cram, *J. Am. Chem. Soc.*, 1991, **113**, 2167.

solutions. Shorter times gave as high as a 60% conversion of **2·3** to bound lactone **2·4**, which on further irradiation went to **2·(CH)$_4$**. Longer irradiation times produced an ^{1}H NMR signal at δ = 1.91, due to free acetylene.[6] Thus cyclobutadiene gave acetylene, just as did cyclobutadiene in an argon matrix at 8 K.[10]

The crystal structure coordinates on which **2·G** of Figure 10.1 is based were taken from those of **2** in **2·(CH$_3$)$_2$NCHO**. Notice the smallness of the channel that connects the inner and outer phase of **2·G**, and the twist of the two hemispheres with respect to one another in this drawing simulating CPK models.

An FTIR spectrum of **2·(CH)$_4$** in CDCl$_3$ gave no carbonyl signals in the 1650–1850 cm^{-1} region. The appearance of a signal at 1233 cm^{-1} provides additional evidence for the formation of cyclobutadiene. This band is assigned to the carbon–carbon stretching frequency which was found at 1240 cm^{-1} in the argon matrix experiment.[10] Host bands obscured the other signals of (CH)$_4$.[6]

A solution of **2·(CH)$_4$** in (CD$_2$)$_4$O was heated to 220 °C for 5 minutes in a sealed NMR tube. The hemicarceplex underwent complete guest exchange to produce **2·(CD$_2$)$_4$O** which had been characterized previously.[2] The ^{1}H NMR spectrum of the solution showed complete loss of the (CH)$_4$ signal and appearance of a new signal at δ = 5.71, which is due to free cyclooctatetraene, **8**. When the tube was opened, the pungent and characteristic odor of **8** was evident. Addition of authentic cyclooctatetraene intensified this 5.71 peak. Thus the liberated (CH)$_4$ must have dimerized to give **7**, which rearranged to **8**.[6] In earlier experiments, the cyclobutadiene dimers were found to rearrange to **8**, the reaction's half-life being *ca.* 20 minutes at 140 °C.[15]

Cyclobutadiene as a guest in **2** dissolved in CDCl$_3$ reacted readily and completely with O$_2$ to give what is probably **2·(Z)-OHCCH=CHCHO**. The reaction was accomplished by bubbling O$_2$ vigorously through a solution of **2·(CH)$_4$** in CDCl$_3$ at 25 °C; the reaction was complete after 30 minutes. The 500 MHz ^{1}H NMR spectrum at 60 °C of the malealdehyde complex provided guest signals as broad singlets at δ = 5.77 and 3.61 that were coupled. The former signal is due to the aldehyde protons being shifted upfield by Δδ ≈ 4 p.p.m. from their usual positions. An examination of CPK models of **2·(Z)-OHCCH=CHCHO** indicates that these protons must be located in the high-shielding polar regions of the host. The position of the signal at δ = 3.60 is typical for an alkene proton of an incarcerated guest. An FTIR spectrum of this new complex contained a broad band at 1696 cm^{-1}, indicative of the presence of an unsaturated aldehyde. The formation of **2·(E)-OHCCH=CHCHO** instead of the (Z) complex cannot be ruled out. An analogy for our reaction, **2·(CH)$_4$ + O$_2$ → 2·(Z)-OHCCH=CHCHO**, is found in the reaction of tetra-*t*-butylcyclobutadiene, **10**, with O$_2$ to give (Z)-*t*BuCOC(*t*Bu)=C(*t*Bu)CO*t*Bu.[16]

The results of this section coupled with those of Section 10.4 provide

---

[15] M. Avram, I.G. Dinulescu, E. Marica, G. Mateescu, E. Sliam, and C.D. Nenitzescu, *Chem. Ber.*, 1964, **97**, 382.

[16] H. Irngartinger, N. Riegler, K-D. Malsch, K-A. Schneider, and G. Maier, *Angew. Chem., Int. Ed. Engl.*, 1980, **19**, 211.

examples of three photochemical, two thermal, and one bimolecular oxidation reactions carried out in the inner phase of hemicarcerand **2**. Cyclobutadiene as a singlet is thermally stable at ambient temperature when protected from reacting with itself by its occupation of the inner phase of **2**. The interesting question arises as to how 200–250 nm light reaches incarcerated α-pyrone, given the strong light-absorbing properties of the host of wave lengths < 300 nm. The shell might act as a photosensitizer for the reaction. We anticipate that many thermally stable but highly reactive species containing bent acetylene, alkenic or aromatic rings, radicals, carbenes, and the like can be prepared and characterized in the inner phase of appropriate hemicarcerands.

10.6 Oxidations and Reductions of Incarcerated Guests

In Section 9.7, host **12** was reported to incarcerate readily guest diphenols **13–16** at high temperatures to produce hemicarceplexes **12·Guest**, stable at temperatures below 100 °C, which were fully characterized.[17] Attempts to incarcerate their corresponding quinones failed due to the instability of these potential guests at the temperatures required for their passage through the portals of **12**. In the same study, both aniline and nitrobenzene were also introduced into the inner phase of **12** at high temperatures to give **17** and **18**, respectively. These carceplexes were also fully characterized.[17]

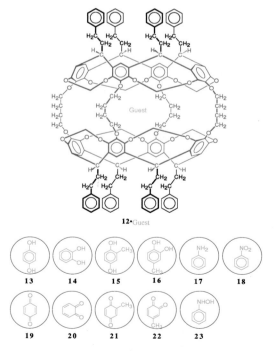

[17] T.A. Robbins, C.B. Knobler, D.R. Bellew, and D.J. Cram, *J. Am. Chem. Soc.*, 1994, **116**, 111.

A second study reported the oxidation of the guest hydroquinones to their corresponding incarcerated quinones **19–22**, as well as the reduction of these hemicarcerand quinones back to their original hemicarcerand hydroquinones. An attempt to reduce incarcerated nitrobenzene (**18**) resulted not in the production of the expected **12**·aniline (**17**), but in **12**·N-phenylhydroxylamine (**23**).[18]

Both $Ce(NH_4)_2(NO_2)_6$–silica-gel–CCl_4[19] and $Tl(O_2CCF_3)$–CCl_4–CH_3OH (two drops of CH_3OH)[20] served as oxidizing agents at 25 °C, and the yields of **19–23** were essentially quantitative. These carceplexes were each characterized and found to be chromatographically stable at room temperature in the dark. When the four free quinones (**19–22**) were heated at reflux in $(CH_2)_4O$ solutions containing two drops of CH_3OH with SmI_2, the original corresponding hydroquinones **13–16** were obtained in essentially quantitative yields. When a solution of incarcerated quinone **19** was shaken with an aqueous solution of Na_2SO_4 for 5 minutes, a 30:70 mixture (^1H NMR spectra) of **13:19** was produced. When the four incarcerated quinones **19–22** were heated at reflux in $(CH_2)_4O$ solutions containing two drops of CH_3OH with SmI,[21,22] the corresponding incarcerated hydroquinones were obtained in essentially quantitative yields. A similar treatment of incarcerated nitrobenzene, **18**, gave not the expected complexed aniline, **17**, but incarcerated N-phenylhydroxylamine, **23**, which was fully characterized.[17] The water produced in these reactions apparently escaped the inner phase through the 26-membered ring portals located in the equatorial–temperate regions of the globe.[17]

The interior of hemicarcerand **12** is too small to accommodate both the reactant and reagent at the same time, and we conclude that the reaction mechanisms all involved electron, proton, and water transfers between the solution phase containing the reagent and the inner phase of **12** containing the reactant. Therefore reaction mechanisms involving coordination of the guest with the reagent are not a necessary condition for these reactions to occur. Particularly noteworthy is the oxidation of the hydroquinones with $Ce(NH_4)_2(NO_2)_6$ adsorbed on silica, since three phases—the inner phase of **12**, the solvent, and the solid silica-gel phase—were all involved.[18] The nine reactions reported in this section coupled with the six thermal and photochemical, one addition–elimination (Section 10.3–10.4), and the acid–base reactions (Section 10.2) suggest that the inner phase of hemicarcerands, tailored to the dimensions of potential transition states, is a unique place to carry out highly specific reactions involving unstable reactants.

10.7 Reductions of the Host of Hemicarceplexes

10.7.1 Octaimine Reductions

Previous sections dealt with transformations of guests carried out with hosts acting as reaction vessels. This section addresses the question of whether host

---

[18] T.A. Robbins and D.J. Cram, *J. Am. Chem. Soc.*, 1993, **115**, 12199.
[19] A. Fischer and G.N. Henderson, *Synthesis*, 1985, 641.
[20] A. McKillop, B.P. Swann, and E.C. Taylor, *Tetrahedron*, 1970, **26**, 4031.
[21] Y. Zhang and R. Lin, *Synth. Commun.*, 1987, 329.
[22] H.B. Kagan, *New J. Chem.*, 1990, **14**, 453.

vessels undergo reactions differently when they contain guests as compared to when they are empty. Accordingly, empty octaimine **24**[23] was reduced to octaamine **25** with $(CH_2)_4O$–NaCN–Ni(OAc)$_2$ at reflux temperature for one hour. The reaction mixture became homogeneous as the reduction proceeded. The octaamine product **25** was obtained in 87% yield and was fully characterized. No reaction occurred in 24 hours at reflux when the Ni(OAc)$_2$ was omitted. The reaction proceeded more slowly and less reproducibly with LiBH$_4$ as the reducing agent. Attempts to incorporate [2.2]paracyclophane into host **25** by the methods used successfully for host **24** failed (see Section 9.2).[24]

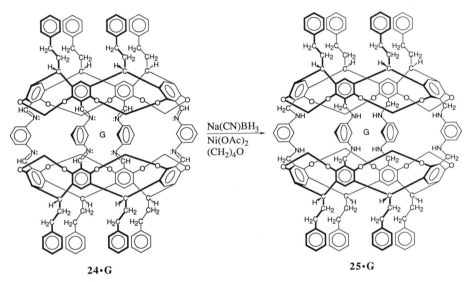

24·G   **25·G**

When **24**·[2.2]paracyclophane was subjected to the same reducing conditions applied successfully to the free host **24**, the reaction proceeded just as readily in comparable yield to give **25**·[2.2]paracyclophane, which was fully characterized. Although CPK models of **25**·[2.2]paracyclophane show the cavity of **25** to be essentially completely filled by the guest, the guest did not appear to inhibit sterically the eight-fold reduction reaction, nor did the guest escape its cavity during the reduction.[24]

10.7.2 Reduction of Hemicarceplexes with Four Acetylenic Bridges

Hemicarcerand **26** was prepared in 2–10% yield from tetrol **27** and (Br)Ts OCH$_2$C≡CCH$_2$OTs(Br) in $(CH_3)_2$SO, $(CH_3)_2$NCHO, or $(CH_3)_2$NCOCH$_3$–Cs$_2$CO$_3$ containing KI as catalyst. When the product was purified by silica-gel chromatography with CH$_2$Cl$_2$ as the mobile phase, the hemicarceplexes initially formed underwent guest-exchange with CH$_2$Cl$_2$, which was easily removed to give free **26** by heating **26**·CH$_2$Cl$_2$ in CCl$_4$, a solvent whose molecules are too

---

[23] M.L.C. Quan and D.J. Cram, *J. Am. Chem. Soc.*, 1991, **113**, 2754.
[24] D.J. Cram and M.L.C. Quan, unpublished results.

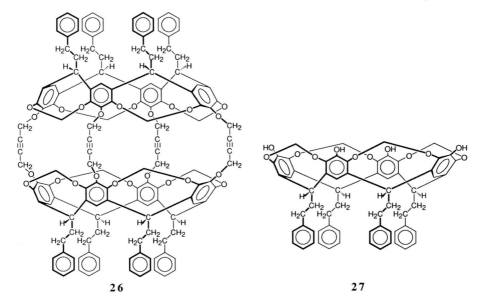

26   **27**

large to enter the inner phase of this host. The following hemicarceplexes were formed by heating **26** or **26**·CH$_2$Cl$_2$ in guest as solvent from 100–130 °C, and were fully characterized: **26**·CHCl$_3$, **26**·CHCl$_2$CHCl$_2$, **26**·CF$_3$C$_6$H$_5$, **26**·1,4-(CH$_3$)$_2$C$_6$H$_4$, and **26**·(S)-(+)-1-bromo-2-methylbutane. Top and side stereoviews of the crystal structure of **26**·CHCl$_2$CHCl$_2$ are found in **28a** and **28b**, respectively. In the latter view, only the four OCH$_2$C≡CCH$_2$O and the guest are included, and the four oxygens in each hemisphere are joined to form two

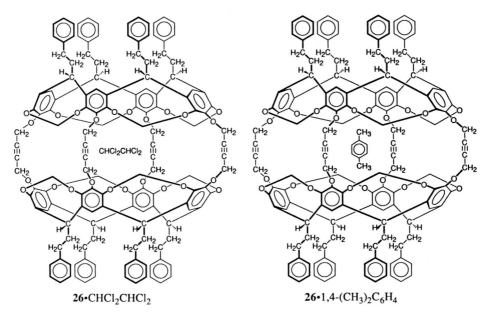

26·CHCl$_2$CHCl$_2$ **26**·1,4-(CH$_3$)$_2$C$_6$H$_4$

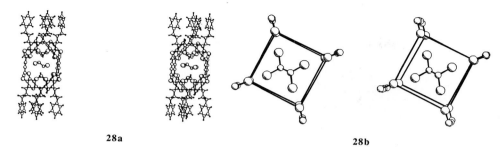

28a   **28b**

parallel squares. Drawing **28b** accentuates the fact that the long axes of the host and guest are noncoincident.[25]

Hemicarceplexes **26**·$CHCl_2CHCl_2$ and **26**·1,4-$(CH_3)_2C_6H_4$ when shaken in benzene solution with H_2–PdC for four hours gave (80% yields) **29**·$CHCl_2CHCl_2$ and **29**·1,4-$(CH_3)_2C_6H_4$, respectively with no loss of guest. The sample of **29**·1,4-$(CH_3)_2C_6H_4$ gave ^{1}H NMR and FAB-MS spectra identical to those of the complex prepared by a different method (**55** of Chart 9.10).[17] Attempts to prepare **29**·$CHCl_2CHCl_2$ by heating empty **29** in $CHCl_2CHCl_2$ at 150 °C for two days failed to give the complex.[17] At least in these examples, hosts of hemicarceplexes undergo chemical reaction without guest modification.[25]

Figure 10.2 provides a dramatic comparison of the differences in appearance and usefulness of CPK *vs.* ball-and-stick models in representing the structure of hemicarcerands. In **26a**, the size and shape of the four entryways connecting the inner and outer phases can be inferred. In contrast, **26b** better traces the connectedness of the atoms composing the host.

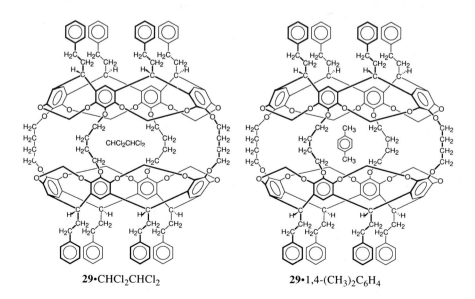

29·$CHCl_2CHCl_2$ **29**·1,4-$(CH_3)_2C_6H_4$

[25] C.N. Eid, Jr., C.B. Knobler, D.A. Gronbeck, and D.J. Cram, unpublished results.

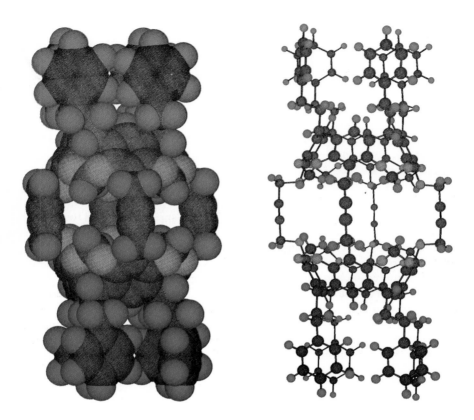

Figure 10.2 *Simulated CPK models colored drawing **26a** based on coordinates taken from the crystal structure of **26**·$CHCl_2CHCl_2$ and colored **26b** ball-and-stick drawing based on the same coordinates. Color code is: C, gray; H, blue; O, red*

We believe the field of host–guest chemistry offers great opportunities for posing and answering fascinating questions of both theoretical and practical nature. We have written this book with the hope of attracting investigators to the field, and providing them with an impression of how molecular modeling can leverage and guide the imagination in the application of the structural theory of organic chemistry to problems rich in novelty, easy to conceive and not difficult to solve.

Subject Index

Ab initio calculation of complex, 9
Acidity of amine salts in hemicarceplexes, 204
Activation free energies,
  complexation–decomplexation of xylyl-bridged hemicarceplexes, 165, 166
  difference between diastereomers, 67, 69
  rotational barriers of incarcerated amides, 203, 204
  rotations of guests in hemicarceplexes, 161
  velcraplex association and dissociation, 129
Amino ester resolving machine, 55
Amino ester salts, catalytic thiolosis of, 67
Analytical determinations of Na^+ and K^+ ions, 48
Aromatic solvent induced shifts ASIS, 146
Aryldiazonium ion in complex, 5, 76, 77
Association constants of,
  alkali metal ion ratios, 41
  bowl-shaped cavitands and acetonitrile, 101
  cavitands of linear guests, 98
  chymotrypsin mimic of guests, 78
  2-cleft cavitands and π-bases, 103
  determination, 8, 34
  enantiomer distribution constants (EDC), 50, 51, 53, 167, 169
  spherand *vs.* podand 41
  triamine host of small compounds, 188
  tris-biacetylenic host of small compounds, 184
  tritosylamide host of small compounds, 188

Bender, M.L., 65

Bergson, G., 59
Billiard ball effect on guests, 138
1, 1′-Binaphthyl corands, 3–7, 50, 53, 55, 57, 59, 61, 62
Bowls, cavitand, 91
  conformation of, 96, 97
  crystal structures, 93–96
  syntheses, 91
  use in syntheses of carcerands, 132
Breslow, R., 65

Carceplexes,
  comparison to other complexes, 147
  expulsion of guest, 138
  guest rotations in, 139, 203
  interior as new phase of matter, 148
Carcerands,
  mechanism of shell closure, 134, 137
  rotations of complexed guests in, 139, 203
  synthesis, 132, 136, 140
Caviplexes, 147
Cavitands, 85
  bowl-shaped, 91–96
  kite-, 108
  synthesis, 87–92
  vase-, 107
  vase–kite conformational interchange, 109
  with one cleft, 103
  with two clefts, 102
Chapman, O.L., 208
Chiral barriers in host, 5
Chiral catalysis, 59
  addition of alkyllithium of aldehydes, 63
  catalytic turnover in, 59
  methacrylate ester polymerization, 61
  Michael additions, 59
  rationalization of, 60
Chiral coraplex catalysis of Michael reaction, 59

Chiral recognition,
    of amino acid esters, 50
    by corands, 49–57
    in guest release, 168
    by hemicarcerands, 167
    in transacylations, 68
Chromophores in spherands, 46
Chromatographic behavior of
    carceplexes, 147
Chromatographic resolution of amino
    acid esters, 55
Chymotrypsin mimic, 77
Clathrates, 147
Closed-surface hydrocarbon sphere, 131
Collet, A., 166, 167, 181, 184, 185
Colorimetric determination of metal
    ions, 46–48
Competitive inhibition, 68, 79
Complementarity, principles of, 39, 40, 98
Constrictive bonding, 165
Corands, 7
    binding free energies of, 34–36
    chiral, 5
    resolution of, 5
Coraplexes, 6
Corey, E.J., 206
CPK molecular models, 2
Cramer, F., 65
Cryptahemispherands, 31–33
    binding free energies, 37
Cryptaplexes, 147
Crystal structures,
    bowls, 12–19
    carceplexes, 142–144
    caviplexes, 88, 93–97, 100, 101, 103,
        111, 138
    cavitand bowls, 93–97
    cavitands, 100, 105
    comparison with molecular models,
        12–19
    corands, 12–19
    coraplexes, 12–19
    cryptahemispherands, 31–33
    cyanospherand, 44
    cyclic urea hemispheraplexes, 73, 74
    hemicarceplexes, 152, 159, 171, 173,
        175, 178, 187, 196–199, 209, 215
    hemispherands, 28–31, 39
    hemispheraplexes, 28–31, 39
    kite-cavitand dimers, 117
    octaamide hemicarceplex, 177
    octaimine hemicarceplex, 175
    octalactone hemicarceplex, 171
    [1.1.1]orthocyclophane
        hemicarceplex, 182

octols, 91
spherands, 25–28, 76
spheraplexes, 25–28, 76
tetraacetylene hemicarceplex, 215
tetramethylenedioxa hemicarceplex,
    196–201
triamine hemicarceplex, 186
tris-biacetylene hemicarceplex, 152
tris-bridged hemicarceplex, 152
tritosylamide hemicarceplex, 186
vase-caviplex, 111
velcrand, 117
velcraplexes, 116–120
xylyl-bridged hemicarceplex, 158–160
Cyanospherands, 44
Cyclic urea corands, 71
Cyclobutadiene stabilized by
    incarceration, 208
Cyclooctatetraene, 210
Cyclotriveratrylene, 86, 87

Dalcanale, E., 111
Decomplexation, rates, 44
Definitions,
    binding forces, 1
    carceplexes, 132
    carcerands, 131
    cavitands, 85
    complementarity, 39
    complex, 1
    constrictive binding, 165
    corands, 7
    coraplexes, 7
    cryptahemispherands, 31
    cyanospherands, 44
    hemicarceplexes, 150
    hemicarcerands, 150
    hemispherands, 28
    hemispheraplexes, 28
    intrinsic binding, 165
    kite-cavitand, 108
    preorganization, 39
    solvophobic effect, 1
    spherands, 20
    vase-cavitand, 107
    velcrands, 115
    velcraplexes, 112
Diastereomeric complexes, 3, 68
Diastereomeric complexes of corands,
    and amino acid esters, 57
    rates of guest exchange, 167
    rates of guest release, 168
Diastereomeric hemicarceplexes, 167
Diastereomeric hosts, 182
Diazomethane protected by host, 206

Diazonium ion complexed, 6, 76
Diederich, F.J., 125
N,N-Dimethylacetamide guest, 144, 194, 198
  chemical shifts of, 157
  effect on R_f value, 146
  infrared spectrum of, 147
  rate constants for complexation, 165
  rate constants for decomplexation, 162
  rotation of, 144, 203
  sensitivity to solvent magnetic properties, 146
N,N-Dimethylformamide guest, 145
  chemical shifts of, 157
  effect on R_f value, 147
  infrared spectrum of, 147
  rotation of, 145, 203
Dimethylsulfoxide guest, 145
  chemical shifts of, 157
  effect on R_f value, 147
  rotation of, 145

Enantiomer distribution constants (EDC),
  determination of, 49
Enantiomer resolution,
  by chromatography, 5
  by complexation, 52
Enthalpy of binding,
  bowl and acetonitrile, 101
  of cavitands, 98, 102
  of velcrands, 124, 127, 128
  of xylyl-bridged hemicarceplexes, 165
Enthalpy-driven complexation of velcrands, 127
Entropy of binding,
  bowl and acetonitrile, 101
  of cavitands, 102
  of velcraplexes, 124, 127, 128
  of xylyl-bridged hemicarceplexes, 165
Entropy-driven complexation of velcrands, 127
Enzyme mimics,
  chymotrypsin, 77
  papain, 66
  serine protease, 77
Erdtman, H., 86, 87

Fast atom bombardment mass spectra (FAS-MS), 133
Free energy of binding, 7–10, 166
  ammonium ions, 34–37
  bowl and acetonitrile, 101
  cavitands of linear guests, 98
  cleft cavitand and π-bases, 103
  corand and hemispherand comparison, 34–36
  corands of amino acid ester salts, 50
  coraplexes of t-butylammonium cyanide, 8–11
  cryptahemispherands of picrates, 34–36
  cyanospherands of picrates, 46
  cyclic urea hemispherands of picrates, 72, 74, 75
  decreasing order of, 41
  determination, 8, 34, 51, 121
  diastereomeric coraplexes of amino acid and ester salts, 50, 51, 53, 60
  effect on pK_a of chromogenic ionophores, 47
  hemispheraplexes of picrates, 34–36
  highly preorganized hosts, 75–77
  intrinsic binding, 165
  metal cations, 34–37
  of octacyanospherand, 44
  partition between constrictive and intrinsic binding, 164–167
  preorganization effects, 42
  spheraplex of picrates, 38, 39
  thiolysis, 67
  three-bridged hemicarceplexes, 155
  triamine host of small compounds, 188
  tris-biacetylene host of small compounds, 184, 185
  tritosylamide host of small compounds, 188
  velcrand dimers, 121, 124
  xylyl-bridged hemicarceplexes, 165
Free energy values for diastereomeric complexes, 51, 53

Guest rotations, 144
Guests, amino acids and ester salts,
  alanine methyl ester perchlorate, 51, 52, 57
  alanine p-nitrophenyl ester perchlorate, 67–69, 79, 80, 82, 83
  alanine perchlorate, 51, 52
  hexefluorophosphate salts, 53
  4-hydroxyphenylalanine methyl ester perchlorate, 57
  4-hydroxyphenylglycine methyl ester perchlorate, 57
  isoleucine p-nitrophenyl ester perchlorate, 67, 68
  methionine methyl ester perchlorate, 51, 52

Guests (*cont.*)
  methionine perchlorate, 51, 52
  phenylalanine methyl ester perchlorate, 51, 52
  phenylalanine *p*-nitrophenyl ester perchlorate, 67, 68
  phenylalanine perchlorate, 51, 52
  phenylglycine methyl ester hexafluorophosphate, 55, 57
  phenylglycine methyl ester perchlorate, 51, 52
  phenylglycine perchlorate, 51–55, 57
  proline *p*-nitrophenyl ester perchlorate, 67
  tryptophane methyl ester perchlorate, 51, 52
  tryptophane perchlorate, 51, 52
  valine, 5
  valine methyl ester perchlorate, 51, 52
  valine perchlorate, 51, 52
  valine *p*-nitrophenyl ester perchlorate, 67–69
Guests, cations,
  alkylammonium ions, 34, 46, 66, 70, 74, 75
  ammonium ion, 34, 46, 70, 75
  aryldiazonium ion, 5, 77
  cesium ion, 32, 34, 46, 70, 75, 133
  lithium ion, 22, 28, 30, 34, 46, 70, 75
  potassium ion, 30, 32, 34, 46, 70, 75
  rubidium ion, 30, 34, 46, 70, 75
  sodium ion, 22, 30, 32, 34, 46, 70, 75
Guests, organic compounds,
  acetone, 100, 101, 195
  acetonitrile, 93, 138, 157, 161, 163, 184, 186, 188
  adamantane, 173
  amantadine, 173
  1-aminobutane, 157, 205
  aniline, 195, 211
  anthracene, 173
  anthraquinone, 173
  argon, 133
  benzene, 96, 100, 157, 184, 195
  1,2,9,10-bis-dehydro[2.2]paracyclophane, 191
  bromobenzene, 101
  1-bromo-2-methylpropane, 167, 214
  2-butanol, 167
  2-butanone, 139, 163, 165, 166
  t-butyl alcohol, 184
  5-*t*-butyl-1, 3-dimethylbenzene, 167
  γ-butyrolactam, 195
  γ-butyrolactone, 195
  carbon disulfide, 97, 98, 156, 187, 188
  camphor, 173
  catechol, 195
  chloroform, 96, 161, 163, 167, 183, 184, 214
  cubane, 184
  cyclobutadiene, 208–210
  cyclohexane, 96
  cyclopentanone, 195
  2-cyclopentenone, 195
  1,4-diacetoxybenzene, 180
  1,2-dehydro[2.2]paracyclophane, 191
  diazomethane, 206
  1,4-dibromobenzene, 195
  dibromomethane, 157, 206
  1,2-dichlorobenzene, 172
  dichloromethane, 87, 96, 100, 111, 157, 170, 173, 183, 184, 186, 214
  2,5-dichlorotoluene, 195
  3,4-dichlorotoluene, 195
  dicyanobenzene, 103
  diethylamine, 157, 205
  1,9-dihydroxy[2.2]paracyclophane, 191
  1,4-diiodobenzene, 195
  2,9-diketo[2.2]paracyclophane, 291
  1,4-dimethoxybenzene, 195
  N,*N*-dimethylacetamide, 141, 143, 144, 146, 151, 157, 160, 163, 165–167, 170, 195
  1,3-dimethyladamantane, 191
  N,*N*-dimethylformamide, 107, 133, 140, 141, 146, 151, 157, 163
  dimethylsulfoxide, 141, 151, 157, 195
  1,3-dinitrobenzene, 103
  ethanol, 139, 188
  ethyl acetate, 161, 165, 166, 195
  ethylbenzene, 161, 168
  4-ethyl[2.2]paracyclophane, 191
  ferrocene, 173
  4-fluorotoluene, 195
  Freon-114, 134, 163
  guanidinium ion, 5
  guanine, 106
  hexachloroacetone, 184
  hexachlorobutadiene, 173
  hexafluorobenzene, 195
  hexamethylenetetramine, 173
  hexamethylphosphoramide, 173
  hydroquinone, 195, 212
  iodobenzene, 195
  iodoethane, 195
  2-iodobutane, 167
  menthol, 173
  methanol, 138, 140, 186, 188
  4-methoxytoluene, 195

4-methylcatechol, 195
2-methylhydroquinone, 195
nitrobenzene, 171, 177, 195, 211
nitrogen, 187, 188
oxygen, 187, 188
[2.2]paracyclophane, 173, 191, 213
[2.2]paracyclophanemonoquinone, 191
[2.3]paracyclophane, 191
[3.3]paracyclophane, 191
3-pentanone, 139
N-phenylhydroxylamine, 212
1-propanol, 163
propylene oxide, 184
propyne, 98
pyridine, 157, 205
α-pyrone, 206
quinone, 212
ruthenocene, 173
1,1,2,2-tetrachloroethane, 161, 170, 184, 214
tetradeuterofuran, 210
tetrahydrofuran, 157, 163
toluene, 93, 96, 100, 161, 163, 165, 166, 183
1,3,5-triethylbenzene, 191
triethyl phosphate, 173
trifluorotoluene, 214
1,3,5-tri-isopropylbenzene, 191
tripropyl phosphate, 173
water, 16, 157, 177–180, 187, 188, 196
1,2-xylene, 195
1,3-xylene, 195
1,4-xylene, 161, 167, 195, 214

Hemicarceplexes, 150
  crystal structure of, 152, 159, 196
  decomplexation of, 152
  proton signals of guests in, 162
  rate constants for decomplexation, 154, 163
Hemicarcerands, 150
  chiral recognition by, 167
  with four portals, 158
  with one portal, 151
  octaamide, 170
  octaimine, 172, 212
  octalactone, 170
  as protecting container, 206
Hemispherands,
  binding free energies, 34–37
  complexation of metal cations, 28, 30
  cyclic urea units in, 71
  preorganization of, 28, 30
  rates of complexation, 43

Highly reactive species protected by host, 206
Hosts,
  acetylene hemicarcerand, 213–216
  biacetylene hemicarcerand, 189–193
  binaphthyl diamines, 63
  binaphthyl hemicarcerand, 167, 168
  bowls, 91–96, 132, 136
  carcerands, 131–136, 140–144, 147
  cavitands, 83–90, 97–106, 132, 136, 138, 141, 148, 151, 173, 177
  corands, 4–19, 34–36, 43, 50–56, 59–62, 66, 68
  crowns, 2, 3, 6, 7, 8, 11, 16, 43
  cryptahemispherands, 31–33, 37, 40, 47, 48
  cryptands, 16, 40, 148
  cyanospherands, 44
  hemicarcerands, 150–169
  hemispherands, 28–31, 40, 44, 70–75, 78–82
  kites, 109, 112, 113, 115
  octaamide hemicarcerand, 176–179
  octaamine hemicarcerand, 213
  octaimine hemicarcerand, 172–176, 212
  octalactone hemicarcerand, 170–172
  octols, 89–91, 114, 171, 214
  podands, 3, 7, 10, 21, 41, 66, 74
  spherands, 20–28, 38, 40, 41–44, 47, 75–77
  tetra-bridged hemicarcerands, 158–169
  tetramethylenedioxa hemicarcerand, 193–201, 211, 212
  triamine hemicarcerand, 185–189
  tris-biacetylene hemicarcerand, 180–184
  tri-bridged hemicarcerands, 150–158, 203–211
  tritosylamide hemicarcerand, 185–189
  vases, 107, 111–113
  velcrands, 115–130
  xylyl hemicarcerand, 158–164
Hosts, chiral, 50
  reorganization of, during complexation, 23
  two chiral elements in, 53
Högberg, A.G.S., 87, 90
Hydroquinone guest, oxidation of, 212

Imidazole unit in transacylation catalyst, 80
Inner phase of carcerands and hemicarcerands, 203, 204

Intrinsic binding, 165
  enthalpy of, 166
  entropy of, 166
  free energy of, 166

Kaiser, E.T., 65
Kellogg, R.M., 87
Kinetics of catalyzed transacylations, 81
Kite-cavitand dimers, 113–115
  crystal structure of, 117
  design of, 113
  dimerization in solution, 120
  enthalpy of association, 121
  entropy of association, 121
  free energy of formation, 121
  hetero-dimers, 121
  homo-dimers, 121

Lehn, J-M. 31, 185
Lindsey, A.S., 86
Lithium ion, see Guests, cations

Maier, G., 208
Masamune, S., 208
Mechanism of chiral transfer in,
  addition of alkyllithium to aldehydes, 63
  anionic polymerization, 61
  Michael reaction, 59
  thiolysis, 67
Mechanism of guest exchange in hemicarceplexes, 162
Mechanism of shell closure for carcerand, 134, 137
Michael addition, chiral catalysis in, 59
Molecular containers, 203
Mukaiyama, T., 64

New phase of matter, 148

Octaacetylene hemicarcerand, 189
Octaamide hemicarcerand, 176
Octaimine hemicarcerand, 172
Octols as cavitand precursors, 88–92
Origins of host–guest chemistry, 1
[1.1.1]Orthocyclophane hemicarcerand, 180
Oxidations of incarcerated guests, 211

Papain mimic, 66
Paracyclophane corands, 12
Pedersen, C.J., 2, 7
Preorganization,
  effect on binding cations, 42
  of hemispherands, 28, 30
  principles of, 39–42, 98
Protecting container for diazomethane guest, 206
Pyridyl corands, 9
α-Pyrone, photochemical conversion of, 206

Quinone guest, reduction of, 212

Racemic host, resolution of, 55
Rate enhancements by complexation determination, 66, 67, 78–80, 81–84
Rates of association–dissociation,
  difference between diastereomers, 67, 68
  four-bridged biacetylene hemicarceplex decomplexation, 191–193
  protonation of incarcerated amines, 205
  spherands and hemispherands, 44
  three-bridged hemicarceplexes, 155
  xylyl-bridged hemicarceplexes, 163, 164
Rate constants for decomplexation of hemicarceplexes, 162, 172
Reactions of incarcerated guests,
  amine protonation, 204, 205
  amide rotations, 203–204
  cyclobutadiene to acetylene, 210
  cyclobutadiene to malealdehyde, 210
  α-pyrone to cyclobutadiene, 208
  oxidations of hydroquinones to quinones, 211
  α-pyrone to photolactone to thermal lactone to α-pyrone, 206, 207
  reduction of nitrobenzene to N-phenylhydroxylamine, 211
  reduction of quinones to hydroquinones, 211
  reductions of hemicarceplex hosts, 212–214
Reductions of incarcerated guests, 211
Reorganization of host during complexation, 12, 17
Resolution of,
  amino acid esters by chromatography, 55
  by transport, 55
  racemic guest, 56
  racemic host, 52
Resolving machine, 56
Resorcinol as starting material, 87
Rotations of guests in hemicarceplexes, 161

Subject Index

Seebach, D., 64
Serine protease mimic, 77
Smithrud, D.B., 125
Sodium ion, see Guests, cations
Specificity in spherand binding of metal cations, 30
Spherands,
  complexing metal cations, 22–24
  crystal structures, 25–28
  preorganization of, 21
  rates of complexation, 41
  specificity of binding, 37
  symmetry properties of, 25
  syntheses of, 22
  templation in synthesis, 22
Spheraplexes, 26, 147
  anti-bridged, 26
  crystal structures, 25
  syn-bridged, 26
Sphere, closed surface hydrocarbon, 131
Solvophobic effect, 1, 127, 129, 130
Stereoisomers of bridged spherands, 24
Structural recognition,
  of ammonium ions, 8
  comparison of corands, 8
  of metal cations, 22
  ratios for metal cations, 40
Symmetry properties of,
  cyanospherand, 44, 45
  spherands, 25

Tabushi, I., 65
Thiolysis, catalysis of, 67
Transacylations catalyzed, 68, 78, 80
Transition states for catalyzed transacylations, 82
Triamine hemicarcerand, 186
Tris-biacetylene hemicarcerand, 180–184
Tritosylamide hemicarcerand, 186

van Hooidonk, C., 65
Vase-kite conformational interconversion, 109
Velcrands, 115
Velcraplexes, 115

Wynberg, H., 59

Yuki, H., 61

Zeolites, 147